Roll-to-Roll Manufacturing

# Roll-to-Roll Manufacturing

## Process Elements and Recent Advances

*Edited by*

*Jehuda Greener, Glen Pearson, and Miko Cakmak*

WILEY

This edition first published 2018


*Registered Offices*
John Wiley & Sons, Inc., 111 River Street, Hoboken, NJ 07030, USA

*Editorial Office*
111 River Street, Hoboken, NJ 07030, USA

For details of our global editorial offices, customer services, and more information about Wiley products visit us at www.wiley.com.

Wiley also publishes its books in a variety of electronic formats and by print-on-demand. Some content that appears in standard print versions of this book may not be available in other formats.



*Library of Congress Cataloging-in-Publication Data*

Names: Greener, Jehuda, editor. | Pearson, Glen Hamilton, 1948– editor. | Cakmak, Mukerrem, editor.
Title: Roll-to-roll manufacturing : process elements and recent advances / edited by Jehuda Greener, Glen Pearson, Miko Cakmak.
Description: Hoboken, NJ : John Wiley & Sons, 2018. | Includes bibliographical references and index. |
Identifiers: LCCN 2017043485 (print) | LCCN 2017052231 (ebook) | ISBN 9781119163800 (pdf) | ISBN 9781119163817 (epub) | ISBN 9781119162209 (cloth)
Subjects: LCSH: Manufacturing processes–Technological innovations. | Flexible electronics–Processing.
Classification: LCC TS183 (ebook) | LCC TS183 .R65 2018 (print) | DDC 670.42–dc23
LC record available at https://lccn.loc.gov/2017043485

Cover Design: Wiley
Cover Images: Courtesy of Miko Cakmak; © Iaremenko Sergii/Shutterstock

Set in 10/12pt Warnock by SPi Global, Pondicherry, India

Printed in the United States of America

10 9 8 7 6 5 4 3 2 1

*For Helena, Maya, Eli, Karen, Emre, and Gulgun*

# Contents

# Preface

Roll-to-roll (R2R) manufacturing is a well-established manufacturing technology platform used for over a century in many mature industries, such as printing, paper consumables and silver-halide photography, to produce two-dimensional, film-like products on a mass scale. The main appeal of this technology and the reason for its widespread use is its high throughput capability combined with a low manufacturing cost. The last two decades have seen a strong resurgence of this technology as it is being extended and adapted in many new technology areas, including microelectronics, display and photovoltaics, in an attempt to leverage some of the tangible benefits of R2R manufacturing, especially its low cost, in a new lineup of innovative products.

This volume mainly aims to review the state-of-the-art and shed a new light on R2R technology, to familiarize a new generation of researchers and practitioners with many aspects of this technology as it is being applied in an ever-growing number of industries and product categories. Some key elements of R2R manufacturing are reviewed by highly experienced experts in the field, with emphasis on practical, hands-on application principles. We also introduce the reader to a number of novel extensions and upgrades of R2R technology, designed to meet new and challenging requirements in the new generation of products.

Although we do not attempt to cover all aspects and possible variants of this widely used manufacturing tool, we sincerely hope that this volume will provide a solid foundation for students, practitioners and researchers already involved in R2R operations or those contemplating this process option in future development programs.

November 2017

*Jehuda Greener,*
*Glen Pearson,*
*Miko Cakmak*

# 1

# Roll-to-Roll Manufacturing: An Overview

*Jehuda Greener*

*J. Greener Consulting, Rochester, NY, USA*

## 1.1 Introduction

Roll-to-roll (R2R) manufacturing is an important manufacturing technology platform widely used for a host of applications and product categories, spanning many industries. These cover the gamut from traditional and mature technologies such as printing and silver halide photography to more novel application areas including flexible microelectronics [1–4], thin flexible batteries [5, 6], photovoltaics [7–10], and display [11–13]. A typical R2R production line, with standard coating, drying, and lamination steps, is depicted in Figure 1.1. This particular line, however, represents only one specialized process layout from within a multitude of manufacturing processes that can be broadly classified as R2R operations. A common thread in all these diverse manufacturing operations is that in all cases relatively thin and flat, film-type two-dimensional (2D) structures are processed continuously on a flexible moving web that is conveyed at some fixed speed between two or more rotating rollers. The web comprises an inert and flexible substrate on which a layer (or layers) of a functional material is applied by some means. The *functional* layer possesses some desired physical/chemical property that has special utility to its intended application. Many types of functional layers are applied in R2R operations reflecting the wide variety of applications utilizing this manufacturing platform. These include chemically sensitized layers used in traditional photography, ink layers used in various printing lines, optically refractive, diffusive, or collimating layers used in optical films for liquid crystal displays, photovoltaic layers used in flexible solar cells, barrier layers used in various packaging applications and magnetic layers used in magnetic tape, to name just a few.

*Roll-to-Roll Manufacturing: Process Elements and Recent Advances*, First Edition.
Edited by Jehuda Greener, Glen Pearson, and Miko Cakmak.

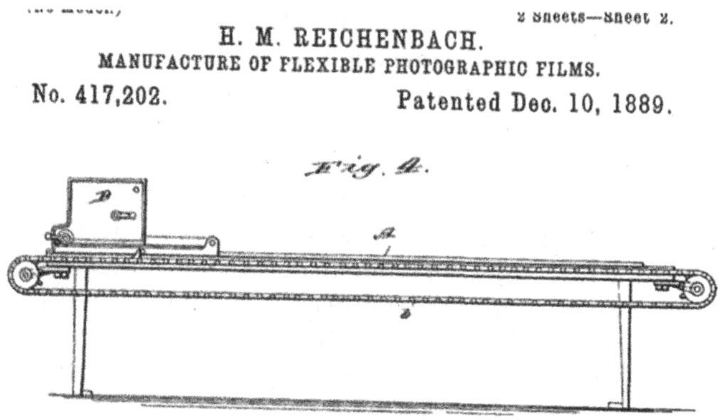


**Figure 1.2** Drawing of a roll-to-roll coating process in Reichenbach's 1889 patent.

reach of the average consumer. In the early 1880s, George Eastman and the newly founded Kodak (later to be named Eastman Kodak) Company developed a novel manufacturing process that facilitated the coating of light-sensitive materials on flexible substrates using a R2R operation that helped launch the mass production of photographic film [30]. The first step in the development of the R2R manufacturing process of photographic film was to coat photographic emulsions on a rolled paper substrate based on key inventions by William Walker and George Eastman [30]. This was followed by replacing paper with a clear cellulose nitrate substrate first proposed and patented by H. Reichenbach, one of Eastman's early collaborators (see Figure 1.2), based on an earlier invention by H. Goodwin of a process for making cellulose nitrate film [31]. These pioneering developments in the manufacturing of photographic film led to the popularization of photography, making it accessible to the average consumer as well as creating a demand for cameras, film processing, and printing equipment and related consumer products. Indeed, many of the early developments and innovations in R2R manufacturing technology in the late 19th Century and first half of the 20th Century were prompted and driven by the fast growth of the photographic and printing industries but especially by the exacting demands of the photographic industry that required precise deposition of up to 24 thin light-sensitive layers on a fast-moving web as well as creating the need for advanced finishing methods and novel substrate technologies [32].

Today, many wide-ranging industries, including printing, paper, packaging, and photography, among others, benefit directly from the operational and cost advantages of the R2R manufacturing technology platform. Many focused attempts to adapt and extend R2R processing practices to various new technology areas, particularly in microelectronics, display, and photovoltaics, are currently underway in many research groups around the world [1–4, 7–13].

Some common operational features of R2R processes are highlighted in Section 1.2 followed by a general discussion of cost and environmental, health, and safety considerations in Sections 1.3 and 1.4.

## 1.2 R2R Operation Overview

R2R operations are as varied and diverse as the markets and product categories they serve, but they share many underlying common features and operational principles. Figure 1.1 represents a typical R2R production line with conventional coating, drying, and lamination steps, but a more generic schematic layout of a R2R operation is shown in Figure 1.3. In this schematic a flexible web is conveyed between two rollers while passing through various process (converting) steps. An unprocessed web, comprising an uncoated or pre-coated substrate, is fed from a supply station (*unwinder*) wherein the raw or partially processed substrate is unwound from a supply roller and fed into the R2R machine (dashed frame). The raw substrate then undergoes a series of consecutive process (converting) steps (S1, S2, ...) while being conveyed at a controlled speed through the R2R machine.

The web, consisting of a substrate and deposited layers, is driven through the R2R machine by the *winder* and its conveyance is facilitated by a number of conveyance rollers or idlers (C) placed along its path. Tension in the conveyed web is controlled by the winder but is typically adjusted by one or more tension rollers (T) distributed within the R2R machine to insure flatness, planarity, and defect-free conveyance throughout the operation. We note that a single R2R line could have more than one unwinder if the operation comprises lamination

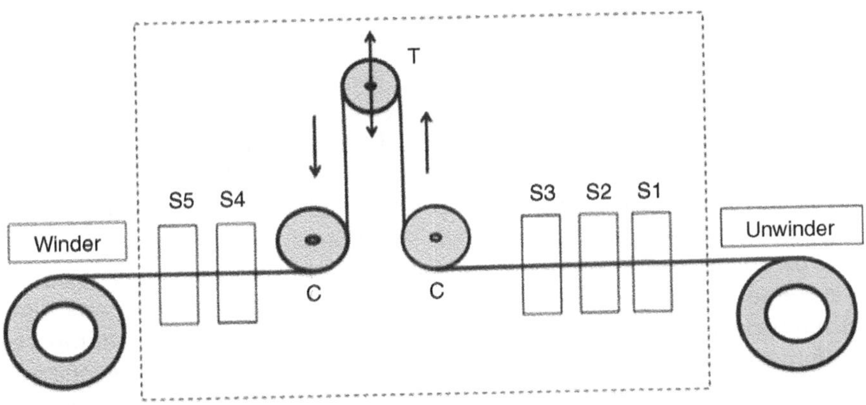


**Figure 1.3** Generic roll-to-roll operation with five converting steps (S1–S5), one tension roller (T), and two conveyance rollers (C).

or interleaving steps, as illustrated in Figure 1.1, but the line is always terminated by a single winder. The only exception here is for the case where the incoming substrate is protected by a sacrificial protective layer ("release liner"), which must be peeled off using a separate winder before the substrate is processed in the R2R machine.

Five converting steps are illustrated in the schematic layout of Figure 1.3, though the number of steps is arbitrary and usually greater than 1. A wide variety of converting steps are used in R2R operations, most of which involve the deposition and posttreatment of thin layers of some functional and ancillary materials on the moving substrate. Indeed, typical converting steps can be categorized by two main classes: (i) *film-forming* steps whereby a thin material layer(s) is deposited on the moving substrate and (ii) *film enhancement* steps whose function is mainly to consolidate, modify, and improve the performance of the deposited layer(s). Examples of common R2R film-forming steps include wet coating [33, 34], vacuum deposition [35, 36], printing [20, 21, 37], solvent casting, extrusion casting, and lamination, while typical film enhancement steps include drying [33, 34, 38], radiation curing [39–41], thermal curing, calendering, micro- or nano-patterning [14, 17, 18], heat treatment, Corona discharge treatment [42], annealing, chemical treatment, cleaning, interleaving, and so on. Some steps may require a special environment (temperature, humidity, vacuum, nitrogen blanket, etc.) so the web moving through the corresponding station must be properly enclosed, which may present some challenges as discussed later. Figure 1.3 underscores the modular nature of R2R operations whereby converting steps (modules) can be added or removed from the production line as dictated by the special requirements of the film product.

As the web is conveyed at a constant speed through all converting steps, it is important to select a speed (equivalent to residence time) that satisfies all process elements to allow an overall robust operation. In this sense we say that the process steps are coupled and the final line speed is constrained by the slowest process step. This is the so-called rate-determining step for the R2R operation. If this is not possible, that is, if one or more steps must operate at very different speeds from the rest, the operation must be split into more than one line, each operating at different speed, or the web needs to run multiple times through the same line at different speeds. Such a "multi-pass" (or "multi-line") operation is, of course, costlier than a single-pass operation and must be avoided if possible. Indeed, as discussed in Section 1.3, the selection of line speed is dictated not only by quality and operational considerations but also by cost considerations. Another factor, aside from line speed, that would require multi-line or multi-pass operation is process environment; for example, addition of vacuum deposition steps to a R2R line. Generally, confining a fast moving web in a vacuum atmosphere may be particularly challenging when combined with atmospheric pressure process steps, which may necessitate splitting the operation into an atmospheric pressure line and a vacuum deposition line.

Line speeds in R2R operations vary widely depending on the type of film being processed, the quality requirements, and the number of converting steps along the R2R line. For mature and commoditized technologies such as printing and various paper products, line speeds of up to 2000 m/min are not uncommon. But for more specialized application areas and advanced technologies with film products having complex and exacting layer structures with tight registration requirements, speeds of less than 10 m/min are often used. As discussed in Section 1.3, cost considerations dictate operating at the highest possible line speed, but product quality and processability will determine the limiting speed at which the line can be operated *robustly*; when operating above this limiting speed, film quality will not meet product specifications *consistently.* Several common rate-limiting effects that define the effective operational range for the film product are various flow instabilities (for wet coating), machine vibrations, drying load, radiation dosage, air entrainment, fluid delivery and pumping capacity (especially for viscous liquids), and deposition time (for vacuum deposition steps.). Thus, identifying the optimal line speed range for a given product that satisfies both cost and quality requirements is key to an effective operation of the R2R line (see Section 1.3 for further discussion on this point).

Another "coupler" for the R2R operation is web tension, which is controlled throughout the line by the winder, but it can be adjusted locally by idler "tension" rollers distributed along the line (e.g., see Figures 1.1 and 1.3). Nonoptimal or nonuniform tension can give rise to various defects, such as wrinkling, coating lines, coated layer thickness nonuniformity, and deformation in the substrate and deposited layers, so it must be carefully monitored and controlled [43]. As the main object of the R2R operation is to produce defect-free films with uniform functional layers and optimal performance, in-line monitoring and control of certain key attributes is often necessary to reduce waste, improve yields, and lower manufacturing cost (see Section 1.3). General web handling considerations for R2R operations are discussed elsewhere in this volume [43].

The processed substrate is finally wound up into a master roll in the winder station on the output end of the R2R machine. The winder is also the driver pulling the web through the machine at the prescribed line speed and tension. The master roll is the final product roll prepared by the R2R operation. When it exceeds a certain size, it must be removed from the winder and "finished" in a separate operation to produce the final product. Finishing typically entails cutting the processed web down to the final product dimensions (singulation), which are often different from the full web dimensions. Other possible finishing steps include polishing, cleaning, quality inspection, packaging, and shipping. Finishing, which can add significantly to manufacturing cost (see Section 1.3), can be performed on-site in close proximity to the R2R machine or off-site closer to the end customer, depending on cost and logistic considerations.

It is common, for logistic reasons, to store the fully processed master roll in a wound (rolled) state for some time before finishing and shipping to customers. Also, if the product is finished off-site, product master rolls are commonly shipped to the finishing site before unwinding and cutting. A prolonged rolled storage, however, can impact product quality in three possible ways:

1) *Curl*: The final product could acquire some curl (curvature) and become non-planar [44].
2) *Core damage*: High winding pressures and stresses within the wound roll could damage the functional layers and produce a defective product if it is wound up under excessively high tension.
3) *Blocking*: The top layer (functional layer side of the film structure) could interact with or adhere to the backing layer on the bottom side of the film when they come in contact during rolled storage, thus damaging the final film product.

It is well known, based on the so-called core-set curl effect [44], that polymeric films tend to take up some curl and deviate from a desired planarity after being stored for some time in a wound (rolled) state. The extent of curling depends primarily on the winding diameter (related to core size), thickness of the film, storage time and conditions, and the mechanical and viscoelastic properties of the web materials and particularly the substrate, which is typically the thickest layer in the film structure. The effect can be significant under some conditions rendering the final product defective in some application areas. Similarly, storing the processed roll under finite tension can produce high compressive stresses within the wound roll [45–47] that could damage the functional layers and negatively impact the performance of the final product. These stresses and the resultant damage are highest near the core hence the term *core damage*. One possible remedy is to minimize winding tension and rolled storage time. However, too low winding tension could make the roll "unstable," leading to "telescoping" (outward sliding of wound layers near the core area) and other problems, thus requiring a judicious choice of winding tension without compromising the quality of the final product. It is also desirable to minimize rolled storage time although it is often dictated by supply chain and logistics considerations. The problem of *blocking*, commonly associated with rolled storage, is not dependent on winding pressure *per se* but can be exacerbated by high winding pressures. This problem is typically resolved either by reformulating one of the layers in direct contact during winding or by adding an inert interleaving layer to physically separate the top and backing layers during rolled storage. All these effects and the corresponding quality and cost implications need to be carefully considered when selecting storage options for a product produced in an R2R operation.

## 1.3 Process Economics

There are two main reasons for the attractiveness and widespread use of R2R processes throughout many segments of the economy, compared with corresponding batch or sheet-to-sheet processes. First, such processes can readily accommodate high-volume, high-throughput operations, thus making them well suited for mass production of 2D film-type products. Second, R2R processes are highly cost effective by virtue of their high throughput and because all process steps in such operations are co-located in a single site and often conducted by a single production machine and fewer operators, unlike batch operations in which all or most process steps are performed independently from one another. The higher throughputs, simplified logistics and operation, and lower capital assets ultimately lead to lower manufacturing cost.

In order to examine the cost structure of the R2R process, it is necessary to define its daily production output, $Q_{R2R}$ (usually expressed in m$^2$/day):

$$Q_{R2R} = Y \cdot S \cdot W \cdot t \tag{1.1}$$

where $Y$ is the yield fraction (<1) for the R2R process, $S$ is the line speed, $W$ is the effective ("coatable") web width that directly relates to machine size and capacity, and $t$ is the effective daily run time (hr/day) for the R2R machine. Overall manufacturing cost, $C_{R2R}$ ($/day), can be expressed by

$$C_{R2R} = C_{AU} \cdot t + C_F + C_M \cdot Q_{R2R} \tag{1.2}$$

$C_{AU}$ is the asset utilization cost (cost to run the R2R machine per hour), while $C_F$ is a fixed daily cost (e.g., asset depreciation, maintenance, utilities, clean room operation, etc.). $C_M$ represents the cost of materials and other consumables per square meter of final product. Thus, the unit manufacturing cost for the R2R process, $UMC_{R2R}$ ($/m$^2$), of the film product takes the form

$$UMC_{R2R} = \frac{C_{R2R}}{Q_{R2R}} = \frac{C_{AU} \cdot t + C_F}{Q_{R2R}} + C_M = \frac{1}{Y \cdot S \cdot W}\left(C_{AU} + \frac{C_F}{t}\right) + C_M \tag{1.3}$$

From inspection of Eq. (1.3), the three key *operational* parameters that directly contribute to cost are yield ($Y$), line speed ($S$), and effective web width ($W$); based on the simple relationship of Eq. (1.3), all three need to be maximized in order to lower $UMC_{R2R}$. However, these variables are generally interdependent—in particular, the yield function depends in some way on line speed and machine width, $Y = f(S, W)$; controlling product quality and uniformity is often more challenging at higher speeds and for wider webs. The effect of line speed on yield and $UMC_{R2R}$ is illustrated schematically in Figure 1.4, where an optimal operational range—the "process window"—is shown. Thus, a key challenge for process engineers is to maximize line speed while maintaining

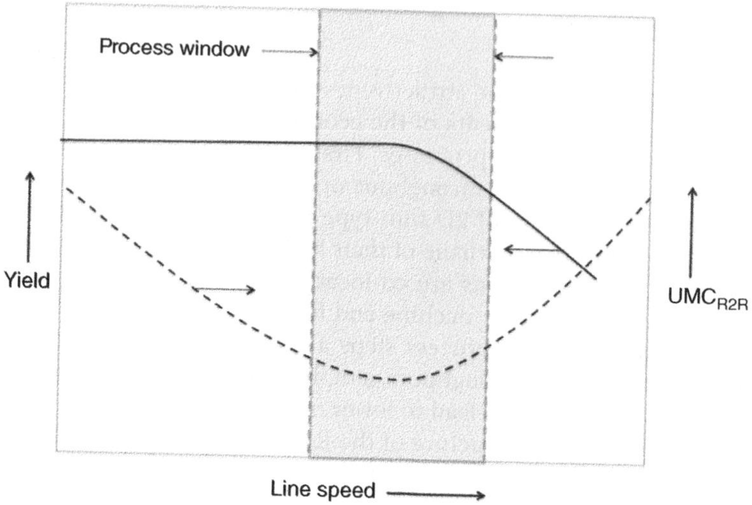


Figure 1.4 Schematic definition of "process window" for a roll-to-roll operation.

acceptable yields and, specifically, identify the "process window" or the effective operational range for a given set of product specifications. Increasing the width of the web (~machine size) is often less practical or feasible because the machine size is tied up to a capital asset, and increasing the size of the machine could lead to high capital expenditure. Aside from the need to acquire a larger and costlier machine, it might also require relocating the manufacturing operation to a larger and more expensive physical structure. Also, controlling product quality and uniformity across wider webs is usually more difficult. The "process window" is generally defined as the actual range for *all* key operational (independent) process variables (e.g., S, temperature, UV dosage, line tension, and the like) that correspond to the highest yields, lowest $UMC_{R2R}$, and, overall, to the most robust operation; the larger the "window," the more robust the operation. It is finally noted that the cost advantage of an R2R operation over a corresponding batch process can be understood in a very simplistic way by assigning a very low effective line speed to the batch process, which would drive the UMC higher based on Eq. (1.3). A batch process could also have higher fixed and asset utilization costs.

If a particular product requires multiple passes through the R2R machine because of large differences in process conditions (especially line speed, environment, line tension, etc.) for the different converting steps, the final UMC for the R2R process becomes

$$UMC_{R2R} = \sum_{i=1}^{n} UMC_{R2R,i} \tag{1.4}$$

where $UMC_{R2R,i}$ is the unit manufacturing cost for the $i$th pass and $n$ is the total number of passes for the entire R2R operation. Each pass has a different yield ($Y_i$), line speed ($S_i$), and material cost ($C_{M,i}$) associated with it, but all other parameters ($W$, $C_{AU}$, $C_F$, $t$) are expected to remain the same for all passes if the same R2R machine is utilized. Thus, by implication, each additional pass adds to the final manufacturing cost associated with the R2R process, and limiting the number of passes should be generally cost advantaged and therefore highly desirable.

The final unit manufacturing cost of the finished product must include not only the cost of the R2R process but also the corresponding finishing costs. Indeed, the total unit manufacturing cost of the final finished product can be written as

$$UMC_p = UMC_{R2R} \cdot A_p + UMC_f \tag{1.5}$$

where $A_p$ is the actual area of the finished (cut) product and $UMC_f$ expresses the total cost of finishing per final cut product. This quantity depends on the particular finishing steps and conditions applied, and it contributes independently to yield losses (1-$Y_f$), which must be combined with yield losses associated with the R2R process (1-$Y$). Indeed, key contributions to manufacturing yield losses come from three main sources: (i) process (R2R)-induced defects (1-$Y$), (ii) finishing losses (1-$Y_f$), and (iii) defective raw materials. Process-induced defects can occur when the R2R process is operated outside its optimal range. For example, as discussed in Section 1.2, excessively high line speeds could induce a variety of disturbances such as mechanical vibrations, air entrainment, flow instabilities, insufficient drying or curing, and so on, leading to a defective product (see Figure 1.4). Likewise, out-of-range, spatially variable temperature or UV dosage and nonoptimal line tension could also induce a variety of defects in the final product. It is, therefore, important to identify optimal ranges, which define the "process window" (e.g., see Figure 1.4), for the key process variables to ensure a robust, defect-free, and cost-effective operation. Thus, continuous monitoring of the key process variables is essential for controlling the process and operating within its optimal range. Monitoring, however, must be conducted *along and across* the full web to ensure that the monitored process variable is uniform throughout the moving web. Variability along the moving web (machine direction (*MD*)) is often associated with different process disturbances compared with variations across the web (transverse direction (*TD*[2])), so changes along both principal directions must be fully monitored and established in order to identify the underlying root cause for the observed variability and/or defect. Variability along or across the web could lead to spatial nonuniformity of

2 In some industries the term cross direction (CD) is used instead of TD.

the film product and corresponding yield losses. Process control requires also full or partial monitoring of key product attributes, which can be done online, that is, directly on the moving web, or offline, by removing some select sections of the processed web and testing independently from the R2R machine. Online monitoring has many advantages, including speed, more complete spatial resolution, and better integration with efficient open- or closed-loop process control strategies. But such monitoring is usually more capital intensive and not always easy to implement.

A common metric used to define process robustness is the process capability index [48, 49]. For a two-sided specification for a key *performance variable* (a dependent variable directly tied to the performance of the final product), this parameter is defined as

$$CI_p = \frac{USL - LSL}{6\sigma} \tag{1.6}$$

where USL is the upper specification limit, LSL is the lower specification limit, and $\sigma$ is the standard deviation or variability of the performance variable. For one-sided specification, the process capability index takes the form

$$CI_{p,upper} = \frac{USL - \mu}{3\sigma} \quad \text{For upper specification limit only} \tag{1.7a}$$

$$CI_{p,lower} = \frac{\mu - LSL}{3\sigma} \quad \text{For lower specification limit only} \tag{1.7b}$$

where $\mu$ is the mean value of the performance variable. If the process mean is not centered between USL and LSL for a two-sided specification, the process capability index of Eq. (1.6) is redefined as follows:

$$CI_{pk} = \min\left(\frac{USL - \mu}{3\sigma}, \frac{\mu - LSL}{3\sigma}\right) \tag{1.8}$$

It is generally recommended that the process capability index falls in the range 1–2 (corresponding to $3\sigma$–$6\sigma$ operation) to assure a well-controlled, robust operation, but the actual target for $CI_p$ should depend on the particular product requirements and cost/yield tolerance.

Losses and waste from finishing steps can also contribute to a lower overall yield and higher manufacturing cost. As noted in Section 1.2, the final processed web, wound up into a roll by the winder, has to be "finished," that is, cut down to the final product dimensions, inspected, packaged, and shipped out to customers. Finishing may also entail other steps such as polishing, cleaning, removal of interleaving layer, and so on, but most of the finishing losses are usually related to the cutting step. There are three main sources of waste associated with cutting:

1) Diagramming losses—only a portion of the linear processed web can be cut into product (e.g., as in the case of a round final product that must be cut from a rectilinear web) and the leftover portion is waste.
2) Edge effects—edges are usually discarded because product quality falls off near the edges of the web (This is actually a R2R process control issue.)
3) Cutting damage—the cutting step, which often involves high stresses and strains [50], can damage some sections of the product, usually near the cut edges, and render the cut product defective. Damage can also be incurred by debris, generated during the cutting step, adhering to the cut film product. Minimizing cutting damage requires selection of appropriate cutting methodology and tools for the product at hand. Many cutting methodologies are known [50, 51], and these should be properly matched to the type of materials being cut, the required cutting precision, the available cutting infrastructure, and the economics of the process.

Another category of finishing losses corresponds to defects associated with rolled storage. As discussed in Section 1.2, rolled storage can induce three types of defects in the final product: high curl and deviations from planarity due to core-set curl [44], core damage associated with high winding pressures [45], and blocking, that is, interaction between the bottom and top layers in the film structure during rolled storage. The potential for yield losses during rolled storage requires careful consideration of storage options during the post-process finishing phase.

The final contribution to yield losses comes from defective raw materials. If any of the raw materials used in the R2R operation (substrate, functional layer components, ancillary layer components, etc.) is defective in some way, or if the materials become defective during storage (e.g., due to inadequate shelf life) before or after R2R processing, they will produce a defective product. These losses can be minimized by close inspection of the incoming materials, proper storage and containment of the materials, and close monitoring of key material properties *before* R2R processing.

## 1.4 Environmental, Health, and Safety Considerations

The cost and overall layout of R2R operations are often dictated by health, safety, environmental, and regulatory considerations. Two elements of R2R operations that are particularly susceptible to environmental challenges are solvent coating and drying steps and some chemical vapor deposition steps involving hazardous gases or plasma. If the solvents used in wet coating are hazardous in liquid or vapor form, special steps must be taken to contain them in their storage and delivery systems and during solvent removal in the drying step to keep emission levels within the mandated limit. In addition, if the solvent vapor poses a potential fire or explosion hazard, its so-called low explosion limit (LEL) must not be exceeded during any steps of the R2R operation, and regulatory mandates must be strictly adhered to [52].

It is often desired to use volatile, low boiling point solvents in wet coating operations because of the ease with which such solvents can be removed from the coated web, which, in turn, corresponds to lower operational and capital costs. The drying step often requires long web stations (large footprint) and relatively high temperatures to ensure effective solvent removal [33, 38] (e.g., see Figure 1.1). The size/length of the drying station and the drying temperatures can be reduced substantially, and/or line speed can be increased with increase in solvent volatility (decrease in boiling point), thus decreasing the overall operational and capital costs. However, such solvents are often environmentally hazardous and more difficult to fully capture in the drying station, which results in higher emissions, so the operational and capital benefits of utilizing volatile organic solvents must be weighed against the potential environmental liabilities and the cost of solvent and vapor containment infrastructure mandated by environmental regulations. Aqueous systems are naturally preferred from an environmental standpoint, but such formulations are not always feasible in some application areas, and removal of moisture by drying can be quite challenging and costly [38].

Another commonly employed approach to get around potential solvent hazards is to replace the solvent-based formulation with a *solvent-less* formulation, such as a radiation-curable (or temperature-curable) resin system, whereby the coated resin is solidified (cured) by exposure to a suitable radiation at a desired dosage [39, 40], which eliminates the need to remove any residual solvent or reactant from the coated web. This approach eliminates the use of solvents and drying steps although various types of radiation could also carry some potential health risks of their own. Indeed, other potential environmental hazards in R2R operations include UV radiation (in a UV curing station), fire resulting from static electricity (when a dry web is rubbing against coated rollers), leaks of hazardous liquids or gases when stored or pumped through long delivery lines, bodily injury when working in close proximity to a fast moving web, and so on. All these potential hazards must be given full consideration when designing a safe and cost-effective R2R operation. For example, fire hazard due to static electricity can be minimized by adding a suitable antistatic layer [25] to the final film structure that can effectively dissipate static charges. Similarly, exposure to UV radiation can be eliminated by using suitable protective gear and by designing the UV station with proper radiation shields.

## 1.5 Summary

R2R operations are widely used in many segments of the economy covering the gamut from commoditized and well-established product categories such as paper consumables, printing, silver halide photography, and packaging films to many new and emerging application areas such as display, photovoltaics, microelectronics, and microfluidics. In all cases a flexible film-type product with a uniform or patterned 2D structure is processed by conveying a web between two or more rotating rollers at a finite line speed while applying various functional and ancillary layers onto a moving substrate. The main attraction of the R2R approach over corresponding batch or sheet-to-sheet processes stems from its low manufacturing cost, simplified logistics, and easy adaptability to mass production operations. Generally, low cost manufacturing requires minimum number of passes through the R2R production line and operating at the highest possible line speed and widest web while meeting product specifications. Effective design of R2R production lines must take into consideration capital and operational costs as well as health, safety, and environmental factors mandated by current regulations, and make them easily adaptable to multiple product iterations and categories. More in-depth discussions of some key aspects and common steps of R2R operations are given in Chapters 2–7 in this volume, followed by reviews of several new application areas and novel extensions of the traditional R2R process in Chapters 8–12.

## References

1 Willmann, J.; Stocker, D.; Dorsam, E. *Org. Electron.* **2014**, 15(7), 1631–1640.
2 Pastorelli, F.; Schmidt, T. M.; Hosel, M.; Søndergaard, R. R.; Jørgensen, M.; Krebs, F. C. *Adv. Eng. Mater.* **2016**, 18(1), 51–55.
3 Yeo, J.; Kim, G.; Hong, S.; Kim, M. S.; Kim, D.; Lee, J.; Lee, H. B.; Kwon, J.; Suh, Y. D.; Kang, H. W.; Sung, H. J.; Choi, J. H.; Hong, W. H.; Ko, J. M.; Lee, S. H.; Choa, S. H.; Ko, S. H. *J. Power Sources* **2014**, 246, 562–568.
4 Hassinen, T.; Ruotsalainen, T.; Laakso, P.; Penttilä, R.; Sandberg, H. G. O. *Thin Solid Films* **2014**, 571(P1), 212–217.
5 Nishide, H.; Oyaizu, K. *Science* **2008**, 319(5864), 737–738.
6 Hu, L.; Wu, H.; La Mantia, F.; Yang, Y.; Cui, Y. *ACS Nano* **2010**, 4(10), 5843–5848.
7 Sondergaard, R.; Hosel, M.; Angmo, D.; Larsen-Olsen, T.; Krebs, F. C. *Mater. Today* **2012**, 15(1–2), 36–49.
8 Galagan, Y. In *Roll-to-Roll Manufacturing: Process Elements and Recent Advances*; Greener, J., Pearson, G. H., Cakmak, M., Eds.; John Wiley & Sons, Inc.: New York, **2017**. Chapter 11.

9 Hu, X.; Chen, L.; Zhang, Y.; Hu, Q.; Yang, J.; Chen, Y. *Chem. Mater.* **2014**, 26(21), 6293–6302.

10 Hwang, K.; Jung, Y. S.; Heo, Y. J.; Scholes, F. H.; Watkins, S. E.; Subbiah, J.; Jones, D. J.; Kim, D. Y.; Vak, D. *Adv. Mater.* **2015**, 27(7), 1241–1247.

11 Gregg, A.; York, L.; Strnad, M. In *Flexible Flat Panel Displays*; Crawford, G. P., Ed.; John Wiley & Sons, Inc.: West Sussex, **2005**; pp 409–462.

12 Montbach, E.; Davis, D. In *Roll-to-Roll Manufacturing: Process Elements and Recent Advances*; Greener, J., Pearson, G. H., Cakmak, M., Eds.; John Wiley & Sons, Inc.: New York, **2017**. Chapter 10.

13 Yeh, Y.-H.; Cheng, C.-C.; Lai, B. C.-M.; Leu, C.-M.; Tseng, Y.-L. *J. Soc. Inf. Disp.* **2013**, 21(1), 34–40.

14 Deng, Y.; Yi, P.; Peng, L.; Lai, X.; Lin, Z. *J. Micromech. Microeng.* **2015**, 25(6), 065004.

15 Wang, M.-W.; Tseng, C.-C. *Opt. Express* **2009**, 17(6), 4718–4725.

16 Whitesides, G. M. *Nature* **2006**, 442, 368–383.

17 Ahn, S. H.; Guo, L. J. *ACS Nano* **2009**, 3(8), 2304–2310.

18 Lim, H.; Choi, K. B.; Kim, G.; Lee, S.; Park, H.; Ryu, J.; Jung, S.; Lee, J. *Microelectron. Eng.* **2014**, 123, 18–22.

19 Hosel, M.; Sondergaard, R. R.; Jorgensen, M.; Krebs, F. C. *Energy Technol.* **2013**, 1(1), 102–107.

20 Sutanto, E.; Alleyne, A. *Mechatronics* **2015**, 31, 243–254.

21 Roth, B.; Søndergaard, R. R.; Krebs, F. C. In *Handbook of Flexible Organic Electronics: Materials, Manufacturing and Applications*; Elsevier: Amsterdam, **2014**.

22 Zhang, Z.; Qiu, J.; Wang, S. *Manuf. Lett.* **2016**, 8, 6–10.

23 Cheng, J. Y.; Mayes, A. M.; Ross, C. A. *Nat. Mater.* **2004**, 3, 823–828.

24 Rimai, D. S., DeMejo, L. P., Mittal, K. L., Eds. *Fundamentals of Adhesion and Interfaces*; VSP BV: Utrecht, **1995**.

25 Jonas, F.; Morrison, J. T. *Synth. Met.* **1997**, 85(1–3), 1397–1398.

26 Holleck, H. *J. Vac. Sci. Technol. A* **1986**, 4(6), 2661.

27 Yalcin, B.; Cakmak, M. In *Roll-to-Roll Manufacturing: Process Elements and Recent Advances*; Greener, J., Pearson, G. H., Cakmak, M., Eds.; John Wiley & Sons, Inc.: New York, **2017**. Chapter 6.

28 Brackley, D.; Marshall, D.; Merz, G.; Miller, E. In *Roll-to-Roll Manufacturing: Process Elements and Recent Advances*; Greener, J., Pearson, G. H., Cakmak, M., Eds.; John Wiley & Sons, Inc.: New York, **2017**. Chapter 8.

29 Meggs, P. B. *A History of Graphic Design*, 3rd ed.; John Wiley & Sons, Inc.: New York, **1998**.

30 Brayer, E. *George Eastman: A Biography*; University of Rochester Press: Rochester, NY, **2006**.

31 Goodwin, H. Photographic pellicle and process of producing same. US 610861, **1887**.

32 Fleischer, C.; Bauer, C. L.; Massa, D. J.; Taylor, J. F. *MRS Bull.* **1996**, 21(7), 14–19.

33 Lightfoot, E. J.; Cohen, E. D. In *Roll-to-Roll Manufacturing: Process Elements and Recent Advances*; Greener, J., Pearson, G. H., Cakmak, M., Eds.; John Wiley & Sons, Inc.: New York, **2017**. Chapter 2.

34 Cohen, E. D.; Gutoff, E. B. *Modern Coating and Drying Technology*; Wiley-VCH: New York, **1992**.

35 Bishop, C. A. In *Roll-to-Roll Manufacturing: Process Elements and Recent Advances*; Greener, J., Pearson, G. H., Cakmak, M., Eds.; John Wiley & Sons, Inc.: New York, **2017**. Chapter 4.

36 Kobayashi, T.; Bando, M.; Kimura, N.; Shimizu, K.; Kadono, K.; Umezu, N.; Miyahara, K.; Hayazaki, S.; Nagai, S.; Mizuguchi, Y.; Murakami, Y.; Hobara, D. *Appl. Phys. Lett.* **2013**, 102(2), 023112.

37 Hosel, M.; Sondergaard, R. R.; Angmo, D.; Krebs, F. C. *Adv. Eng. Mater.* **2013**, 15(10), 995–1001.

38 Shams Es-Haghi, S.; Cakmak, M. In *Roll-to-Roll Manufacturing: Process Elements and Recent Advances*; Greener, J., Pearson, G. H., Cakmak, M., Eds.; John Wiley & Sons, Inc.: New York, **2017**. Chapter 3.

39 Decker, C. *Polym. Int.* **1998**, 45(2), 133–141.

40 Drobny, J. G. *Radiation Technology for Polymers*, 2nd ed.; CRC Press, Taylor & Francis Group, LLC: Boca Raton, **2010**.

41 Fouassier, J. P.; Rabek, J. F. *Radiation Curing in Polymer Science and Technology: Practical Aspects and Applications (Vol. IV)*; Elsevier Applied Science: London and New York, **1993**.

42 Chang, J. S.; Lawless, P. A.; Yamamoto, T. *IEEE Trans. Plasma Sci.* **1991**, 19(6), 1152–1166.

43 Roisum, D. R.; Guzman, G.; Shams Es-Haghi, S. In *Roll-to-Roll Manufacturing: Process Elements and Recent Advances*; Greener, J., Pearson, G. H., Cakmak, M., Eds.; John Wiley & Sons, Inc.: New York, **2017**. Chapter 5.

44 Greener, J. In *Roll-to-Roll Manufacturing: Process Elements and Recent Advances*; Greener, J., Pearson, G. H., Cakmak, M., Eds.; John Wiley & Sons, Inc.: New York, **2017**. Chapter 7.

45 Yagoda, H. P. *J. Appl. Mech.* **1980**, 47(4), 847–854.

46 Good, J. K.; Wu, Z.; Fikes, M. W. R. *J. Appl. Mech.* **1994**, 61(1), 182–185.

47 Koc, H.; Knittel, D.; de Mathelin, M.; Abba, G. *IEEE Trans. Control* **2002**, 10(2), 197–208.

48 Kotz, S.; Johnson, N. L. *Process Capability Indices*; Chapman & Hall: London, **1993**.

49 Henderson, G. R. *Six Sigma Quality Improvement with Minitab*; John Wiley & Sons, Ltd: Chichester, **2011**.

50 Hertzberg, R. W.; Vinci, R. P.; Hertzberg, J. L. *Deformation and Fracture Mechanics of Engineering Materials*, 5th ed.; John Wiley & Sons, Inc.: New York, **2013**.

51 Meehan, R. R.; Burns, S. J. Exp. Mechanics **1998**, 38(2), 103–109.

52 Flammability Limit https://en.wikipedia.org/wiki/Flammability_limit (Accessed on October 19, 2017).

# 2

# Coating and Solidification

*E. J. Lightfoot*[1] *and E. D. Cohen*[2]

[1] *DuPont Photovoltaic and Advanced Materials, Yerkes Research Laboratory, Buffalo, NY, USA*
[2] *Edward D. Cohen Consulting, Inc, Fountain Hills, AZ, USA*

Coating technology is well over a century old; however, the development of modern coating technology has largely occurred over the past 50 years. This includes the evolution of high speed coaters for commodity applications, flexible coaters for niche applications, simultaneous multilayer coaters for creating unique structures, high precision multiple-zone dryers, and contamination-free coaters for demanding applications. Fundamental advances in understanding of coating, drying and defect formation mechanisms, coating hardware design and fabrication, process support equipment, online quality coating weight measurement, and surface inspection systems have led to drastic improvements in coating quality and productivity over the past decade. This chapter covers the coating, drying, and curing processes along with defect management.

## 2.1 Coating Methods

There are roughly thirty coating methods in use with thousands of variations on these [1–11]. Only the techniques most commonly used in the converting and advanced materials industries are discussed here (slot die, extrusion, reverse roll, Mayer rod, and gravure).

The coating applicator distributes the coating across the sheet, ensures that the liquid wets the web passing under the coating head, and controls the thickness of each layer of coating. Some coating methods apply a single layer at a time; other methods apply multiple layers simultaneously. Detailed discussions of the fundamental fluid mechanics of coating flows are available elsewhere [6].

*Roll-to-Roll Manufacturing: Process Elements and Recent Advances*, First Edition.
Edited by Jehuda Greener, Glen Pearson, and Miko Cakmak.

However, some knowledge of the fundamentals of coating is necessary to understand how to choose coating methods; therefore, the fundamentals are covered first, hardware second, and selection criteria last.

### 2.1.1 Coating Fundamentals

#### 2.1.1.1 Wetting

The first requirement for coating is that the liquid coating must wet the substrate at the interface. Static wetting is well understood—a drop placed on a surface will either spread over the surface to its maximum extent or assume a definite contact angle. The contact angle is defined as shown in Figure 2.1 with the convention that if the contact angle is zero, the liquid completely wets the substrate; if the contact angle approaches 180°, the liquid balls up on the surface.

If the surface tension of the liquid is lower than the surface energy of the solid, the liquid wets the solid; if the surface tension is greater than the surface energy of the solid, the liquid will retract to give a measurable contact angle. Surface tensions of liquids are generally measured with a Dunouy ring or Wilhelmy plate. The surface energy of the solid is more difficult to determine. The Zisman plot graphs the cosine of the contact angle against the surface tension of various liquids. The critical surface tension is found by extrapolating to $\mathrm{Cos}(\theta) = 1$ (perfect wetting). The most common practice in industry is the use of "dyne pens"—markers containing test solutions specified in ASTM D2578-04a. ASTM D7490-13 breaks the surface energy into polar and non-polar components (numerous correlations have been proposed to make this separation [12]).

One significant complication to static wetting is hysteresis. If the substrate is tilted and the drop starts to move across the surface, the advancing contact angle is different from the receding contact angle. This indicates the surface is not homogeneous—there are areas on the surface that the liquid would like to wet and others it would prefer not to wet. The areas that do not wet result in a high advancing contact angle; those that do not de-wet result in a low receding contact angle. This hysteresis brings up the most significant issue: coating is a dynamic process and it requires dynamic not static wetting.

Dynamic wetting is a complex phenomenon that remains a field of active research; thus, it is beyond the scope of this overview to discuss it in detail.

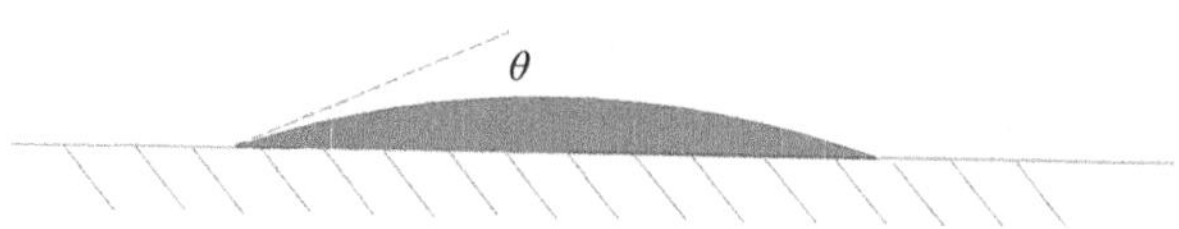


**Figure 2.1** Definition of the contact angle.

The difficulty in understanding dynamic wetting is increased by the complexity of technological approaches used to enhance wetting (e.g., the use of surfactants and cosolvents in formulations, substrate pretreatment, electrostatic assist, and fluid mechanical assists such as the inertial assist in curtain coating). Because the maximum speed of dynamic wetting scales with the inverse of the viscosity of the liquid, dynamic wetting is the fundamental limit on coating speed. Several classes of theories of dynamic wetting have been proposed. Snoeijer [13] and Bonn [14] have given recent reviews of the literature of dynamic wetting. Blake [15] compares the success of the kinetic–hydrodynamic theories, theories that combine hydrodynamic and molecular approaches, and the nonequilibrium thermodynamic/hydrodynamic model of Shikhmurzaef. All of these models have been used successfully in some cases, and Blake gives examples of each model providing a better fit to data than the others.

The basic mechanism of dynamic wetting failure was first elucidated by Deryagin and Levi [16]. As the moving web passes through by the coating head, the dynamic contact angle is increased as the liquid is pulled downstream. If the dynamic contact angle reaches 180°, the meniscus is no longer capable of supporting a force to exclude air from being entrained under the coating and bubbles form. However, air entrainment is preceded by a corrugation of the dynamic contact line. Although many patterns are possible, the famous patterns are triangular pockets of air that may shed small bubbles from their tips [17]. These bubbles may collapse from conjoining pressure [18] and may remain visible in a microscope [19].

The industrial approach to dynamic wetting remains largely empirical and proprietary. One of the more interesting empirical observations comes from Blake and Morley [20]: "It is well known that coating speeds are limited due to the onset of air entrainment. These limiting coating speeds have been found to be highest on solid surfaces of intermediate wettability." Common practice in the converting industry is to ignore the difference between static and dynamic wetting, and various rules of thumb are cited such as keeping the surface tension of the liquid five to six dynes per centimeter below that of the solid. Such heuristics have the advantage of simplicity and offer closure on a subject of discomfort to many. Unfortunately, they provide little guidance in the optimization of coating speed.

Electrostatic assist uses an electric field to attract the liquid coating to the substrate. This is common with gravure printing (and coating) in which it helps empty the cells [21]; it is less commonly used in slot die and slide coating to attract the coating bead to the surface. Curtain coating uses the momentum of a falling curtain to exclude the air from the surface ("hydrodynamic assist") and produces the highest coating speed on continuous webs such as film and foil. Because air can escape through porous substrates like ordinary paper and nonwovens, coating speeds on these substrates are higher than on film by a factor of two or more.

#### 2.1.1.2 Coating Distribution

The second requirement for successful coating is that the coating applicator must be capable of spreading the liquid across the substrate. The third requirement is that the thickness of each layer must be controlled to target.

It is helpful to divide coating methods into three classes: pre-metered, self-metered, and doctored coating. In pre-metered coating, the liquid is pumped at a controlled rate through a cavity with a narrow exit slot that distributes the coating across the sheet. In self-metered coating, the properties of the fluid itself determines the thickness (cross sheet distribution is usually accomplished in a bath). With doctored coating, an excess of coating is spread across the surface, and the excess is "doctored" off by mechanical means.

The greatest challenge in coating is producing uniform thickness. Coating flows tend to break into alternating thin and thick streaks called ribbing or—in the extreme—to break into coated and uncoated streaks called rivulets. Linear stability analysis has been used to predict when a flow that is producing a smooth coating becomes unstable with respect to ribbing. However, the picture of stable and unstable regions in the coating window can be misleading. There are also regions of the coating window in some processes in which either ribbing or uniform coating is possible. An example would be a die coater that produces stable ribbing flow, but, if a shim is drawn across the slot by a skilled operator, can be converted to a uniform flow that also remains stable. This indicates a physical bifurcation in operating states, not an instability (in this case, the two states involve the meniscus being pinned at different points on the coating head).

Machine direction variations in coating weight are different from the transverse direction variations. There are always disturbances that translate into machine direction variations present from building vibrations, variations in web speed and pumping rate as the web passes through the coating head. The coating process will have some sensitivity to these disturbances. Periodic disturbances may be attenuated or grow to manifest themselves as periodic flows such as the barring instability seen in bead coating at low speeds.

Each coating techniques is capable of producing uniform coatings under certain conditions, yet each is prone to certain defects that are characteristic of that technique. While some defects can be virtually eliminated, the endemic defects remain present on the Pareto list of defects. For example, gravure coatings tends to produce patterns that replicate the engraving pattern on the gravure roll; slot die coating is prone to streaks caused by particles lodged in the slot.

#### 2.1.1.3 The Coating Window

The region on the operability diagram that gives uniform coating is called the coating window. Finding a technique that is in a robust portion of its coating window is a primary goal in selecting a coating technique.

There is always a maximum speed above which dynamic wetting failure occurs. A given technique may be limited by coating defects before it encounters wetting failure under some conditions; however, in most cases, the highest operable speed under any condition will be limited by dynamic wetting of the substrate. The most common fluid mechanical limit is ribbing (or rivulets) with many other defects being specific to a given method. A third common limit to coating speed is drying. In solvent coating the drying limit is often the safety limit of not producing an explosive gas mixture in the oven or the maximum rate of solvent removal by pollution abatement devices. Limitations on the residual solvent level are common in solvent as well as aqueous coating.

The term "window" conjures up images of a two-dimensional rectangle. This makes it hard to resist assigning simple attributes to a coating technique such as "slot die coating works up to 1000 fpm." Unfortunately, the coating window is a volume in a multidimensional space that is generally far from rectilinear. Slot die coatings have been made at speeds over 1000 fpm; however, most coating formulations cannot be applied to most substrates at anything like that speed with a generic setup.

It is convenient to measure the coating window in simple experiments in two physical variables at a time. Because of the importance of productivity, coating speed is almost always one of the key variables. The other tends to be a convenient process variable such as vacuum setting in slot die coating, roll speeds in gravure, and reverse roll coating. It is important to use a mapping technique that allows for discontinuous variations in the process response and hysteresis.

The general procedure for mapping a coating window is to begin with the initial formulation candidate, set up the coater to run in a nominally operable speed, and then vary the other process setting to define the range of stable operation. For example, a slot die coater may exhibit ribbing at low vacuum and weeping into the vacuum box at high vacuum; thus, a window of vacuum versus line speed may be mapped as follows. Lower the vacuum until ribbing occurs, and then raise it until ribbing disappears. In this case, the coating may stay stable to a significantly lower vacuum, and then it will change back to a uniform coating. In such cases one must exercise judgment to decide how much, if any, intervention (e.g., "shimming the gap") to use in reestablishing stable flow. Having mapped the span of the low vacuum limit, the vacuum is then increased until weeping occurs, and a similar procedure is used to determine the vacuum latitude (i.e., the limits for increasing and decreasing vacuum). This procedure is repeated at various line speeds until the useful section of the coating window has been mapped.

Statistically designed experiments (often called DOEs) are widely used in industry. Measuring the coating window is one case where the normal statistical designs must be used with great care. DOEs are appropriate for measuring continuous state variables (variables that depend on the settings, not how one got to those settings). In the example given earlier, a given vacuum might produce either "good" or "bad" coating depending on how one got to that vacuum level. Setting up a regular array of points run in random order can produce irreproducible but not effectively randomized results and will not locate the edge of the coating window with any precision.

The coating window defines the region in the operating window that produces more or less uniform and stable coating; however, that does not mean that all points in the coating window give equally attractive coatings. Coating quality is usually not constant throughout the operable region. For example, some portions of the coating window may be more or less sensitive to vibrations or speed fluctuations in the web transport system. For pre-metered coatings, every point in the window gives the same coating weight; for self-metered and doctored coating methods, the coating weight varies within the window. Standard statistical designs produce smooth fits (typically quadratic) to a response surface that can be very effective for mapping the coating weight and quality with the caveat that all states in the DOE must fall within the coating window.

The art of managing coating windows is not in generating the initial window but in determining how the shape of the coating window changes with formulation (i.e., physical properties) and other process settings. For example, the generalized coating windows for slide coating should be described in terms of dimensionless bead height, capillary number, Weber number (or Reynolds numbers), Stokes number, dimensionless pressure, and dimensionless electrostatic force (for electrostatic assist) [22]. The coating window also varies with geometry (e.g., slide angle), the number of layers, ratio of layer thicknesses, and ratios of viscosity from layer to layer. Clearly such a window is too complex to be illustrated. However, it is often possible to correlate the onset of flow features using a subset of these numbers. For example, Schweizer [23] was able to correlate a critical capillary number for the onset of vortices in the region above the slide as a function of Reynolds number and a physical property number ($Po = Ca^3St/Re^2$).

### 2.1.2 Coating Hardware

#### 2.1.2.1 Pre-metered Coating

In pre-metered coating, the liquid is pumped to a die cavity that distributes the coating liquid across the web before it emerges from a slot and is applied to a web. The average coating weight is controlled by flow rate; all other degrees of freedom are used to optimize quality. Pre-metered coating with in-line injection is preferred for coating reactive systems. The most common pre-metered coating methods are slot die (for nominally liquid coating) and extrusion coating (for coating molten polymers).

#### 2.1.2.1.1 Slot Die Coating

Eastman Kodak [24] developed slot die coating. Slot dies can be constructed with multiple cavities to produce simultaneous multilayer coatings. There is no theoretical limit to the number of layers in slot die coating, but the usual practice is to construct the die with each cavity having its own exit slot that impinges directly onto the web. Geometric constraints make slot coaters with more than three layers rare.

There are a number of variations of slot die coaters in use today. A slot die pressed against a rubber roll is sometimes referred to as contact coating to distinguish it from a die deployed near a steel roll (with or without a vacuum box) that is sometimes called proximity die coating or proximity coating. Slot dies that drop the coating onto the web from a distance are a form of curtain coating. This has occasionally been called extrusion coating; however, that term can be confusing. Slot dies can also be deployed against a tensioned web in what is increasingly known as tensioned web over slot die (TWOSD) coaters. TWOSD and contact die coating can produce very thin (single- or multilayer) coatings.

Dies can be designed to produce uniform coating at one viscosity or optimized for a range of properties. Slot die coating can be one of the more robust methods for dealing with non-Newtonian effects (e.g., shear thinning coatings) if properly designed. Dies for general-purpose pilot lines are often designed with large cavities and relatively tight exit slots. Dies for wide applicability often use active control elements that deflect one lip of the die to control the coating weight profile. This maximizes the range of coatings they can process, but this may not be the optimal design for production. Once the characteristics of the final coating are known, cavities designed for a specific class of coating can offer additional features. For example, there are companies that specialize in dies that produce a minimum shear rate at the wall to avoid gel formation in unstable coatings. The operability of a slot die varies greatly with the die to web gap, the angle between the die face and the web, and changes in the die to web angle (e.g., alternately diverging and converging channels [25]).

#### 2.1.2.1.2 Extrusion Coating

Extrusion coating [1, 5] can be thought of as a variation on die curtain coating where the coating liquid is a polymer melted in an extruder (usually a single-screw extruder). The high viscosity of molten polymers precludes the use of edge guides, and the use of metering pumps is uncommon. Mixing in polymers is problematic; thus, in-line injection is not naturally paired with extrusion coating as it would be with the low viscosity die coating. Without edge guides, the flow in the curtain necks in and produces edge beads that are generally allowed to extend beyond the edge of the substrate, slit off, and—ideally—recycled. Start-up losses in extrusion coating can be significant; however, extrusion coating does not require drying and the productivity is high. Thus, if the demand for the coated product makes extrusion coating a viable option, it is generally the most economical choice. Multilayer extrusion coating is also possible although the common practice is to use co-extrusion technology rather than the multiple manifolds commonly found in slot die coating; however, both approaches are used. Seven-layer co-extrusions are not uncommon. There are reports of thousands of "micro-layers" being co-extruded to produce optical effects.

The feed to an extrusion coater is quite different from that to a low viscosity liquid coater. An extruder (normally a single screw) is used to melt the thermoplastic resin and produce the pressure needed to drive the melt through a screenpack (a filter made by various layers of wire mesh supported by a "breaker plate") and the die. The die is fairly typical of flat film extrusion. The melt exits the die and falls into the laminating nip (consisting of one rubber covered roll and one cooled steel roll called the chill roll). An auxiliary cooling roll is used to cool the rubber roll. If a single substrate is being coated, then it passes around the pressure roll to allow intimate contact between the melt and the chill roll. If lamination is occurring, then one substrate (the less thermally conductive substrate if they are different) passes around the pressure roll and the other passes around the chill roll. The melt is usually brought in contact with the substrate slightly before the nip point.

Die cavities are most commonly of a "T-bar" or coat hanger design. T-bar dies use a large cavity to minimize the pressure drop in the cavity and taper to a narrow slot that provides the controlling pressure drop. Although slots defined by constant height "lands" are common, there are vendors who specialize in producing variable height lands. Coat hanger cavities have a primary distribution chamber consisting of two tapered channels that are angled down and across the face of the die. These feed a triangular region (known as a preland) that is designed to equalize the resistance to flow so that the flow at the bottom of the preland is uniform. A secondary distribution channel (to equalize pressure and give more flexibility for changing resin or line speed) may be cut straight across the die face at the exit of the preland. The (primary) land is a section of the die face that is nominally flat and designed to create enough pressure drop to control the flow exiting the die. Most coating dies are of so-called clamshell construction: two mating sides (one with a flexible lip, the other a rigid lip) bolt together to form the die cavity. Dies are mostly electrically heated (hot oil is less common).

The die may contain one or two means of adjusting the coating weight lane by lane. Adjusting bolts on the lips of one half the die (which is machined to enhance die bolt adjustments) can deflect the land on that side of the die, raising or lowering the pressure drop in that region of the exit slot. Internal choker bars controlled by bolts may be used to adjust the cavity as well. Die bolts may be used without choker bars, but choker bars are not used without die bolts. The use of computer control for the die bolts has grown to where it is now the norm.

To prevent the melt from sticking to the pressure roll, endless belts of Teflon® PTFE impregnated fiberglass or Kevlar® polyaramide (Teflon is a registered trademark of Chemours Inc. and Kevlar is a registered trademark of E. I. du Pont de Nemours and Co.) fiber cloth are wrapped around the pressure roll (and the auxiliary cooling roll). These release cleanly from most extruded thermoplastics.

To improve adhesion of the extruded film to the substrate, adhesion-promoting "primers" are usually applied to the web before the laminator. Priming can be electrostatic (corona treatment), chemical (i.e., wet coating), flame treatment, or ozone treatment. Coating weights are controlled by the line and extruder speeds, but in many cases, the chill roll capacity limits the maximum thickness that can be obtained. Extrusion coating lines can operate at speeds up to 1000 m/min in some cases, although a more typical operation would be in the range of 150–250 m/min. Coating weights are typically in the range of 10–75 $g/m^2$.

#### 2.1.2.2 Self-Metered Coating

##### *2.1.2.2.1 Reverse Roll Coating*

Reverse roll coating is a versatile coating method and can give uniform, defect-free coatings (12–1200 μ wet thickness) over a wide range of coating speeds and fluid viscosities. However, reverse roll coating is unusual in having a split coating window: it is stable at low speed and high speed, but not intermediate speed. Comments about the attributes of reverse roll coating should be interpreted in the context of which operable region is being considered. The low speed window is relatively robust for non-Newtonian behavior; the high speed window less so. Reverse roll is more commonly used for smooth substrates than rough. It is also a relatively expensive technique requiring precision rolls and bearings as well as robust mechanical design to minimize any deflections or vibrations that will impact coating uniformity.

There are several configurations of reverse roll coating. The common element that defines reverse roll coating is an excess of coating fluid on the applicator roll that is metered (doctored) off of the applicator by a reverse-turning metering roll, and the remaining fluid is transferred to the web traveling in the reverse direction. The two-roll coater, referred to as a levelon coater or doctor roll coater, can be thought of as a simplified reverse roll configuration; however, most reverse roll coaters have three or more rolls as shown in Figure 2.2.

The thickness of the film on the applicator roll is determined by controlling the speed ratio defined as the ratio of the metering roll to the applicator roll speeds. The final film thickness is equal to the thickness of the film on the applicator roll times the wipe ratio that is defined as the speed of the applicator roll divided by the speed of the web. Figure 2.3 shows how the thickness of the metered coating on the applicator roll varies as a function of the gap, speed ratio, and capillary number based on the applicator roll speed.

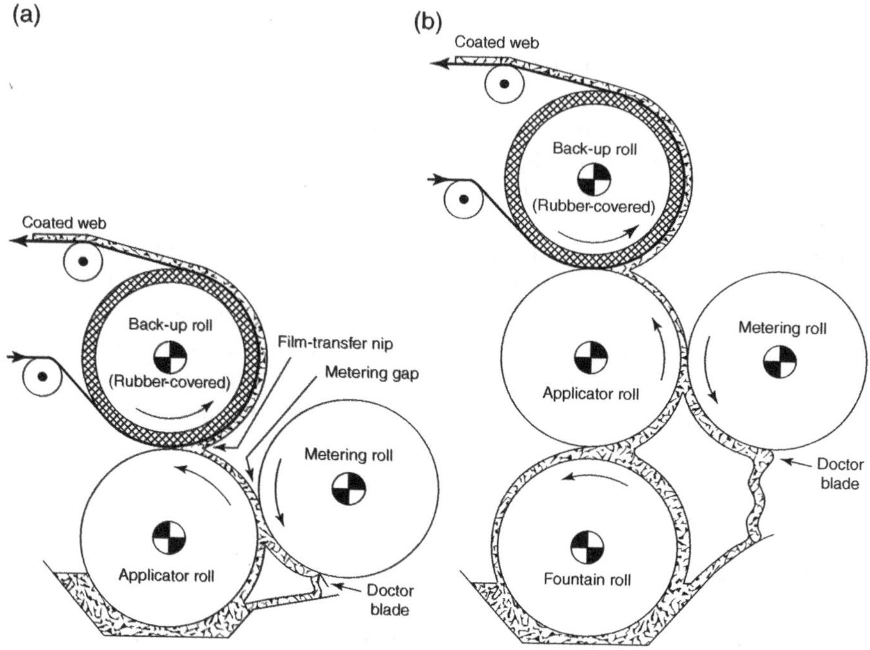


**Figure 2.2** Pan-fed reverse roll coaters: (a) three roll and (b) four roll [4]. © John Wiley & Sons.

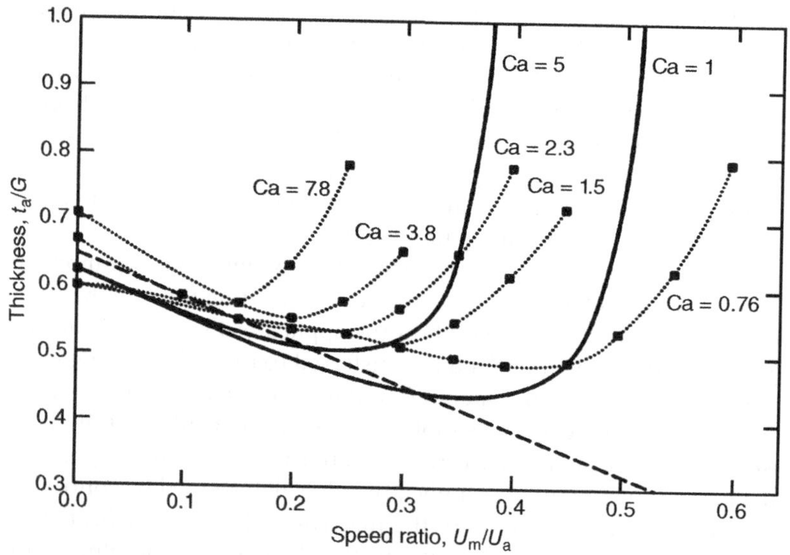


**Figure 2.3** Reverse roll, metered film thickness on the applicator roll divided by gap, $G$, as a function of the ratio of the metering roll speed $U_m$ to applicator roll speed $U_a$ for various capillary numbers based on $U_a$. (—) represents theoretical values; experimental ones; and (- - -) is the lubrication model. Final thickness is applicator roll thickness times the wipe ratio [4]. © John Wiley & Sons.

Low speed reverse roll coating is quite robust to defect formation; however as the capillary number, defined in terms of the speed of the applicator roll

$$\mathrm{Ca} = \frac{\mu U_{\mathrm{a}}}{\sigma}$$

increases, ribbing starts to appear at a relatively low speed ratio. Another defect, called cascade or seashore, tends to form at moderate to high capillary numbers and high speed ratio. The onset of cascade defect can be predicted to engineering accuracy for gaps of 25–250 μ and capillary numbers from 0.2 to 4 by the correlation [26]

$$\frac{U_{\mathrm{m}}}{U_{\mathrm{a}}} = 0.43\mathrm{Ca}^{-0.46}$$

For intermediate Ca, there is no region free of ribbing and cascade. As the speed is increased, a narrow window of operability opens up. The onset of cascade is a relatively sharp transition; the transition to ribbing is more gradual. An operability diagram, showing the region of stable flow as well as the regions where these defects form, is given in Figure 2.4. The principal advantage of reverse roll coating is that conditions can be adjusted to give a stable, defect-free coating at high coating speeds. Using precision bearings, reverse roll coaters can lay down a coating with a tolerance of about ±2%.

#### 2.1.2.2.2 Mayer Rod Coating

The Mayer rod coater is one of the earliest coating methods [27] and is still widely used. There are many names for this process: Mayer bar (and the alternate spellings Meyer bar and Meyer rod), coating rods, applicator rod, wire-wound rod, equalizer bars, rolled bars, and doctor rods. It is used to coat a wide variety of products at moderate speeds with varying degrees of precision. The defining feature of the Mayer rod coater is a pool of liquid on the web that is doctored off by a rod. The prototypical Mayer rod is a wire-wound rod as shown in Figure 2.5. However, a variety of designs from smooth rods to rods with machined corrugation are used as well. An excess of liquid is created by a pre-wetting section (usually a kiss coater although dip coating is also used). The rod is held against the web (which may be supported by a roll or under tension) and is commonly rotated so that the surface of the rod moves in the opposite direction of the web. The rods themselves are rather flexible; thus, they require a rigid support to prevent deflection. Numerous holder designs have been patented and sold [2].

Control of the wet coating thickness is achieved by varying the cross-sectional area of the cavities. For wire-wound rods, the cavity cross section per unit width is 10.7% of the wire diameter. Rods are marked with the wire size or

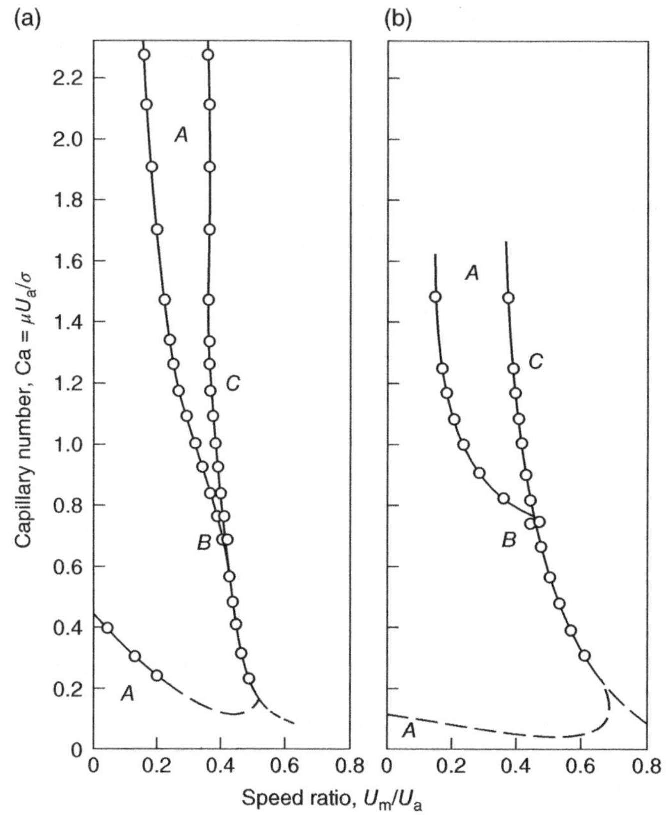


**Figure 2.4** Operability diagram for reverse roll coating [4], where *A*, represents a stable coatings area; *B*, ribbing; and *C*, cascade; for (a) a gap $G = 750\,\mu$ and (b) a gap $G = 250\,\mu$. © John Wiley & Sons.

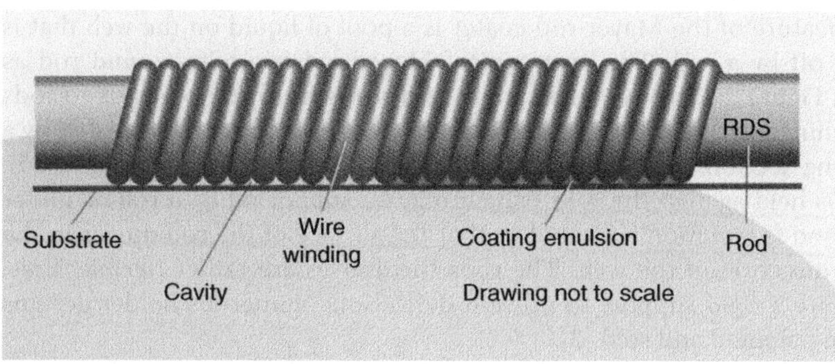


**Figure 2.5** Mayer rod details. (Courtesy of RD Specialties.)

equivalent (for machined rods) that is nominally the projected volume times ten; thus a #24 rod (English units) might be expected to give a 2.4 mil thick wet coating. However, the wet coverage is always less than the theoretical value, sometimes as low as one half of this. The actual thickness depends on the fluid rheology, the substrate speed, the web tension when coating against an unsupported substrate, and the direction of rotation.

The strength and weakness of Mayer rod coating is that the rods are consumable: subject to wear but easy to change and inexpensive. It is usually more cost-effective to replace a rod than to clean it when dirty. Rods can be rewound without incurring the risk of yield loss from ineffective cleaning. Rotation increases rod life by causing even wear and preventing particles from getting caught under the rod and causing streaks. The coating rods are compact and simple, but they wear rapidly when used with abrasive fluids.

There are two defects usually associated with wire rod coating. First, coatings leave the rod with a corrugated surface reflecting the geometry of the cavities. If the coating is not self-leveling or the line speed is too high to allow complete leveling before the coating is solidified, the dry coating will retain a fine line structure that is an artifact of the rod geometry. Smoothing rods may be used to smooth out these defects (although smoothing rods often produce hydrodynamic ribbing in place of the wire marks). Second, rod coaters are subject to particles getting caught in the wires, leading to streaks. As mentioned earlier, the rods are usually rotated to minimize this defect. Formed, gapped, and double wound rods create cavities that are less prone to clogging. There are rod holder designs that incorporate doctor blades to further reduce the tendency of particles to get caught in the rod [2].

Mayer rod coating gives good coating quality with reasonable uniformity (generally better than 10%) for low viscosity (easy leveling coatings). It is easy to use and flexible. Rods can be changed quickly, making Mayer rod coating particularly attractive for short-run coatings. The disadvantages are a low viscosity range (typically not much over 1000 cP) and low wet thickness. The height of the corrugations in the coating surface is proportional to thickness, and the rate of leveling is inversely proportional to viscosity times thickness. Although Mayer rod coating has been reported at speeds as high as 1000 fpm, it is more commonly used at lower speeds. The need for leveling creates inverse relationships between maximum viscosity and maximum line speed as well as maximum thickness and maximum line speed.

#### 2.1.2.2.3 Gravure Coaters

The term gravure coating refers to a family of coating methods tied together by the use of an engraved roll that delivers a controlled amount of liquid coating to the coating nip. The engraved rolls are filled by immersion in a bath. A doctor blade wipes off excess coating above the cells. Thus, in one way, gravure can be thought of as a doctored coating method. However, the transfer of the coating from the gravure cylinder to the web can be thought of as a self-metered mechanism, and gravure is sometimes referred to as a hybrid coating method.

Various configurations of gravure coating can apply wet layers from <1 to 400 μm thick. The coated thickness with gravure coating is substrate dependent: open (permeable) substrates such as paper can accept much thicker coatings than impermeable substrates like foil and film. Gravure coating has been used to apply coatings with low shear viscosities from less than one to thousands of cP. However, one should remember that most coatings that are quoted as having high viscosity are shear thinning and the high shear viscosity range is much narrower.

The nomenclature for gravure coaters can be confusing but is well established. What is now called direct gravure coating uses gravure rolls that match the speed of the web. Some call all engraved rolls with uniform patterns "anilox rolls," while the term "gravure roll" is reserved for rolls engraved with an image for printing. Many others coat uniform layers with "gravure rolls." In differential gravure coating the gravure roll is moving in the same direction as the web but is driven at a different speed. Offset gravure uses a rubber roll between the gravure roll and the web with all rolls turning at the same surface speed. This is also referred to as "indirect gravure." In reverse gravure, the gravure roll is in direct contact with the web but is driven in the opposite direction from the web. Combinations such as differential offset gravure also exist. For differential and reverse gravure, the relative speeds of the rolls are usually expressed as a gravure ratio defined as the ratio of the specified roll to the line speed (expressed as a decimal, percent, or percent over speed). These configurations are illustrated in Figure 2.6.

The gravure configurations discussed earlier use a heavily loaded backup roll (sometimes called an impression roll) to achieve good transfer to the web. Booth [2] quotes the usual force at 2,000–20,000 N/m, while Weiss [11] claims that most gravure heads are operated at a constant displacement not a constant force. In kiss coating (also known as tensioned web or unsupported web coating), the backup roll is eliminated. This results in lower coating weights. The gravure roll in kiss gravure can rotate in either the forward (sometimes called "direct") direction or the reverse direction.

Microgravure coating is a kiss gravure coating with a 20–50 mm diameter gravure roll [28]. To prevent bending these small diameter rolls, this technology is practiced with minimal deflection of the web as it passes above the roll. This leads to a narrow meniscus with a sharp angle of divergence, which is claimed to eliminate transverse direction defects as well as minimize the impact of the coating solution on the web.

The final major variation on gravure coating, pressurized gravure, is defined in terms of how the engraved roll is wetted. As shown in Figure 2.7 the engraved roll can be wetted through immersion in an open pan, through contact with a wetted fountain roll (i.e., immersed in a pan) or a fountain die, or by passing

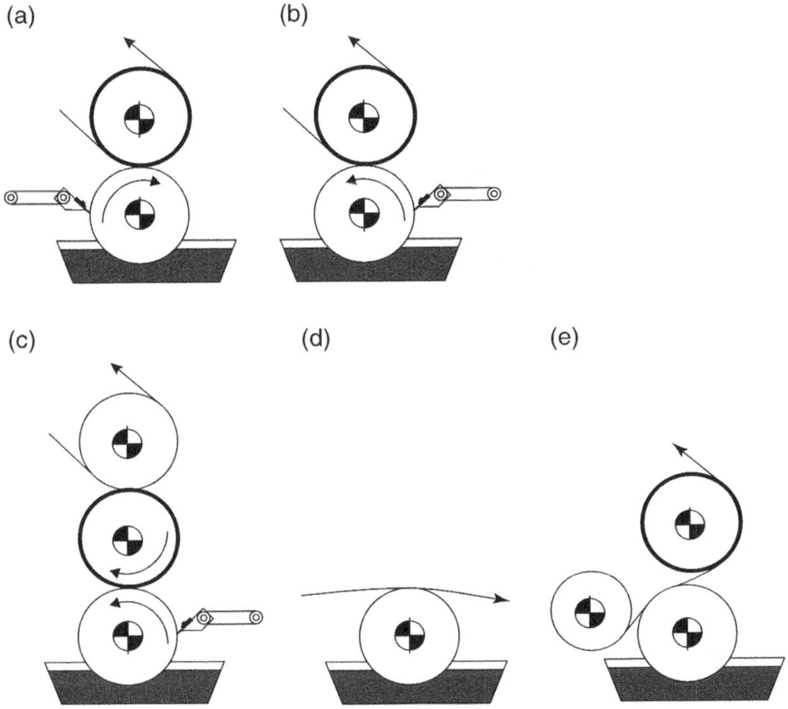


**Figure 2.6** Common configurations of gravure coating: (a) direct (and differential), (b) reverse, (c) offset, and (d, e) two implementations of kiss (tensioned web) gravure (the engraved roll may rotate in either direction in kiss coating).

through a closed chamber (sometimes called a doctor box). The enclosed chamber may be a relatively large cavity or machined to be close fitting or have baffles to optimize the cell filling [29]. Open pan coating is limited in speed by air entrainment in the pan and incomplete filling of the cells that leads to gravure bands (see Figure 2.8); fountain rolls were widely used to extend the maximum speed for gravure coating up to 1200 fpm for favorable coatings [2], but fountain rolls have largely been replaced by "pressurized gravure."

There are two fundamental variations on pressurized gravure: the closed chamber may be pressurized (leading to a leakage of extra coating material under the doctor blade), or the chamber can be configured for countercurrent flow along the roll face to maximize filling of the cells in the gravure roll [29].

Once the cells have been filled, the excess coating is doctored from the surface of the engraved roll by the "doctor blade." Doctor blades can be mounted in either a "trailing" or "reverse" angle as shown in Figure 2.9. Closed head (and pressurized) gravure chambers use one trailing angle and one reverse angle doctor blade to contain the liquid (leading to the alternate name "doctor box").

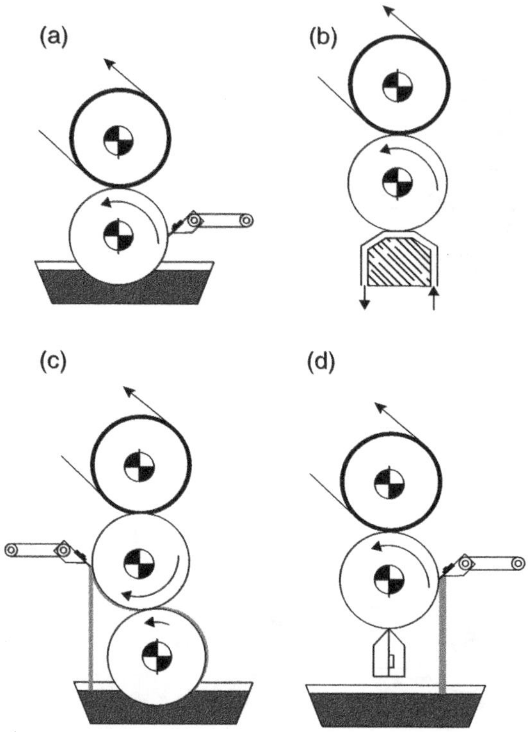


**Figure 2.7** Wetting options for gravure coating: (a) open pan, (b) closed head (or pressurized), (c) fountain roll, and (d) fountain die.

**Figure 2.8** Gravure bands (vertical = MD) and doctor blade chatter (dark horizontal lines just below the midpoint of the picture) [29]. © John Wiley & Sons.

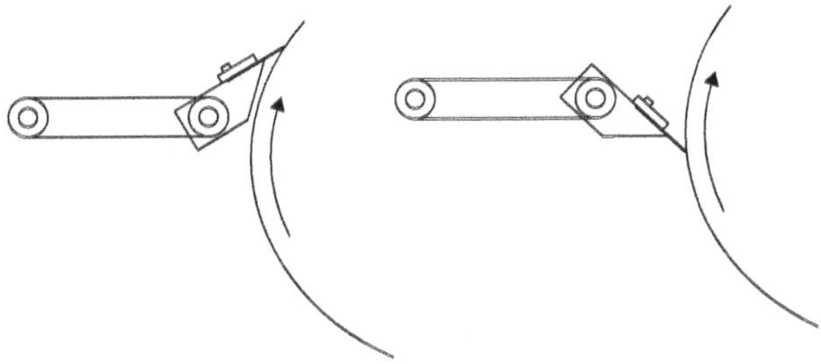

Trailing angle
Spreading angle
Wiping angle
"Conventional"

Reverse angle
Plowing angle
Chiseling angle
Shearing angle
Cleaning angle
"Doctoring"

**Figure 2.9** Doctor blade configurations.

There are several significant differences between trailing angle and reverse angle wiping. Trailing angle blades often need a high external load to maintain good wiping; reverse angle blades are relatively lightly loaded because friction plus hydrodynamics push the blade against the surface. Lubrication pressure can cause trailing angle blades to lift, suddenly flooding the nip with excess coating [30]. This has relatively little effect in direct gravure; however, in reverse gravure and kiss gravure, this leads to a sudden increase in solvent level in the oven that can create an explosion hazard if the solvent is combustible.

Doctor blades may be made from stainless steel, blued mild steel, or plastic (often referred to as "laminate" blades). Steel doctor blades vary in thickness from roughly 0.005″ to 0.025″. Laminate blades are available in thicknesses from 0.005″ up about 0.125″. Thinner blades are generally supported in the doctor blade holder by "backer blades" (usually made of spring steel). There are numerous designs of doctor blade tips being sold—beveled, lamellar, square cut, ceramic coated, etc.—which are largely sold on promises of extended blade or roll life. Doctor blades may be oscillated 6–50 mm in the transverse direction to give even wear and to dislodge particles or bubbles trapped under the tip of the blade that cause streaks. Improper design of the oscillation system can lead to doctor blade chatter shown in Figure 2.8. Trailing angle blades may operate in either a stiff blade regime or a bent blade regime [31].

There are two common measures of the angle the doctor blade makes to the gravure roll: the "set angle" and the "contact" angle. The set angle is the apparent angle between the doctor blade as set in its holder (without being pressed against the roll) and a normal angle to the roll. The contact angle is the actual angle the blade makes with the roll surface after it has been deflected by the externally applied load. At zero loading, the contact angle is 90° minus the set angle. Various vendors recommend different contact angles for different doctoring configurations. Recommendations for trailing angle generally fall between 55° and 70°, while recommendations for reverse angle generally fall between 30° and 40°. The sealing blades on the upstream side of a closed head are much lower angle—10–20°—since they are not intended to doctor excess liquid from the roll.

The most common construction of gravure roll is a steel roll plated with a soft copper layer that is mechanically (or electromechanically) engraved and then chrome plated. Laser-engraved ceramics are used where the need for longer life justifies the added cost of a ceramic roll. The use of chemically etched rolls (also known as "conventional" etching) rather than mechanical engraving has become less common largely because of seam lines that occur as an artifact of the photoresist imaging process. The process for manufacturing microgravure rolls is significantly different than the traditional engraving process and produces gravure rolls with extremely tight tolerances. Both trihelical and quadrangular patterns have been used; however, trihelical patterns seem to be the more widely used for microgravure.

The common mechanically engraved cell patterns are illustrated in Figure 2.10. Patterns are specified by a combination of a line count and a geometrical abbreviation. The cell pitch or line count is the number of cells per inch or centimeter measured perpendicular to the pattern and usually ranges from 4 to 160/cm. Engraving patterns can be divided into closed and open cells. Pyramidal cells have a pointed bottom; quadrangular cells have a flat bottom (to improve emptying). Hexagonal cells can be arranged in a pattern that has either a 30° or 60° angle (depending on whether a side or an apex is the leading feature of the hexagon). The trihelical pattern consists of continuous grooves with a triangular cross section that spiral around the roll, usually (but not always) at a 45° angle. The RotoFlow patterns are pyramidal and quadrangular patterns that have small channels connecting them to provide a slight flow to mitigate cell plugging. The QCH (or PQCH) pattern consists of hexagonal cells connected with a channel of either triangular or trapezoidal cross section. Note that the illustrations in Figure 2.10 are somewhat idealized: in practice it is virtually impossible to create sharp corners with engraving. Furthermore, engraving tools are subject to wear; thus, most features of "triangular" cross section have rounded bottoms and edges. Coincidentally, this ameliorates the issues with sharp corners that have been discussed in the academic literature [32].

The dimple up (also known as pins up or crosshatched diamond) pattern is the most open of the patterns but is less common largely due to difficulties in doctoring and a tendency to wear quickly. Open cell patterns are generally preferred for pressurized gravure coating.

The primary factor controlling the coating weight for gravure coating is the total volume of cells engraved in the face of the roll per unit area. This is usually

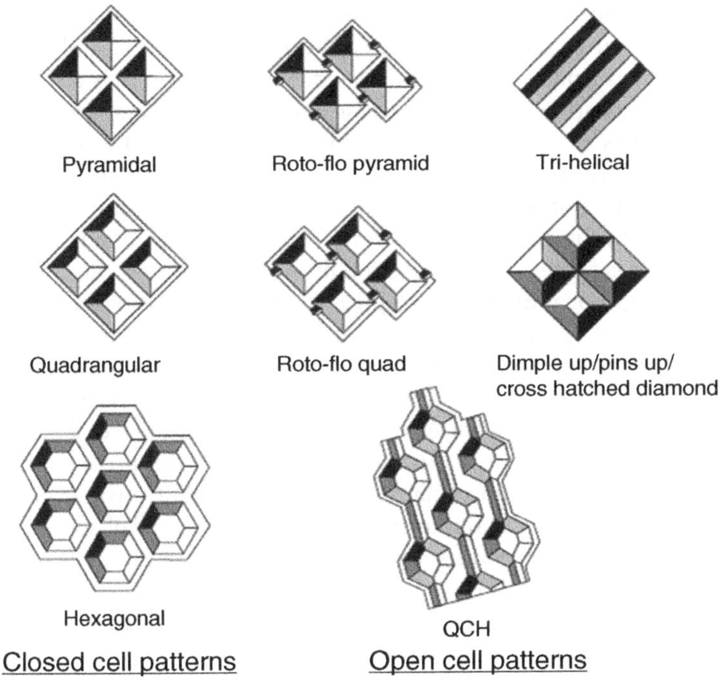


**Figure 2.10** Common mechanical engraving patterns for gravure coating.

expressed in the hybrid unit of billions of cubic microns per square inch. For direct and differential gravure coating, coating weight is limited by the amount of liquid that can pass through the coating nip, and the final thickness is determined by a film splitting flow at the downstream separation point. In contrast, reverse gravure can lead to a flooded nip with the excess coating being transferred to the web. Thus, the mechanisms for controlling coating weight can be quite different for these different techniques.

For direct gravure, coarse coating weight control is generally accomplished by varying the cell volume with finer coating weight control affected by changing the concentration of the coating. Backup roll pressure and doctoring have minor effects on direct gravure coating. For reverse and differential gravure, the roll speeds are set independently, and fine coating weight control is normally affected by varying gravure roll speed. The ratio of roll speeds is characterized by the "gravure ratio," which is defined as the ratio of the absolute value of the surface speed of the gravure roll to the web speed (often expressed as a percentage). This definition of gravure ratio is virtually universal—the prominent exception being certain academic studies that have chosen to express the roll speeds in terms of a "speed ratio" that is the inverse of the gravure ratio (which has apparently been chosen to suggest a parallel to reverse roll coating). Differential offset gravure offers the possibility of varying two ratios, and these are usually expressed as a percentage of line speed. Nip force generally also has a significant effect on coating weight for reverse and differential gravure.

Two limiting models of the wet coating thickness are used to correlate coating weight in gravure coating. The constant "pickout" model assumes that a more or less constant fraction of the liquid delivered to the coating nip is transferred to the web; the constant "transfer efficiency" model assumes the wet coating thickness is some constant fraction of the cell volume (volume per unit area can be thought of as an average depth). However, the use of the terms "transfer efficiency" and "pickout" is not universal. For direct gravure, these two models are indistinguishable. For differential and reverse gravure, the two models predict different behaviors because the gravure roll and web can move at different speeds. Constant transfer efficiency predicts that the wet coating thickness is independent of gravure ratio, while constant pickout predicts that the wet thickness is directly proportional to the gravure ratio. Constant pickout, constant transfer efficiency, and intermediate behaviors are all observed in practice. Given the range of behaviors seen with gravure coating, the engraving pattern selection is done empirically.

The prototypical defect in gravure coating is a partial replication of the engraving pattern. Gravure pattern is worse in direct gravure than in reverse or differential gravure and is exacerbated by low ratios, large spacing between cells, and high viscosity or yield stress. Smoothing bars or rolls (generally counter-rotating in a free span) can also be tried although they often produce ribbing.

Flashing and multiple line pickout are common defects in direct gravure coating with trihelical rolls [33]. Flashing occurs when one or more grooves in the roll do not pick out correctly from the roll, leaving an uncoated diagonal line on the web. Multiple line pickout occurs when the coating from more than one groove in the roll merges to form one heavy diagonal streak. Multiple line pickout is caused by the natural wavelength for ribbing on a forward (smooth) roll coater, matching some multiple of spacing of trihelical grooves.

Like all roll coating processes, gravure coating is subject to film splitting defects at the downstream meniscus. These include ribbing, and if the coating is sufficiently elastic, it may lead to filamentation and misting (if the filaments break to produce droplets).

All configurations except kiss gravure use a rubber backup or transfer roll. Elastomeric rolls tend to produce what are often called "footprints" or backup roll repeaters. The common process for manufacturing rubber rolls is to wind a thin sheet of rubber around the roll in what is often called a window shading process, vulcanizing this, and then grinding it to final shape. This process produces a distinct seam line. The common alternative to the window shading process is a casting process (usually used for urethane rolls) that does not exhibit seam lines. However soft spots, hard spots, inclusions in the rubber, and mechanical damage will often lead to repeating artifacts in the coated film.

### 2.1.3 Selecting a Coating Method

Experience attests to a great reluctance for industry to change coating methods once a product line has been commercialized. Thus, it is of critical importance to be sure the coating method is well chosen or that the coater is designed for rapid changes of coating methods (i.e., cartridge coating heads). Multi-method semiworks coaters are available to evaluate various methods during scale-up.

There is no single coating method that can coat the entire range of process and product properties that might be encountered. There are a number of guides available to help in the early stages of choosing a coating method [3, 4, 34–37]. However, these should only be used for prescreening generic methods for practical evaluation: no guide (including this one) contains enough information to make an intelligent choice of a specific coating configuration. Each generic technique exists in many variations that typically have different operability. Specific coating methods should be selected based on hands on evaluation with the intended coating and substrate followed by evaluation by the target customers for the product. Some of the basic attributes of generic coating methods are listed in Table 2.1.

The principal substrate factors influencing the choice of a coating method are permeability, tolerance for thickness variations, and tolerance to the presence of nonplanarity defects.

Open porous materials (e.g., paper, fabric) are permeable to the coating and prone to "strike-through" (the coating soaks through and wets the backup roll). This can make doctored coating methods problematic for permeable webs since they pre-wet the web. However, short-dwell flexible blade coaters are the mainstay of the paper coating industry: while they may only be operable at high speed, high speed brings economic advantage. Clearly one should be careful with but not discount doctored coating methods for permeable web applications. Permeable materials are much easier to wet and coat at higher speed. Ideally, the web to be coated is flat and of uniform thickness. However, just as every coating method has endemic defects, every substrate manufacturing process has its own signature and a corresponding Pareto list of defects. The norm in the industry is to qualify substrates by production line (not merely by vendor). All webs have some variations in thickness (gauge variation), and a tendency of some portions to be longer or shorter than others (a planarity defect known as buckle). Edge buckle (baggy edges or droop) are common in tenter frame made film; center buckle (baggy centers) are common in blown film. If the buckle occurs in a narrow lane, it is called a buckle lane, a baggy lane, a gauge band, and so on. It is common to apply tension to the web to hold all lanes in tension before passing through a coating head. However, if the buckle is severe (most commonly a

**Table 2.1** Attributes of some common coating techniques.

| Method | Tolerant to thickness variations | Cockle tolerant | Buckle tolerant | Low shear viscosity (cP) | No. of layers | Wet thickness (μm) | Coating accuracy | Max. speed(m/min) | Intermittent/ patterned |
|---|---|---|---|---|---|---|---|---|---|
| Mayer rod | Yes | Marginal | No | 20–1,000 | 1 | 5–50 | Moderate | 250 | No |
| Dip | Yes | Ok | Yes | 20–1,000 | 1 | 5–100 | Precision | 150 | No |
| Air knife | Yes | Ok | Marginal | 5–500 | 1 | 2–40 | Good | 500 | No |
| Forward roll | No | No | Marginal | 20–1,000 | 1 | 10–200 | Moderate | 150 | No |
| Reverse roll | Yes | Marginal | No | 100–50,000 | 1 | 5–400 | Precision | 400 | No |
| Knife over roll | No | No | Marginal | 100–50,000 | 1 | 25–750 | Moderate | 150 | No |
| Knife over blanket | Marginal | No | No | 100–50,000 | 1 | 25–750 | Moderate | 150 | No |
| Flexible blade | Yes | No | No | 500–40,000 | 1 | 25–750 | Good | 1500 | No |
| Slot | Yes | Marginal | Marginal | 5–20,000 | 1–3 | 15–250 | Precision | 500 | Ok for patches |
| Extrusion | Yes | Marginal | Marginal | 50,000–5,000,000 | 1–15 | 15–1000 | Good | 700 | No |
| Slide | Yes | Marginal | Marginal | 5–500 | 1–18 | 15–250 | Precision | 200 | No |
| Curtain | Yes | Yes | Yes | 5–500 | 1–18 | 2–500 | Precision | 400 | No |
| Gravure, direct | Yes | Ok | No | 1–5,000 | 1 | 1–25 | Good | 700 | Excellent |
| Gravure, offset | Yes | Ok | No | 100–50,000 | 1 | 1–400 | Good | 300 | Flexography |

| | | | | | | | | | |
|---|---|---|---|---|---|---|---|---|---|
| Gravure, reverse (including pressurized) | Yes | Ok | No | 1–5,000 | 1 | 1–400 | Good | 1500 | No |
| Microgravure | Marginal | No | No | 1–4,000 | 1 | 1–40 | Precision | 100 | No |
| Comma | No | No | Marginal | 1000–30,000 | 1 | 20–350 | Moderate | 300 | No |
| Nip roll (Dahlgren) | Yes | Yes | No | 100–500 | 1 | 50–200 | Moderate | 600 | No |
| Dip and scrape | No | Ok | No | 25–500 | 1 | 50–450 | Moderate | 200 | No |
| Spray | Yes | Yes | Yes | 10–300 | 1 | 50–400 | Poor | 100 | No |

buckle lane), wrinkles will pass through the coating head. If the coating head has a nip, the bagginess will gather behind the nip until it reaches a critical value and then pass through the nip as an intermittent crease. This is generally considered unacceptable, and these techniques are listed as not buckle tolerant. If the coating head has a gap, the wrinkles will pass through—probably producing a coating disturbance—but without being turned into creases. These techniques are listed as marginal for tolerating buckle defects. Cockle, as opposed to buckle, refers to bumps or undulations in the web that are fixed in location: the bump goes through as it reaches the coating head without being gathered behind a nip. Cockle can be a small bump or troughs that run in the machine or transverse (cross) directions. Cockle, like buckle, can usually be managed with tension and spreading devices; however, severe cockle reduces the gap in a coater, creating thin spots or streaks in doctored coating. Techniques that produce coating defects from cockle are listed as intolerant to cockle.

Most applications require uniform, continuous coatings. However, an increasing number of advanced material applications require intermittent or patterned coating. Some of the traditional coating methods are readily adaptable to printing patterns (direct gravure becomes gravure printing if the engraving is patterned, offset gravure becomes flexography if the rubber roll is patterned, etc.). Other traditional coating methods can readily apply stripes and patches. These are noted in Table 2.1.

The following procedure is recommended for selecting a specific coating method and configuration:

- Start by establishing the requirements for the product and process (including the nature of the substrate).
- Match the capabilities of the generic coating methods shown in Table 2.1 to these requirements considering that:
  - Virtually all industry standard guides (including this) use low shear apparent viscosity (Brookfield viscometer): do not take the limits too literally or as an entitlement.
  - Speed ranges are approximate numbers dependent on other parameters (coating weight, rheology etc.). Quoted speeds are for film coating (c.f. permeable webs) and should not be taken as an entitlement or with precision.
  - Operability will vary from one variation of a generic technique to another (consult the vendor or references for additional guidance).
- Compare the coating requirements with the coating method capabilities and select candidates for evaluation.
- Run semiworks experiments to verify suitability of each candidate. These should begin with mapping the coating window and identifying the limiting defects, include attempts to eliminate or delay the limiting defects (e.g., extend the maximum speed before encountering defects) as well as statistically designed experiments to determine product quality.

- Use the most promising method for semiworks production and customer sampling/market development. If production data do not meet the end user requirements for this application, then another method or configuration should be evaluated.

## 2.2 Drying and Curing

This section discusses the hardware and issues associated with turning the liquid coating into a dry material that can be wound into a roll without blocking.

The coating process requires the coated layer to be in a liquid state at the time of coating. However, the coated material must be wound up, and that requires the liquid coating to be solidified at some point between the coating head and the windup. The most common way to do this is drying; however, hot melt coatings require only cooling, electron beam (e-beam) coatings require only irradiation to initiate a chemical reaction, and ultraviolet (UV)-curable coatings may require drying but in some cases can be solidified by curing alone.

The common unit operations in the solidification section are as follows:

Heat transfer mechanisms
- Conduction—Widely used in substrate manufacture (both paper and plastic film), hot melt coating, and extrusion coating
- Convection—Most widely used method in the converting industry because it provides both heat and mass transfer
- Radiation—Always present to some degree, primary mechanism for infrared (IR) drying, microwave drying, and radio-frequency drying

Curing
- Thermally induced
- UV initiated
- E-beam initiated

Mass transfer (i.e., solvent removal)
- Safe handling of solvent vapors
- Minimize human exposure to chemicals
- Prevent explosive mixtures

Web transport—Conveying the web from the coating head through the solidification section

### 2.2.1 Principles

#### 2.2.1.1 Choice of Solidification Method

The first choice to be made is the mechanism by which the liquid coating is converted to a solid or semisolid material (e.g., pressure-sensitive adhesives are dried to become sticky and contained between a release a liner and the facestock but never become solid).

If the product can be coated neat (i.e., without solvents or diluents that require drying), both capital cost and operating cost can be expected to be lower than if a solvent or diluent is used. If the product can be made by hot melt or extrusion coating, this is usually the lowest-cost option. Extrusion coating is a higher throughput process than hot melt coating: if the volumes and product warrant extrusion coating, that is generally the most economical choice. If the product is amenable to neat coating and can be solidified by e-beam curing or UV curing without drying, these are generally more economical options than drying.

Drying is generally more expensive than the simple cooling or radiation-induced crosslinking because it involves both heat and mass transfer. Aqueous drying requires fans and ductwork to convey the evaporated water out of the building; flammable solvent drying also requires air handling equipment but brings the added costs of environmental controls and mitigating the hazard of vapor phase explosions. That being said, drying is by far the most flexible solidification technique, and most high value in use "advanced materials" are dried. Drying requires both heating (to offset the heat of vaporization of the solvent or diluent) and convection to remove the vapor from the dryer.

There are some applications where the coating can be partially solidified in a dryer with minimal mass transfer. One well-known, though largely obsolete, example is the chilling of gelatin-based coatings used in the photographic film industry (warm gelatin solutions are set by cooling below the melting point of the gelatin). Where rapid heating to set the coating is required, IR, microwave, and radio-frequency heating (often coupled with convective drying in so-called "hybrid" dryers) may be attractive options.

#### 2.2.1.2 Coupled Mass and Energy Transfer

Evaporation is always enhanced by convection and the moist or solvent laden air needs to be conveyed from the vicinity of the coating into the air handling system. In most cases convective drying is turbulent although laminar flow air can be used for very fluid coatings.[1]

Because heat transfer is more or less independent of solvent while mass transfer rates vary from solvent to solvent, most drying hardware is specified by energy transfer capabilities (i.e., heat transfer coefficient). Drying air is commonly heated electrically, by steam, hot oil, or direct-fired gas. The most common specification offered on a dryer is the total heat load that can be met by these heating systems. This specification is useful in determining the maximum amount of air that can be heated to the design temperature but is not directly applicable in predicting the drying rate that can be obtained. The dryer will also have a permissible temperature range dictated by safety concerns such

1 One exception to this is "gap drying" [38].

as maximum skin temperature, material constraints, and the ability of mechanical tolerances to accommodate thermal expansion. Solvent drying is limited by the generation of explosive mixtures of solvent in air. These determine the maximum usable heat supplied to the oven.

Quoted heat transfer coefficients h are usually defined by

$$\frac{Q}{A} = h\Delta T$$

where $Q/A$ is the heat flux per unit area and $\Delta T$ the difference in temperature between the coating and the incoming air. Similarly the mass transfer coefficient at low mass transfer rates is defined by

$$N = kA\frac{y}{1-y}$$

where $N$ is the drying rate (expressed in units of moles per time) and $y$ is the mole fraction of solvent in the air in the vicinity of the coating.[2]

The fundamental correlation between heat transfer coefficient $h$ and mass transfer coefficient $k$ is the Chilton–Colburn analogy [39]:

$$\frac{h}{k} = C_{\mathrm{p}}\left(\frac{\kappa}{\rho C_{\mathrm{p}} D_{\mathrm{ab}}}\right)^{\frac{2}{3}} = C_{\mathrm{p}}\left(\frac{\alpha}{D_{\mathrm{ab}}}\right)^{\frac{2}{3}} = C_{\mathrm{p}}\mathrm{Le}^{\frac{2}{3}}$$

where $\kappa$ is the thermal conductivity, $\rho$ the density, $C_{\mathrm{p}}$ the heat capacity of the gas stream at constant pressure, $D_{\mathrm{ab}}$ the diffusivity of the solvent in the gas stream (either air or an "inert" mixture of nitrogen and oxygen that is sufficiently depleted in oxygen to preclude an explosion), and $\alpha$ the thermal diffusivity of the gas stream. The ratio of the thermal diffusivity to the mass diffusivity Le is the Lewis number.

Accurate modeling of drying is far more complex than is practical in most situations. However, crude models of drying can give valuable physical insight. The simplest model of steady-state drying requires the heat flux to balance the heat of vaporization times the mass flux:

$$hA\Delta T \approx \Delta H_{\mathrm{Vap}} kA\frac{y}{1-y}$$

This ignores sensible heating of both the air and the coating but is close enough to the heat and mass balances in a wet bulb thermometer to lead many

---

2 The drying rate can be expressed in mass rather than molar units, but the units must be consistent throughout all calculations. The reader is referred to the standard texts on mass transfer for details.

practitioners of aqueous drying to use psychometric charts to model drying. Ignoring sensible heating is a limitation worthy of further discussion.

One obvious difference between drying and a wet bulb thermometer is that the wet bulb thermometer is a relatively compact body in a large uniform airflow. Air in a web drying oven is blown from nozzles that are much smaller than the area they serve to ventilate. The wet bulb thermometer operates at constant temperature, so sensible heating of the thermometer is not significant; sensible heating of the web and coating can be very important in the dryer. Furthermore, the air between the nozzles is somewhat cooled from the supply temperature and is also laden with more solvent than the feed condition; thus the driving forces for both heat and mass transfer are reduced in the area between the nozzles. The local heat transfer coefficient (defined with respect to the initial driving force) decays asymptotically as the inverse of the distance from the slot. This suggest that defining heat transfer coefficients in terms of a log mean driving force would be more accurate; however, that is not the norm in the industry.

Boundary layer theory predicts the Nusselt number scales with the two thirds power of velocity. While the well-known Martin correlation for heat transfer coefficients in the presence of a strongly impeded return path exhibits just such a dependence [40], other correlations such as the correlation of Saad et al. [41] for unimpeded airflow have a velocity (Reynolds number) exponent significantly greater than two thirds. One explanation for this is that boundary layer theory holds in the area between the nozzles, but the assumption of boundary layer theory that pressure varies slowly along the solid–gas interface does not hold under the slot (the region of peak heat transfer) [42]. The Martin correlation gives a lower limit (for impeded return flow); the Saad correlation gives an upper limit (for unimpeded return). These two limiting correlations for heat transfer coefficients are given in Figure 2.11.

Another limitation on the psychrometric model is that the web in a closed oven is subject to a significant heat flux from thermal radiation. For ideally absorbing planar webs enclosed in a box, the heat flux is given by the Stefan–Boltzmann law:

$$Q_{\mathrm{Rad}} = F_{12}\sigma\left(T_{\mathrm{Wall}}^4 - T_{\mathrm{Web}}^4\right)$$

where $F_{12}$ is the view factor and $\sigma$ is the Stefan–Boltzmann constant [43]. Inside the box, the view factor from each side approaches one; thus the total heat flux from incidental radiation is roughly $2\sigma(T_{\mathrm{Wall}}^4 - T_{\mathrm{Web}}^4)$. Obviously, the magnitude of this incidental heat flux depends on the operating temperature; however, ignoring incident radiation is frequently the single largest error in the simple psychrometric model of drying and the use of the Chilton–Colburn analogy (the heat flux is significantly larger than for pure convection, while there is no corresponding boost in mass transfer).

Saad correlation (slot nozzles only)

$$Nu_w = 0.14Re_w^{0.775}(w/X_p)0.314(w/H)^{0.286}$$

| | |
|---|---|
| Nomenclature | Discharge coefficient |
| $C_d$ | Specific heat at constant pressure |
| $C_p$ | Hole diameter |
| $D$ | Open area fraction |
| $f$ | Average heat transfer coefficient |
| $h$ | Nozzle to plate clearance |
| $H$ | Thermal conductivity of air (at $T_f$) |
| $\kappa$ | Nusselt number based on hole diameter |
| $Nu_d = hd/\kappa$ | Nusselt number based on slot width |
| $Nu_w = hw/\kappa$ | Prandtl number (properties evaluated at $T_f$) |
| $Pr = mC_p/\kappa$ | Reynolds number based on hole diameter |
| $Re_d = \rho V_{eff} d/\mu$ | Reynolds number based on slot width |
| $Re_w = \rho V_{eff} w/\mu$ | Temperature of incoming air |
| $T_{air}$ | Average air temperature |
| $T_f = (T_{air} + T_{plt})/2$ | Temperature at the impingement point (plate) |
| $T_{plt}$ | Velocity for calculating effective Reynolds number |
| $V_{eff}$ | Exit velocity |
| $V$ | Slot width |
| $w$ | Nozzle to nozzle centerline pitch |
| $x_p$ | Dynamic viscosity of air (at $T_f$) |
| $\mu$ | Density of air (at $T_f$) |
| $\rho$ | |

Martin correlations for impingement nozzle heat transfer

| Nozzle type | Range of applicability | Correlation |
|---|---|---|
| Multiple slot jets | $750 < Re_w < 20{,}000$<br>$0.008 < f < 2.5 f_o$<br>$2 < H/w < 80$ | $\frac{Nu_W}{Pr^{0.42}} = \frac{2}{6} \cdot f_o^{\frac{3}{4}} \cdot \left[\frac{4 \cdot Re_W}{\frac{f}{f_o} + \frac{f_o}{f}}\right]^{\frac{2}{3}}$<br>$f_o = \left[60 + 4 \cdot \left[\frac{H}{(2 \cdot W)} - 2\right]^2\right]^{-\frac{1}{2}}$<br>$f = \frac{W}{x_p}$ |
| Multiple round jets | $2000 < Re_d < 100{,}000$<br>$0.004 < f < 0.04$<br>$2 < H/d < 12$ | $\frac{Nu_d}{Pr^{0.42}} = K \cdot \sqrt{f} \cdot \left[\frac{1 - 2.2 \cdot \sqrt{f}}{1 + 0.2 \cdot \left(\frac{H}{d} - 6\right) \cdot \sqrt{f}}\right] \cdot Re_d^{\frac{2}{3}}$<br>$K = \left[1 + \left(\frac{H}{0.6 \cdot d} \cdot \sqrt{f}\right)^6\right]^{-0.05}$<br>$f = \frac{\pi}{2 \cdot \sqrt{3}} \cdot \left(\frac{d}{x_p}\right)^2$ |

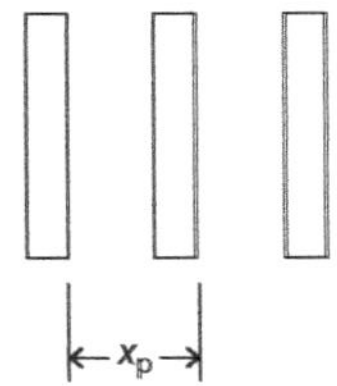


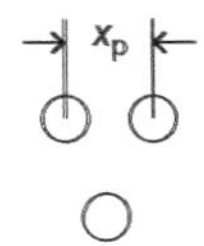


Note: If $C_d < 1$ then $w_{eff} = w\,C_d$

$d_{eff} = d\,C_d 0.5$

$V_{eff} = V/C_d$

$f_{eff} = f\,C_d$

**Figure 2.11** Comparison of the Martin and Saad correlations for convective heat transfer coefficients.

Additional corrections for high rate drying may apply. The expression for low rate mass transfer becomes infinite as the web temperature approaches the boiling point of the solvent. In practice, when the web approaches the boiling point, convection reduces the rate of drying. This is rarely encountered in practice (it is quite hard to boil water without the use of IR, microwave, or RF heating, and it is quite dangerous to dry organic solvents at high rates). There are several different models for high rate mass transfer, and the reader is referred to the standard references on mass transfer for further details [43].

The most common complication of drying is the use of mixed solvents in coating formulations. Multicomponent drying is significantly more difficult to model than single component. Unless the solvent blend forms an azeotrope, either a thermodynamic azeotrope or a dynamic azeotrope [44], there will be no constant rate period as discussed in the following text.

#### 2.2.1.3 Infrared Drying

IR drying is relatively common in the paper and printing industries. While less common in the converting and advanced materials industries, it is still a viable option in many cases.

The Stefan–Boltzmann law cited earlier applies to black bodies (ideally absorbing/emitting—the absorbance and emissivity being equal by Kirchhoff's law). While this is sufficient to gain insight into incidental IR heat transfer, IR drying requires a bit more care. There is a standard theory for gray bodies [43]; however, real webs have complex absorption spectra that do not correspond to the gray body theory. This has led to the common practice of measuring the IR absorption of a material, choosing IR sources that emit strongly in the wavelength region of maximum absorption, and generating basic data on a pilot line. The peak wavelength of emission of a hot body depends primarily on the temperature of the emitter. Thus, IR emitters are typically divided into three temperature ranges near IR (short wave), middle IR (medium wave), and far IR (long wave).

The fundamental limitation on IR drying is that radiation increases heat transfer into the coating without enhancing external mass transfer. Thus, convective mass transport is needed in addition to IR. This combines all of the complexity of convection with IR. There are also significant safety considerations for IR with solvents (the temperatures of the IR panels exceeds the autoignition point of most solvents). Pure IR heating is useful for the heat-up period (described in the following text under drying curve). Hybrid hot air/IR drying gives very high heat and mass fluxes if the coating absorption can be matched to an appropriate emitter (again, most commonly in aqueous systems).

#### 2.2.1.4 UV Curing

UV curing is a common option for crosslinking coatings. Conceptually, UV curing is simple: a neat coating (either coated at 100% solids or solvent coated and dried) is exposed to UV light that excites a photoinitiator that in turn initiates a crosslinking reaction. Oxygen is an inhibitor to UV curing and much UV curing is done in a nitrogen-purged atmosphere. However, in many cases it is possible to cure in air by exposing the coating to a strong enough UV source that the initiator consumes the oxygen and the reaction is carried out before additional oxygen can diffuse into the coating. UV curing requires a photoinitiator to absorb the light and create a free radical that initiates the reaction. In practice, UV curing is complex—minimum controls include monitoring oxygen levels in the box, control of the total UV dose, and the focus of the light. Coatings must be formulated for UV curing and initiators are often unstable.

#### 2.2.1.5 E-Beam Curing

E-beam curing is a specialized but attractive means of crosslinking a certain coating by exposing the coating to a beam of free electrons accelerated to high speed. E-beam curing units are available with accelerators in the millions of volts; however, e-beam curing for coating applications is typically in the hundreds of kilovolts.

While e-beam curing could be used on dried solvent-borne coatings, the utility comes from its ability to cure solvent-free coatings at high speed (up to 500 m/min), and the general practice is to use e-beam without solvent, thus eliminating the need for a dryer. This greatly reduces capital equipment costs for high speed coating and drying. Because e-beam cured systems are solvent-free, they also eliminate many of the costs associated with solvent drying including the variable cost of solvent. The capital cost of safety and pollution abatement equipment is greatly reduced, as is the overhead in managing emissions. Control of e-beam cure hardware is mostly electrical and is more responsive than control of most drying hardware (dielectric drying being the exception) that involves thermal inertia. E-beam curing reactions are generally extremely fast and are fully reacted off the coater. Crosslinked coatings tend to be insensitive to chemical exposure and can be abrasion resistant since they are 100% solids. E-beam coatings are free from residual solvents (that must diffuse out of dried coatings). While they may contain unreacted monomers, it is easier to get to complete reaction than to complete dryness. The penetration depth of electrons is proportional to the energy. E-beam systems that produce millions of volts can be used to crosslink internal adhesive layers in rather thick laminates.

#### 2.2.1.6 Dielectric Drying

There are two other techniques for getting the energy into the coating that should be mentioned: radio-frequency drying and microwave drying. Although well known and commercially available, neither is common. Both depend on the coating—generally the solvent—strongly absorbing the emitted energy. In both cases water has a strong absorption and both techniques are primarily found in aqueous coating. One advantage of dielectric drying is that if the energy is absorbed by the solvent but not the dried coatings, these techniques can be used to equalize the solvent level across the sheet: if there are thick lanes that are still wet, these will absorb more energy than areas of the coating that are already dry. This can also be used to minimize the upper temperature the coating and web are exposed to in the falling rate zone discussed in the following text.

#### 2.2.1.7 The Drying Curve

Drying is well known to proceed through sequential periods. In the "heat-up" period, the temperature either rises or falls until the evaporative cooling balances the heat flux. During the heat-up period, sensible heating of the coating and substrate is significant. Once the heating and evaporative cooling are balanced, the coating enters the "constant rate" period. Constant rate drying requires the mole fraction of solvent in the air just above the coating ($y$) to be independent of the concentration of solvent in the coating, and since the web temperature is constant, sensible heating of the solid phase is not significant (thus the analogy to the wet bulb thermometer appears to be appropriate). For porous solids—including porous coatings (e.g., clay coatings on paper)—as the water content decreases below a critical value, the rate at which water can be wicked to the surface rather than convective heating limits the rate of mass transfer, and the vapor phase concentration starts to drop. Evaporative cooling no longer balances the heat flux and the temperature of the coated substrate approaches the dry bulb temperature of the air. This stage is called the "falling rate" period and sensible heating of the substrate is again important. Dried coatings may also be passed through humidification or conditioning zones where a controlled amount of moisture is added back to the coating.

The sequence of heat-up, constant rate, falling rate, and rehumidification seems intuitive for porous materials. It is less obvious why solution coatings that are not porous should follow this same sequence. There are four mechanisms that can cause a nonporous coating to exhibit constant rate drying. First, some continuous coatings phase-separate on drying (e.g., once set, the gelatin-based coatings used in the photographic industry expel water when heated). Second, the vapor pressure of solvents over simple polymer solutions in good solvents does not change appreciably until the volume fraction of dissolved polymer is on the order of 50%. Third, mixed solvent systems can give true azeotropic mixtures. Finally, so-called "selective drying" of mixed solvents can create a dynamic azeotrope [44]. The falling rate period in continuous coatings is most commonly triggered by solution thermodynamics (in particular the entropic effects of polymeric solutions modeled by Flory–Huggins theory). The critical solvent content that divides the constant and falling rate periods is the surface concentration. The concentration profiles depend on diffusion and drying history in a complex way. However, the transition from constant rate to falling rate in nonporous coatings is governed by interfacial thermodynamics.

It is common to build drying ovens in "zones." From a practical standpoint, the dryer must be broken into pieces that can be shipped by truck. From a processing standpoint, building the dryer in zones allows processing conditions to be optimized for each period of the drying curve. For high temperature drying, IR or hybrid hot air IR is often used to minimize the heat-up period. Since the web temperature is reduced by evaporative cooling in the constant rate period, constant rate zones can run at higher air temperatures than the later zones, provided the solvent level does not approach the lower explosive limit (LEL) and the coating does not enter the falling rate period in the zone. The transition from constant rate to falling rate generally happens within a zone (often called the transition zone); thus, it is common to install IR pyrometers at the end of each zone to monitor the exit temperature. The first zone in which the temperature rises above what would be expected from constant rate drying is the transition zone, the zones that come after that are generally falling rate zones. However, there is one common exception to that rule: mixtures of high boiling and low boiling solvents can exhibit multiple constant rate periods (a low boiling constant rate period followed by a transition to a higher boiling constant rate period). Conditioning zones (most commonly rehumidification zones to control the moisture content in aqueous coatings) are also common and may require different control hardware than drying zones (e.g., moisture sensors and typically more humidification capacity to enable two-sided humidity control).

Solvent levels in the dryer should be monitored continuously. For aqueous drying, that can be done with dew point sensors (which are tied into humidity control schemes). For organic solvents, the solvent level must be monitored to avoid vapor phase explosions in the dryer. Fire codes vary, but the guidance offered by NFPA 86 is a good reference point: the average solvent level in the dryer must be below 25% of the LEL by design unless there is continuous monitoring of the solvent level. With continuous monitoring, the average solvent level monitored at the point of highest solvent level in the oven must remain below 50% of the LEL. Given the difficulty in accurately predicting the maximum solvent load sent to the dryer, continuous monitoring is highly recommended. Since the solvent level is referenced to the LEL, it is common to express solvent vapor concentrations as a percent of LEL. Safety policies should include a maximum %LEL allowed in the dryer (reaching the maximum %LEL will trigger an interlock) along with alarm levels. The most common instrument for monitoring LEL level is the Control Instruments PrevEx flammability analyzer. These should be installed in the hot air stream—installing in a side stream risks the possibility of solvent condensing in the line giving an artificially low reading and increases the response time of the sensor. Solvent ovens should also be vented to direct the exhaust from an explosion to a safe spot.

In North America, oven venting should follow NFPA 68. In general, the best guidance for designing a safe solvent oven is offered by a reputable vendor familiar with all applicable local codes.

Although the popularity has waned, another option for mitigating the hazards of vapor phase explosions is the use of inert gas dryers. These dryers use nitrogen-enriched (or oxygen-depleted) air, sometimes loosely referred to as nitrogen, but typical solvents cease to burn if the oxygen level is below a threshold (on the order of 10%). To minimize the cost of makeup nitrogen, most inert gas ovens use solvent condensation versus combustion. Note that the use of oxygen-depleted air introduces a significant asphyxiation hazard to the area.

## 2.2.2 Hardware

### 2.2.2.1 Conduction

The most common method for heat transfer by conduction is the heated or chilled roll. Extrusion coating requires a quench drum to solidify the coating, and most hot melt coaters use cooling rolls as well. Chill rolls are also commonly found in ordinary coating and drying lines to bring the web temperature to ambient prior to winding (to avoid thermal stresses on the wound roll as it cools). The most common design for a chill roll is a double-walled spiral channel. Typical heat transfer coefficients for rollers are much higher than for convective heating or cooling. Table 2.2 gives typical ranges of heat transfer coefficients and drying rates for various unit operations.

There are limitations on how much heat can be taken out of a coating by conduction. First, there are geometric limitations on how much the web can wrap around the roller. Second, in most cases, the rollers cannot contact the wet side of the web; thus only one roller can be used. Third, the surface temperature of the roll is generally limited by the dew point of the air (to avoid condensation). Finally, as the material cools, it tends to shrink (and conversely expands on heating). When the thermally induced shear stress exceeds the traction pinning the sheet to the roll, the web can lose contact with the roll (and may be scratched as it shrinks).

“Steam cans” are common in the paper industry, but conductive heating is uncommon in the converting and advanced materials industries. The most common use of heated rollers is in lamination. Rolls may be heated by steam, hot oil, or electricity. The most uniform roll temperatures are obtained by induction heating with heat pipes to equilibrate temperatures.

### 2.2.2.2 Convection

There are four general types of airflow in convective dryers: laminar flow, parallel turbulent flow, oblique turbulent flow (often opposed oblique flows that produce a “pressure pad”), and turbulent impingement (flow normal to the web). Turbulent flow produces the most efficient drying; laminar flow is

**Table 2.2** Approximate ranges of heat transfer coefficients and drying rate for various unit operations.

| Mode | Examples | HTC (BTU/h·ft²·F) | Max flux (per side[a]) (BTU/h·ft²) | Comments |
|---|---|---|---|---|
| Conduction | Steam can | 100–180 | 50,000 | Web typically pressed on by fabric |
| | Thermal fluid cylinder | 50–100 | 40,000 | Affected by air film gap at high web speeds |
| | Chill roller | 70–130 | 30,000 | |
| Convection | Hot box—Festoon | 1–10 | 5,000 | HTC depends on web speed |
| | Floater | 8–30 | 20,000 | HTC depends on web flotation limits |
| | Impingement | 5–60 | 25,000 | HTC depends on air jet intensity |
| Radiation | Thermal fluid panel | | 2,500 | Long wave lengths, ok for water |
| | Ceramic panels—MW | | 40,000 | Medium wave, ok for water |
| | SW tungsten lamps | | 25,000 | Short wave, less efficient for water |
| | Halogen lamps | | 40,000 | Near IR wave length, ok for pigments, not efficient for water |
| Microwave/radio frequency | Thick materials | | 40,000 (* total-internal heating) | Depends on "loss factor" for material—water absorbs well |
| | Ceiling tiles | | | |
| | Selective heating | | | |
| | Glue beads, heat seals | | | |

[a] While most of the numbers are per side, microwave and radio frequency heating are internal, not per side.

reserved for very delicate coatings such as optical coatings. Parallel turbulent flow produces the lowest pressure spike of the turbulent configurations, while impingement produces the highest. As discussed earlier, the pressure spike under the nozzle seems to be critical in achieving high nozzle efficiency because it overcomes the boundary layer limitation of the heat transfer coefficient varying with the 2/3 power of slot velocity. However, higher pressure spikes on the wet side of the coating increase the probability of disturbing the wet coating (either causing mottle, blow around, or, in the extreme, doctoring the coating off the web in the oven). Usually heat transfer efficiency is desirable and this is sometimes treated as an objective of optimization; inducing defects through airflow is considered a constraint on the optimization.

There are two subclasses to impingement nozzles: slot nozzles (that ideally create a uniform curtain of turbulent air impinging on the web) and hole nozzles (that use an array of holes to create circular jets that impinge on the surface). These are shown in Figure 2.12. Hole nozzles are slightly more efficient for heat transfer but tend to produce machine direction lanes (visible as diffuse lines) in the dried film.

There are six well-known ways to convey the web through an oven (four of which are common in current practice). Stick dryers convey the web by hanging it from sticks that are held in a moving track [11]. These were once common and were known as festoon dryers because the film hung in "festoons" between the sticks. Two-side roll support ovens (in which the web passes over alternating dry side rollers and wet side air bearings) have largely replaced stick dryers for slow drying and are currently sold as festoon dryers although the web is taut between the rolls. The webs can be supported by a vacuum belt or vacuum rolls. Vacuum rolls are commonly used for tension isolation prior to the dryer but

**Figure 2.12** Comparison of slot (top) and hole (bottom) impingement nozzles. Note that the flow is normal to the web on both sides. *Source:* Courtesy of B&W MEGTEC Inc.

are uncommon in the dryer itself. One side roll support (commonly using tendency-driven rolls) arranged in an arch or spiral configuration can be paired with impingement nozzles or tangential flow nozzles for high rate drying or for more delicate webs, a more tangential flow shown in Figure 2.13. Single-side floatation nozzles work by producing high parallel flow regions over the nozzle and low flow regions in the open space between nozzles, leading to alternating suction toward the nozzle (in the high flow areas) and away from the nozzles in the low flow return areas (see air foil illustration; Figure 2.14). Because this pressure field can be modeled with Bernoulli's equation, one-side floatation nozzles are often referred to as Bernoulli nozzles. Two-side floatation uses opposed

**Figure 2.13** Roll support for delicate and extensible webs. *Source:* Courtesy of Polytype Converting.

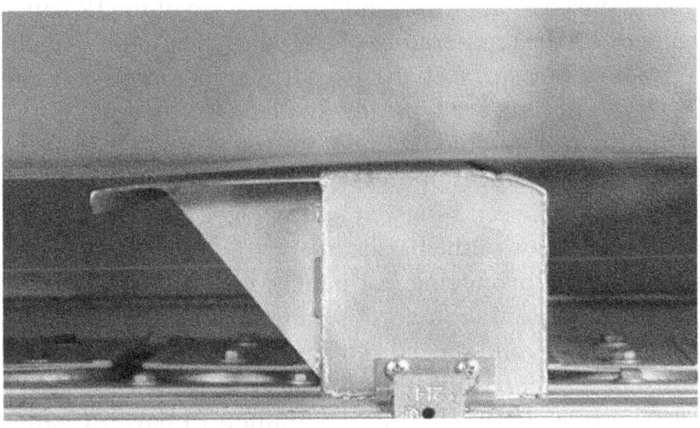

**Figure 2.14** One-sided air floatation nozzle, also known as airfoils or Bernoulli bar. *Source:* Courtesy of B&W MEGTEC Inc.

**Figure 2.15** Two-sided air floatation nozzles also known as air bar or Coanda bars. *Source:* Courtesy of B&W MEGTEC Inc.

oblique flow nozzles on both sides of the web to produce an undulating pattern or "sine wave" in the web. The common nozzles use opposing jets to create a stagnant region between the jets that produces a pressure pad; alternating pressure pads above and below the web create the sine wave. Because these nozzles use the Coanda effect (jets tending to follow a curved surface) to steer the opposing jets, they are often referred to as Coanda bars (see Figure 2.15).

#### 2.2.2.3 IR Drying

As noted earlier, the efficacy of IR drying depends on producing a good match between the absorption of the coating or web and the emission of the IR panel. The peak emission depends on the temperature of the panel; a number of technologies are used to produce hot surfaces that vary in temperature between roughly 500 and 2300°C. IR emitters are typically divided into three temperature ranges: near IR (short wave), middle IR (medium wave), and far IR (long wave); however, definitions of the ranges of these categories vary, [7, 8, 11] and short wave can be subdivided between tungsten and halogen lamps. Satas [7] provides a comprehensive overview of the hardware.

There are four basic modes of producing hot panels: heating with thermal fluids, combustion of natural gas (either in porous panels or catalytic combustion), electrical resistance heating, and specialized lamps (e.g., quartz halogen). For preheating and post-curing operations, IR heating can be done with a large panel. However, for drying operations, it is more common to have IR lamps mounted in a convection nozzle—either between the slots of a Coanda nozzle as shown in Figure 2.16 or in the planar surfaces of a Bernoulli nozzle.

**Figure 2.16** Hybrid IR/Coanda bar nozzle. *Source:* Courtesy of B&W MEGTEC Inc.

There are several operational and safety features of IR heating that warrant mention. First, the rapid heating of solvent coatings without sufficient dilution air can produce large clouds of explosive vapors. Clearly that requires isolation of solvent from ignition sources and careful air management (thus the common use of hybrid hot air/IR heating). Second, since the hot panels needed for effective IR heating are often capable of igniting the solvent and often the web, the IR dryer should be designed to avoid direct contact of the hot panel with the web both in normal transport mode and in the case of a web break. This is frequently accomplished with quartz windows (that must be kept clean). Third, the high temperatures of IR heating can cause partial degradation of solvents and or sublimation of decomposition products in the coating, leading to tarry deposits in the oven and emissions issues. Tars reduce the efficiency of heat transfer, leading to maintenance issues as well as the obvious safety, health, and environmental issues. IR heating has a tendency to melt the web when the line is stopped; thus, it is imperative to work with a vendor that can either provide rapid shutdown of the IR or install shields that can be deployed to protect the web.

Despite the operational concerns noted earlier, IR heating can produce extremely high heat fluxes and hybrid IR/convection dryers can produce among the highest drying rates known as indicated in Table 2.2.

## 2.3 Defect Management

### 2.3.1 Characterizing Defects

The first step in eliminating defects is to characterize the defect (not to name it). There are a wide variety of analytical techniques that are in routine use to characterize defects:

- Optical and digital microscopy
- Scanning electron microscopy
- IR microscopy
- Energy-dispersive X-ray diffraction
- Electron spectroscopy for chemical analysis
- X-ray fluorescence spectroscopy
- Fourier transform IR spectroscopy

Defect mapping shows the location and frequency of defects in both the machine and transverse directions. This map can then help locate where in the process the defect is created. Chatter is another defect for which defect mapping is helpful (the measured frequency of the chatter is compared with the frequencies of all components in the line to pinpoint the source of vibrations).

### 2.3.2 Defect Naming

A defect must have a unique name in order to track its occurrence. Several efforts have been made to standardize names [45, 46]; however, colloquial names based on visual appearance remain the norm. It is important that the defects be given the correct name when consulting references that suggest root causes and corrective actions to eliminate the defect. If the name is not accurate, it can result in focusing efforts in the wrong direction. One should avoid naming a defect by choosing the root cause or corrective action associated with it.

### 2.3.3 Online Defect Characterization Systems

Two categories of properties need to be characterized to ensure product quality: gross variations in coating weight and the presence of defects in the coating. Visual inspection and offline testing have largely been replaced by online monitoring and inspection systems.

There are two general types of weight (thickness) measurement systems: total weight/subtraction gauges and direct coating weight gauges. Total weight gauges compare the total weight under the sensor head before and after coating. These may work by beta transmission, gamma backscatter, or X-ray transmission.

Direct weight gauges detect the presence of the coating only. These commonly work by IR absorption, beta backscatter, or X-ray fluorescence.

There are two basic types of inspection systems in use. Laser systems use a laser beam to scan the substrate and detect defects by deflection of the laser beam. Camera systems capture images of the defect. Laser systems can capture more kinds of defects; camera systems are less expensive and can record images in both color and gray scale. Both types of systems can be used for bright-field (direct, line-of-sight) transmission, dark-field (off-axis) transmission, and reflection. Laser systems offer polarity shift detection as well.

### 2.3.4 Defect Troubleshooting

Coating quality has leapt ahead in the past decade from fundamental understanding of the coating process and improved instrumentation. Nonetheless, defects still occur. There is a dichotomy of thinking in defect management. Some believe in defining a standard process with few if any deviations from standard operating conditions. Others accept variations and routinely troubleshoot the process. Regardless of the approach, the majority of defects arise from the following 10 deficiencies.

#### 2.3.4.1 Contamination

Contaminants and impurities in raw materials lead to several types of defects. Particles that hang up in the coating head lead to streaks; particles that pass through the coating head lead to discrete (spot) defects. Hard particles such as metallic impurities in the coating can be eliminated by filtration. Softer particulates such as gels can deform to pass through filters.

Ambient dirt, grease, or oil contaminant can lead to repellent spots (craters). These can be controlled by keeping the coater and the coating room clean; however, many cleaning products (including cleaning rags and pads) can become sources of contamination. Dryers are particularly prone to airborne dirt drawn in through the filters, bits of dried coating that have been splattered on the walls, shards of broken film left in the oven, and on occasion spores from mold that has grown on air filters.

#### 2.3.4.2 Substrate Deficiencies

If the surface tension of the substrate is higher than expected, the solution will not uniformly wet the substrate, leading to poor coating quality. Contaminants on the substrate surface will lead to physical defects such as spots or craters. Web cleaners can be effective in removing some but not all contaminants.

Thickness variations in the substrate will lead to coating weight variation in a doctored coating process. Nonplanarities (that may or may not be caused by thickness variations) affect coating uniformity in many processes as noted in Table 2.1.

Static charge generated on the substrate surface attracts dirt to the web, and since it results in an uncontrolled electrostatic assist to wetting, it can leave static marks in the coated product. It can also lead to fire if flammable solvents are present.

#### 2.3.4.3 Liquid Coating Quality

If the coating solution or dispersion is not uniformly mixed, it may lead to coating defects due to agglomerates, gel particles, nonuniform composition, and irreproducible rheology. Temperature variations also affect rheology. Dissolved air and bubbles can cause streaks or holes in the coating. Aqueous coatings are particularly susceptible to bacterial and fungal growth.

Many liquid coatings are not stable as they age. Dispersions and emulsions agglomerate and may not be redispersible. Polymers will drop out of solution if raised above or below consolute points, and some solutions will crosslink with age. Storage requirements should be defined at scale-up.

#### 2.3.4.4 Unsuitable Coating Method

Operating close to the coating window boundaries often leads to poor quality (e.g., streak sensitivity) and irreproducibility. The most common reason for use of an unsuitable coating method is a desire to use an existing coater for a new product that does not work well on the old technology.

#### 2.3.4.5 Inadequate Design of Coating Line Equipment

Each coating line has many hardware components. Inadequate design or injudicious compromises made in the design to any of the component or section will result in defects or yield losses.

#### 2.3.4.6 Deterioration of Coating Line Equipment

Maintenance of all mechanical tolerances and alignment on coating heads, pumps, drives, bearings, couplings, and rolls are critical. A sharp edge is required on the essential edge of die, curtain, and knife coaters. However, these can be rounded with time. Knicks lead to streaks. To minimize the effect of wear, the edges should be closely monitored on a regular basis. Once the shape has changed, the hardware should be returned to the manufacturer for refurbishing.

#### 2.3.4.7 Drying-Induced Defects

More defects occur in the dryer than any other place in the coater. These can come from airflow disturbing the wet coating, contamination, residual stresses, and changes in density or surface tension with heating or drying, scratches, and rubs. One pitfall of dryer control is that dampers are generally the least reproducible control elements on the coating line; thus monitoring of airflows is imperative.

#### 2.3.4.8 Variations in Web Handling System

Variations in web speed or tension can induce mechanical chatter, overstretch the web, or cause the web to bounce, leading to air bar rubs. The most common root causes of speed and tension fluctuations are drive systems (especially couplings), bad bearings, out-of-round rolls, insufficient tension isolation from zone to zone, and vacuum roll sticking.

#### 2.3.4.9 Inadequate Operating Procedures and Training

A modern coater can have hundreds to thousands of parameters to control and each product can have unique settings. Standard operating conditions should be defined for all of these (many can be programmed into the control hardware). Adherence to procedures for all process equipment and products is needed to ensure optimal operation.

#### 2.3.4.10 Key Variables Not Properly Controlled

One can debate which variables are key variables; experience dictates that all coating process variables should be considered important and controlled to minimize defect formation. Some say of coating "the Devil is in the details"; others say, "God is in the details." This dichotomy may not be theological as much as a reflection of how much attention has been paid to the details. Those in the former group are more likely to live in a culture of troubleshooting; those in the latter group are more likely to run a standard process.

## References

1 Bezigian, T., *Extrusion Coating Manual*, 4th Ed., Atlanta: TAPPI Press (**1999**).
2 Booth, G. L., *Coating Equipment and Processes*, New York: Lockwood Publishing Co, Inc. (**1970**).
3 Cohen, E. D. and Lightfoot, E. J., Coating Processes. In: Kirk, R. E. and Othmer, D. F. (Eds.) *Encyclopedia of Chemical Technology*, New York: John Wiley & Sons, Inc. (**2011**).
4 Cohen, E. D. and Gutoff, E. B., *Modern Coating and Drying Technology*, New York: VCH Publishers (**1992**).
5 Gregory, B. H., *Extrusion Coating: A Process Manual*, Victoria: Trafford Publishing (**2007**).
6 Kistler, S. F. and Schweizer, P., *Liquid Film Coating: Scientific Principles and Their Technological Implications*, London: Chapman & Hall (**1996**).
7 Satas, D. (Ed.), *Web Processing and Converting Technology and Equipment*, New York: Van Nostrand Reinhold (**1984**).
8 Shepherd, F., *Modern Coating Technology Systems*, Barnet: Emap Maclaren, Ltd (**1995**).

9 Tracton, A. A., *Coating Technology Handbook*, 3rd Ed., New York: Marcel Dekker (**2005**).
10 Wagner, J. R., *Multilayer Flexible Packaging*, 2nd Ed., San Diego: Elsevier (**2016**).
11 Weiss, H. L., *Coating and Laminating Machines*, Milwaukee: Converting Technology Co. (**1977**).
12 Adamson, A. W. and Gast, A. P., *Physical Chemistry of Surfaces*, 6th Ed., New York: Wiley Interscience (**1997**).
13 Snoeijer, J. H. and Andreotti, B., Moving contact lines: Scales, regimes, and dynamical transitions. *Annu. Rev. Fluid Mech.* 45, 269–292 (**2013**).
14 Bonn, D., Eggers, J., Indekeu, J., Meunier, J., and Rolle, E., Wetting and spreading. *Rev. Mod. Phys.* 81, 739–805 (**2009**).
15 Blake, T. D., *J. Colloid Interface Sci.* 299(1), 1–13 (**2006**).
16 Deryagin, B. V. and Levi, S. M., *Film Coating Theory*, London: Focal Press (**1964**).
17 Burley, R. and Kennedy, B. S., A study of the dynamic wetting behaviour of polyester tapes. *Br. Polym. J.* 8, 140–143 (**1976**).
18 Teletzke, G. F., Davis, H. T., and Scriven, L. E., How liquids spread on solids. *Chem. Eng. Commun.* 55(1–6), 41–82 (**1987**).
19 Mues, W., Hens, J., and Boiy, L., Observation of a dynamic wetting process using Laser-Doppler velocimetry. *AICHE J.* 35(9), 1521–1526 (**1989**).
20 Blake, T. D. and Morley, S. D., Coating processes. US Patent 5,792,515 A (**1998**).
21 Rutherford, B. (Ed.), *Gravure: Process and Technology*, Sewickley, PA: Graphic Arts Technical Foundation (**1997**).
22 Hens, J. and Van Abbenyen, W., Slide Coating. In: Kistler, S. F. and Schweitzer, P. M. (Eds.) *Liquid Film Coating: Scientific Principles and Their Technological Implications*, London: Chapman & Hall (**1997**).
23 Schweizer, P. M., Visualization of coating flows. *J. Fluid Mech.* 193, 285–302 (**1988**).
24 Beguin, A. E., Method of coating strip material. US Patent 2681294 A (**1954**).
25 O'Brien, W. G., Beveled edge metered bead extrusion coating apparatus. US Patent 4,445,458 (**1984**).
26 Coyle, D. J., Knife and Roll Coating. In: Kistler, S. F. and Schweitzer, P. M. (Eds.) *Liquid Film Coating: Scientific Principles and Their Technological Implications*, London: Chapman & Hall (**1997**).
27 Mayer, C. W., Coating machine. US Patent 1043021 A (**1912**).
28 Iwasaki, T., Gravure coating device and method. US Patent 4,948635 A (**1990**).
29 Lightfoot, E. J., Pressurized gravure coating. 11th International Coating Science and Technology Symposium, Minneapolis, Minnesota, September 23–25, (**2002**).
30 Patel, R. and Benkreira, H., Gravure roll coating of Newtonian liquids. *Chem. Eng. Sci.* 46(3), 751–756 (**1991**).

31 Pranckh, F. R. and Coyle, D. J., Elastohydrodynamic Coating Systems. In: Kistler, S. F. and Schweitzer, P. M. (Eds.) *Liquid Film Coating: Scientific Principles and Their Technological Implications*, London: Chapman & Hall (**1997**).
32 Dodds, S., Carvalho, M., and Kumar, S., Stretching liquid bridges with moving contact lines: The role of inertia. *Phys. Fluids* 23, 092101 (**2011**).
33 Pulkrabek, W. W. and Munter, J. D., Knurl roll design for stable rotogravure coating. *Chem. Eng. Sci.* 38, 1309–1314 (**1983**).
34 Booth, G. L., Take your choice of coating methods, Part 1. *Mod. Plast.* 36(1), 90–95 (**1958**).
35 Booth, G. L., Take your choice of coating methods, Part 2. *Mod. Plast.* 36(2), 90–99 (**1958**).
36 Jacobs, R. J., Fundamentals to Consider in Selecting Coating Methods, published in five parts, Paper Film and Foil Converter February through July (**1963**).
37 Scriven, L. E., Bringing fundamentals to bear on choosing coating method. Symposium on Coating Technologies for Imaging Materials, The Society for Imaging Science and Technology, Springfield, VA, May 12–17, (**1991**).
38 Yapel, R. A., Huelsman, G. L., Milbourn, T. M., and Kolb, W. B., Gap drying with insulation layer between substrate and heated platen. US 6134808 A (**2000**).
39 Chilton, T. H. and Colburn, A. P., Mass transfer (absorption) coefficients prediction from data on heat transfer and fluid friction. *Ind. Eng. Chem.* 26, 1183–1187 (**1934**).
40 Martin, H., Heat and Mass Transfer Between Impinging Gas Jets and Solid Surfaces. In: Hartnert, J. P. and Irvine, T. F. (Eds.) *Advances in Heat Transfer*, Vol. 13, New York: Academic Press (**1977**): pp. 1–60.
41 Saad, N. R., Mujumdar, A. S., and Douglas, W. J. M., Heat transfer under multiple turbulent slot jets impinging on a flat plate. *Drying* 80, 422–430 (**1980**), Vol. I. New York: Hemisphere Publishing Corp.
42 Lightfoot, E. N. and Lightfoot, E. J., Mass Transfer. In: Ruthven, D. M. (Ed.) *Encyclopedia of Separation Technology*, Vol. 2, New York: John Wiley & Sons, Inc. (**1997**): pp. 1138–1147.
43 Bird, R. B., Stewart, W. E., and Lightfoot, E. N., Jr., *Transport Phenomena*, New York: John Wiley & Sons, Inc. (**1960**).
44 Riede, T. and Schlűnder, E., Selective evaporation of a ternary mixture containing one nonvolatile component with regard to drying processes. *Chem. Eng. Process.* 28, 151–163 (**1990**).
45 Smith, R. D. (Ed.), *Ultimate Web and Roll Troubleshooting Guide*, Atlanta: TAPPI Press (**2013**).
46 Cohen, E. D. and Gutoff, E. B., *Coating and Drying Defects: Troubleshooting Operating Problems*, 2nd Ed., New York: John Wiley & Sons, Inc. (**2006**).

# 3

# Drying of Polymer Solutions: Modeling and Real-Time Tracking of the Process

*S. Shams Es-haghi*[1,2] *and Miko Cakmak*[1,2,3]

[1] *Birck Nanotechnology Center, Purdue University, West Lafayette, Indiana, USA*
[2] *School of Materials Engineering, Purdue University, West Lafayette, Indiana, USA*
[3] *School of Mechanical Engineering, Purdue University, West Lafayette, Indiana, USA*

## 3.1 Introduction

Drying of solvent-cast or coated polymer films is a key polymer processing step in roll-to-roll (R2R) operations with significant relevance to various coating technologies [1], membrane technologies [2, 3], polymer film manufacturing [1, 4], and production of various microelectronic devices [5, 6].

The rapid industrial growth of liquid crystal display applications [7] and the advent of novel technologies such as flexible electronics have prompted the development of novel drying processes of polymer solutions. Hence, the coating of polymer solutions, first developed more than a century ago to supply the demand of the photographic industry, is getting increasingly important for the production of high-quality thin polymer films [1].

The ability of solvent casting technology to produce polymer films with high flatness, uniform thickness, good dimensional stability, and extremely low haze makes it an ideal processing operation for the production of polymer films that are used in many optical applications [1, 4]. Moreover, solvent casting is an excellent technique for the production of films from thermally sensitive polymers [1, 4] and production of high temperature-resistant films that their melt processing is difficult or impossible such as amorphous polymers with high glass transition temperature, for example, polyimides [8, 9].

Solvent casting and coating of polymers are flexible processes that can be combined with other processing operations, such as electrospinning, in order to produce hybrid thin polymer films with embedded nanofibers to gain desired functional properties [5, 10]. Other modifications can be achieved by

*Roll-to-Roll Manufacturing: Process Elements and Recent Advances*, First Edition.
Edited by Jehuda Greener, Glen Pearson, and Miko Cakmak.

developing a casting line equipped with electric, magnetic, and thermal fields to control the alignment of the particles within a film [11].

Apart from the technological importance of casting and coating of polymer solutions, intriguing physical phenomena might occur during evaporation of the solvent that raise important questions in soft matter physics. Skin formation [12, 13], stress development [14, 15], surface roughening [16], phase separation [17], pattern formation, and Marangoni instability [18, 19] are usually observed in evaporating polymer solutions depending on the processing conditions and physicochemical properties of the system. Marangoni instability can be more serious in the evaporation of solvent from a droplet of a polymer solution, which is important in inkjet printing technology [20]. These phenomena often lead to undesired effects, for example, wrinkles, buckling, cracks [21], and formation of bubbles inside the film [22]. A deep understanding of the physics underlying the drying of polymer solutions will help one to have control over the process and to prevent undesirable effects that could change the quality and performance of the final film product. Among the aforementioned physical phenomena, skin formation and stress development play crucial roles in the quality of a cast polymer film. In effect, most of the undesired effects mentioned previously are caused by skin formation and stress development. Moreover, stress development in the system can result in birefringence [14] and interesting phenomenon like optical gradient across the thickness during evaporation of solvent [23].

The temporal evolution of concentration and temperature during the drying process that strongly depends on the physicochemical characteristics of the system, initial and boundary conditions considerably affect the final properties of the polymer film [12, 13]. In effect, physical phenomena such as skin formation and surface roughening can be triggered or suppressed by changing the processing conditions [24, 25]. The phenomenon like phase separation induced by evaporation of the solvent [17] highly depends on the phase behavior of the system in question. As will be mentioned later in this chapter, the polymer solutions that are usually used for industrial applications often contain more than two components, for example, ternary polymer solutions. In such systems, the transport phenomena are more complicated, and, therefore, a good understanding of the physics of polymer solutions especially diffusion and phase behavior is of crucial importance.

Owing to the complexity of the drying process of polymer solutions—also bearing in mind that numerous factors determine the quality of a solvent-borne polymeric film—a basic understanding of the drying process will be achieved only by real-time monitoring of the process under various processing conditions. In fact, the real-time study will help one to unravel the complex interaction between processing conditions and final properties.

This chapter is organized as follows: The second section summarizes the theoretical works regarding the modeling of the drying process. Governing equations formulated for the description of the temporal evolution of concentration and temperature during the drying process of polymer solutions will be discussed. In the third section, a novel real-time measurement apparatus developed recently for tracking thickness, weight, birefringence, and surface temperature during the drying process is described, and real-time experimental observations made on several polymer solutions under various processing conditions are presented. The chapter closes with concluding remarks.

## 3.2 Modeling of the Drying Process

Film formation induced by evaporation of solvent is a moving boundary problem in that the position of the evaporating boundary is not known a priori and must be solved as part of the problem [26]. The objective of a drying model for solvent casting or coating process is the formulation of a set of governing equations that describe the kinetics of drying and also capture the temporal evolution of concentration and temperature in the system after applying initial and boundary conditions under which the drying process is operated. A model will help one to simulate the evaporation of the solvent and predict the kinetics of drying under different material properties and processing conditions without performing further experiments.

An evaporating polymer solution is a nonequilibrium system in which mass diffusion and heat conduction take place. In some cases such as film formation from solutions of crosslinking polymers, chemical reactions occur between functional groups of the components during solvent evaporation that makes the problem more complicated [27].

The models developed for the drying process of polymer solutions rest on some assumptions. A reasonable assumption can simplify the problem; however, simplicity is obtained at the cost of reducing the accuracy of the model. Therefore, great care must be exercised in making assumptions. After model development, a key step before the simulation is to consider functions for transport coefficients, for example, mass diffusion coefficients and thermal conductivity for the system. The transport coefficients are in general functions of concentration and temperature. A transport coefficient may be assumed constant or estimated by some functions proposed based on experimental data or physical models.

In the following, the development of models proposed for the drying process of polymer solutions is summarized.

Roehner [28] modeled solvent removal from thin polymer films cast on a moving web in direct contact with a carrier gas stream. In the model, it was assumed that the steady-state condition exists at each point along the moving web in the fixed coordinate frame. Moreover, the change in the film thickness was neglected. This assumption may be relevant at the point where most of the solvent evaporated; however, it cannot describe the whole drying process. Knowing the fact that an evaporating polymer solution is an open system that exchanges heat and mass with its surrounding, drying must be modeled as a transient mass and heat diffusion problem.

The diffusion coefficient in Roehner's model was in the form

$$D = D_0 \exp\left(\frac{-\mathrm{Ed}}{RT}\right)\exp(\beta\phi_S), \tag{3.1}$$

where $D_0$ and $\beta$ are constants and Ed, $R$, $T$, and $\phi_S$ are activation energy of the diffusion, gas constant, absolute temperature, and weight fraction of the solvent, respectively. Equation (3.1) was proposed by Kokes and Long in their study of diffusion of vapors of some organic solvents into polyvinyl acetate films [29]. They found that for low concentrations of solvent, the data follow Eq. (3.1). Moreover, in their paper $\phi_S$ was introduced as the volume fraction of the solvent in the polymer.

Drying of a paint was modeled by Blandin et al. [30] in which drying was assumed to be an isothermal mass diffusion problem. The governing equation for this problem was

$$\frac{\partial C}{\partial t} = \frac{\partial}{\partial x}\left(D\frac{\partial C}{\partial x}\right), \tag{3.2}$$

where $C$ is the concentration of solvent in the coating and $D$ is the diffusion coefficient, which was measured from the slope of the plot of the solvent remained in the coating as a function of the square root of time and was expressed as

$$D = D_0 \exp\left(\frac{-\alpha}{T}\right)\exp\left(\frac{-\beta}{C}\right), \tag{3.3}$$

where $D_0$, $\alpha$, and $\beta$ are constants. The same procedure was followed by Ion and Vergnaud [31] to model the drying process of a paint.

Tu and Drake [32] developed a model for mass and heat transfer during the drying of a coating containing a nonvolatile solute and a volatile solvent. A schematic representation of the one-dimensional (1D) diffusion problem modeled by Tu and Drake is shown in Figure 3.1. As shown in Figure 3.1, the solution with a uniform initial thickness $h_0$ is cast on a substrate that is kept at a constant temperature equal to the initial temperature of the solution. Uniform concentration and temperature profiles were considered for the initial conditions.

In this model, it was assumed that drying takes place in two stages, evaporation-limited and diffusion-limited stages. At the evaporation-limited stage, sufficient solvent is available at the surface. The solvent evaporates with the

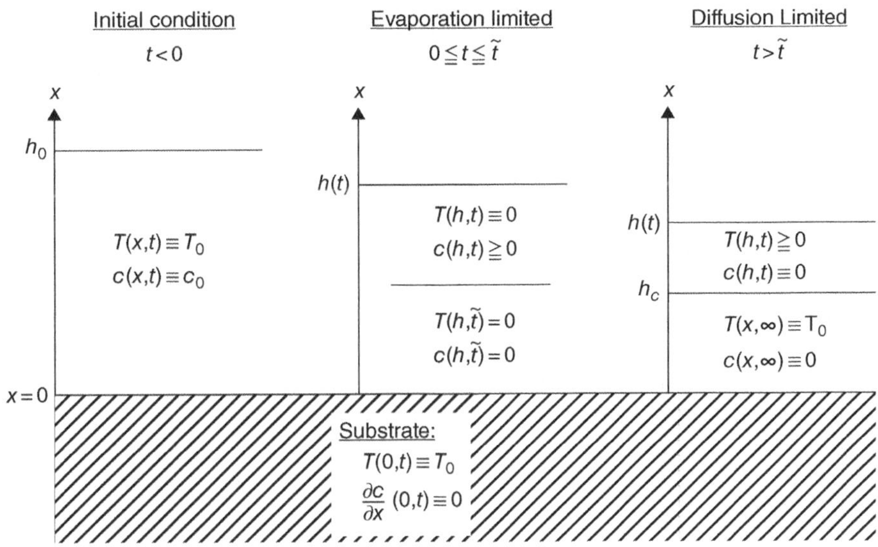


**Figure 3.1** Schematic representation of the one-dimensional mass and heat diffusion during evaporation from a coating. *Source:* Reproduced by permission from Ref. [32].

maximum possible evaporating flux that is associated with $T(h, t) = 0$ in Figure 3.1. The first stage continues until the time $\tilde{t}$ when the concentration at the surface becomes zero. Then, the diffusion-limited stage begins in which the diffusion process cannot transfer solvent to the surface and the temperature at the surface increases. At the end, the final coating reaches the initial temperature, and the concentration of the solvent becomes zero. The governing equations for mass and heat diffusion are given by

$$\frac{\partial c}{\partial t} = \frac{\partial}{\partial x}\left(D\frac{\partial c}{\partial x}\right), \tag{3.4}$$

$$Q\frac{\partial T}{\partial t} = \frac{\partial}{\partial x}\left(k\frac{\partial T}{\partial x}\right) + A(x, t), \tag{3.5}$$

for $0 \leq x \leq h(t)$ and $t \geq 0$. $c$, $D$, $T$, $Q$, $k$, and $A(x, t)$ are volumetric concentration of the solvent, diffusion coefficient, temperature, heat capacity of the solution, thermal conductivity of the solution, and rate of heat generation in the solution, respectively. In this model, the variation of thickness during evaporation of the solvent was taken into account, and the problem was solved numerically as a moving boundary problem. The location of the evaporating boundary was derived by the conservation of mass of nonvolatile solute in the coating.

For numerical simulation, Tu and Drake [32] considered that the material parameters $D$, $Q$, and $k$ are constant and heat generation inside the film was neglected, that is, $A(x, t) = 0$. Therefore, the governing equations were expressed as

$$\frac{\partial c}{\partial t} = D\frac{\partial^2 c}{\partial x^2}, \tag{3.6}$$

$$\frac{\partial T}{\partial t} = \frac{k}{Q}\frac{\partial^2 T}{\partial x^2}. \tag{3.7}$$

The initial conditions, boundary condition at the substrate and constraints for evaporation-limited and diffusion-limited stages, are shown in Figure 3.1. In the evaporation-limited stage, the boundary condition at the moving boundary, $h(t)$, was given by

$$L\rho D\left(\frac{\partial c}{\partial x}\right)_{x=h(t)} = k\left(\frac{\partial T}{\partial x}\right)_{x=h(t)}, \quad T(h, t) = 0, \tag{3.8}$$

where $L$ and $\rho$ are latent heat of vaporization of the solvent and density of the solvent, respectively. The moving boundary was determined by

$$\left[1 - c(t)\right]\frac{dh}{dt} = \frac{k}{L\rho}\left(\frac{\partial T}{\partial x}\right)_{x=h(t)}, \tag{3.9}$$

with the initial condition $h(0) = h_0$.

At $t = \tilde{t}$ when the diffusion-limited stage begins, for $t > \tilde{t}$, the concentration at the surface will be $c(h, t) = 0$, and the equation for the moving boundary will change into

$$\frac{dh}{dt} = D\left(\frac{\partial c}{\partial x}\right)_{x=h(t)}. \tag{3.10}$$

Figure 3.2 shows the temperature and concentration profiles when dimensionless quantities $\alpha = k/QD$ and $\gamma = kT_0/L\rho D$ are 5 and 0.5, respectively.

A model for drying of a polymer solution in a drying chamber was developed by Guerrier et al. [33]. The polymer solution that was called varnish was coated on an aluminum substrate. Then the substrate went through the dryer where perpendicularly it was exposed to hot air jets through a set of nozzles. The typical range of the substrate velocity was between 0.8 and 5 m/s, and the drying took place in a few seconds. The varnish was prepared by dissolving a mixture of several copolymers in an organic solvent. During the course of evaporation, no reaction occurred. Schematic representation of the drying process is shown in Figure 3.3.

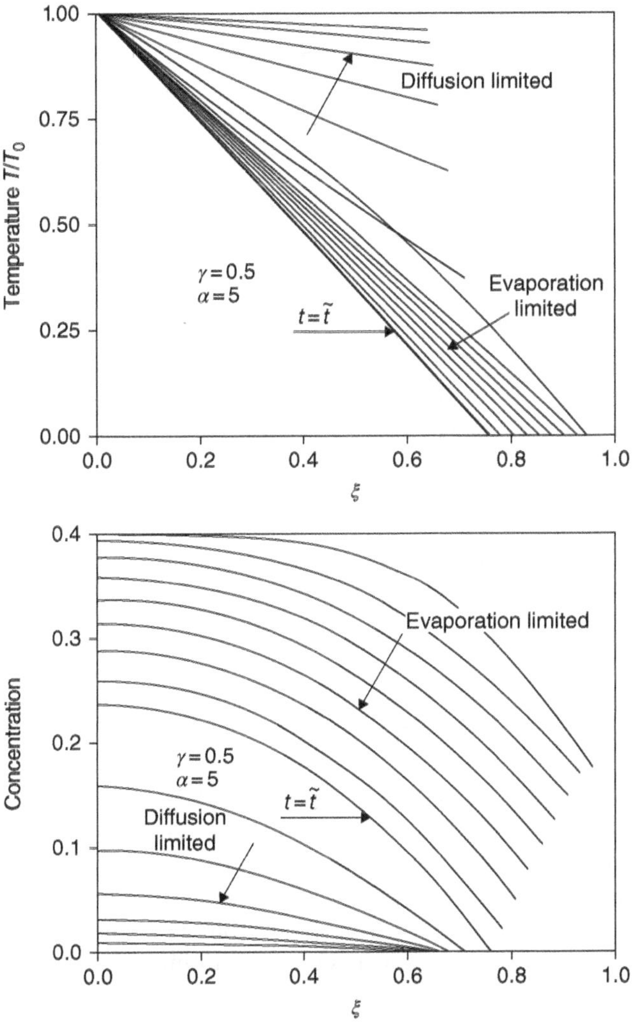


**Figure 3.2** Temperature and solvent concentration profiles for evaporation-limited and diffusion-limited stages. *Source:* Reproduced by permission from Ref. [32].

As shown in Figure 3.3, the drying process was considered as 1D mass and heat diffusion problem. The governing equations for mass and heat diffusion were formulated as

$$\frac{\partial \rho_S}{\partial t} - \frac{\partial}{\partial z}\left( D_{SP}\left(\omega_S, T\right)\frac{\partial \rho_S}{\partial z} \right) = 0, \tag{3.11}$$

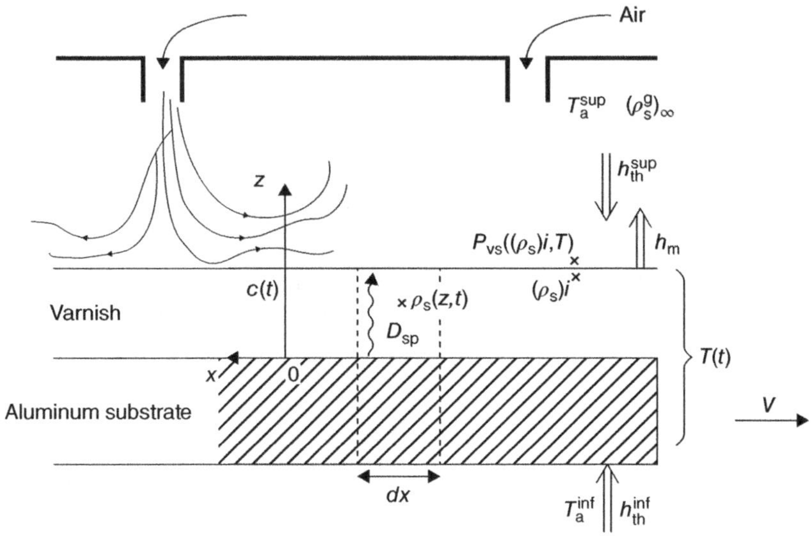


**Figure 3.3** Schematic representation of the drying process of a polymer solution in a drying chamber. *Source:* Reproduced by permission from Ref. [33].

$$\left(\rho^{al}c_P^{al}e^{al}+\overline{\rho c_P e(t)}\right)\frac{dT}{dt}=h_{th}^{sup}\left(T_a^{sup}-T\right)+h_{th}^{inf}\left(T_a^{inf}-T\right)-L_V\Phi_m, \quad (3.12)$$

where $\rho_S$, $D_{SP}$, $\omega_S$, $T$, $\rho^{al}c_P^{al}e^{al}$, $\overline{\rho c_P e(t)}$, $L_V$, and $\Phi_m$ are the local partial density of the solvent, mutual diffusion coefficient, weight fraction of the solvent, temperature, substrate thermal capacity per unit area of the sheet, varnish thermal capacity, vaporization heat, and evaporating solvent flux, respectively. $h_{th}^{sup}$ and $h_{th}^{inf}$ are overall exchange coefficients for the heat transfer on the varnished and non-varnished sides of the substrate, respectively. Evaporating solvent flux, $\Phi_m$, was given by [33]

$$\Phi_m = h_m\left[\frac{(P_{VS})_{interface}M_S}{RT}-\left(\rho_S^g\right)_\infty\right], \quad (3.13)$$

where $(P_{VS})_{interface}$, $h_m$, $M_S$, $R$, and $\left(\rho_S^g\right)_\infty$ are saturating vapor pressure of the polymer solution, solvent mass transfer coefficient at the surface, solvent molar mass, gas constant, and solvent vapor concentration in air far from the surface. Guerrier et al. [33] used the experimental data of diffusion coefficient in poly(*n*-butyl methacrylate)/methyl ethyl ketone solutions obtained by Hwang and Cohen [34] for their model. It should be noted that the polymer solution used in their study was a mixture of several copolymers, and therefore, the mixture was a multicomponent system. However, the problem was treated like a mass diffusion in a binary system, and one mutual diffusion coefficient was used for the mass transfer.

In the formulation of heat transfer in the problem as shown by Eq. (3.12), the gradient of the temperature inside the varnish was neglected. In other words, Guerrier et al. [33] assumed that the temperature does not change across the thickness of the varnish. Therefore, Eq. (3.12) expresses the temperature of the varnish as a function of time only. However, Eq. (3.11) can capture the variation of concentration of solvent as a function of both time and position.

Figure 3.4 shows the concentration profiles of the solvent within the varnish during the course of the drying process. The profiles are the results of numerical simulation of the model developed by Guerrier et al. [33].

Price and Cairncross [35] developed a drying model to optimize the drying process of polymer solution coatings. The governing equations for mass and heat diffusion were expressed as

$$\frac{\partial c_S}{\partial t} = \frac{\partial}{\partial x}\left(D\frac{\partial c_S}{\partial x}\right), \tag{3.14}$$

$$\rho_c \hat{C}p_c \frac{\partial T}{\partial t} = \frac{\partial}{\partial x}\left(\kappa_c \frac{\partial T}{\partial x}\right), \tag{3.15}$$

where $c_S$, $t$, $T$, and $D$ are the solvent concentration in polymer phase, time, temperature, and mutual diffusion coefficient, respectively. $\rho_c$, $\hat{C}p_c$, and $\kappa_c$ are

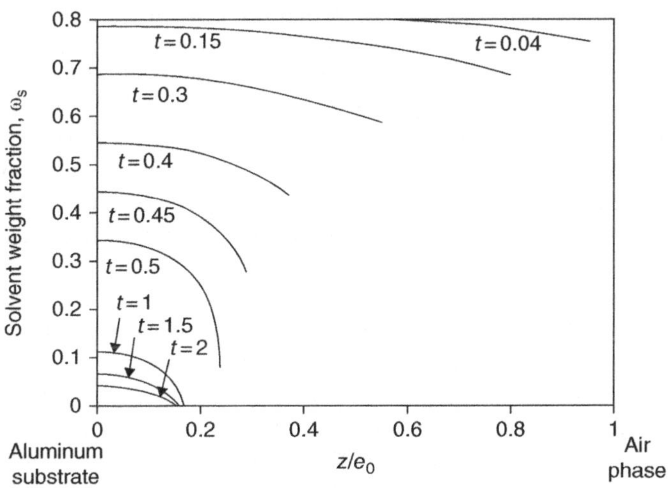


**Figure 3.4** Solvent weight fraction profiles during evaporation of solvent at various times (in seconds). *Source:* Reproduced by permission from Ref. [33].

density, specific heat, and thermal conductivity of the coating, respectively. Price and Cairncross [35] used the Vrentas–Duda free volume theory [36, 37] to describe concentration and temperature dependence of mutual diffusion coefficient that is expressed as

$$D = D_0 (1-\varphi)^2 (1-2\chi\varphi) \exp\left(-\frac{E}{RT}\right) \exp\left(-\frac{\omega_S \hat{V}_S^* + \omega_P \xi \hat{V}_P^*}{\hat{V}_{FH}/\gamma}\right), \tag{3.16}$$

where $\varphi$ is the volume fraction of solvent; $\chi$, Flory–Huggins interaction parameter; $D_0$, pre-exponential factor; $E$, activation energy; $R$, gas constant; $T$, absolute temperature; $\omega_S$ and $\omega_P$, weight fractions of solvent and polymer; $\hat{V}_S^*$ and $\hat{V}_P^*$, the solvent- and polymer-specific critical hole free volumes; $\xi$, the ratio of the solvent and polymer critical molar jumping unit volumes; and $\hat{V}_{FH}/\gamma$, the free volume available for diffusion and given by

$$\frac{\hat{V}_{FH}}{\gamma} = \frac{K_{11}}{\gamma} \omega_S (K_{21} + T - T_{gS}) + \frac{K_{12}}{\gamma} \omega_P (K_{22} + T - T_{gP}), \tag{3.17}$$

where $K_{11}/\gamma$ and $K_{21} - T_{gS}$ are free volume parameters for the solvent and $K_{12}/\gamma$ and $K_{22} - T_{gP}$ are free volume parameters for the polymer. It is important to note that the Vrentas–Duda mutual diffusion coefficient needs numerous parameters. Fourteen independent parameters are necessary to use this model, and these parameters are not usually available in the literature for different polymer solutions [38]. Moreover, the diffusion coefficient given by Eq. (3.16) does not satisfy the asymptotic limit at infinite dilution when the concentration of polymer goes to zero [39]. In Eq. (3.16) when the concentration of the polymer goes to zero, mutual diffusion coefficient becomes zero. However, the mutual diffusion coefficient must become a finite value at this limit [40].

Although in the formulation of governing equations the gradient of the temperature inside the polymer coating was considered during the course of drying, Price and Cairncross [35] reported only the results of their simulation for concentration profiles at different time intervals. Figure 3.5 shows the concentration profiles of the solvent during the drying of poly(vinyl acetate)/toluene solution under low airflow condition.

The models mentioned previously were presented in a chronological order to show the development of the modeling of the drying process of polymer solutions. As one can see, these models developed in a similar framework, that is, writing the mass and energy balance for the system. The difference between the models comes from the assumptions made to simplify the model and the boundary conditions that are associated with different operational conditions. Similar studies on the drying process of polymer solutions were performed by Robinson et al. [41], Price et al. [42], Alsoy and Duda [43], Saure et al. [44], Verros and Malamataris [45], and Ramesh and Duda [46]. In all of these models,

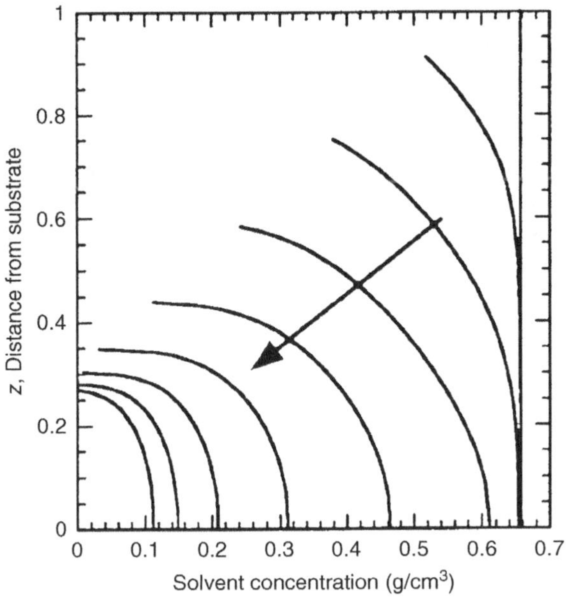


**Figure 3.5** Concentration profiles of the solvent during the drying of poly(vinyl acetate)/toluene coating under low airflow. Profiles are shown in 10 s increments starting at $t = 0$ when the concentration is uniform. *Source:* Reproduced with permission from Ref. [35].

considering the general case of non-isothermal mass diffusion, the dimensionless governing equations for mass and heat diffusion are expressed as [47]

$$\frac{\partial \varphi}{\partial \tilde{t}} = \frac{\partial}{\partial \xi}\left(\tilde{D}\frac{\partial \varphi}{\partial \xi}\right), \tag{3.18}$$

$$\frac{\partial \theta}{\partial \tilde{t}} = \frac{\partial}{\partial \xi}\left(\tilde{K}\frac{\partial \theta}{\partial \xi}\right), \tag{3.19}$$

where $\varphi$, $\theta$, $\tilde{D}$, and $\tilde{K}$ are volume fraction of the solvent, dimensionless temperature, dimensionless diffusion coefficient, and dimensionless thermal conductivity, respectively. The coupling between mass and heat transfer presented by the set of Eqs. (3.18) and (3.19) originates from the temperature dependence of diffusion coefficient or concentration dependence of thermal conductivity. A model represented by Eqs. (3.18) and (3.19) for diffusion phenomena cannot capture the so-called Ludwig–Soret effect or mass diffusion driven by a temperature gradient [48]. Moreover, there is no criterion in such a formulation to be sure that the model is thermodynamically consistent. Mass and heat diffusion are dissipative phenomena and contribute to the production of entropy in the system. In order to develop a thermodynamically consistent model for heat and mass transfer in polymer solutions, the balance of entropy, that is, the local mathematical expression for the second law of thermodynamics, must be considered in the formulation of the governing equations. Apart from the necessity of considering entropy balance, the drying problem of a polymer solution is a nonequilibrium problem and should be modeled in the framework of nonequilibrium thermodynamics. 3D governing equations for mass and heat diffusion derived in the framework of classical irreversible thermodynamics for a binary polymer solution are expressed as [47, 49]

$$\rho_1 \frac{d\varphi}{dt} = \nabla \cdot \left\{ \alpha \left( \frac{\partial \mu}{\partial \varphi} \right)_T \nabla \varphi + \left[ \alpha \left( \frac{\partial \mu}{\partial T} \right)_\varphi + \beta \right] \nabla T \right\}, \tag{3.20}$$

$$\rho c_{\mathrm{P}} \frac{dT}{dt} = -\nabla \cdot \left\{ \left( \mu + \frac{\beta}{\alpha} T \right) J_1 - \left( \gamma - \frac{\beta^2}{\alpha} T \right) \nabla T \right\}, \tag{3.21}$$

where $\rho_1$, is the mass density of the solvent; $\varphi$, volume fraction of the solvent; $t$, time; $T$, absolute temperature; $\rho$, mass density of polymer solution; $c_{\mathrm{P}}$, isobaric specific heat capacity; $J_1$, non-convective mass flux of the solvent; and $\mu$, the exchange chemical potential of the components and is given by

$$\mu = \frac{\mu_1}{M_1} - \frac{\rho_2}{\rho_1} \frac{\mu_2}{M_2}, \tag{3.22}$$

where $\mu_1$ and $\mu_2$ are the chemical potentials of the solvent and polymer, $M_1$ and $M_2$ are the molar mass of the solvent and polymer, and $\rho_2$ is the mass density of the polymer. $\alpha$, $\beta$, and $\gamma$ are phenomenological coefficients that are restricted by constraints

$$\alpha > 0, \gamma > 0, |\beta| < \sqrt{\alpha \gamma T^{-1}}, \tag{3.23}$$

to satisfy the positive definiteness of entropy production in the system.

In this formulation, the transport coefficients are derived in terms of thermodynamic variables. $\alpha(\partial\mu/\partial\varphi)_T$, $\alpha(\partial\mu/\partial T)_\varphi + \beta$, and $\gamma - \beta^2 T/\alpha$ are mutual diffusion coefficient, thermal diffusion coefficient, and thermal conductivity, respectively. The governing equations derived in the framework of irreversible thermodynamics indicate that the temperature gradient contributes to the mass diffusion. Moreover, the mass flux of the solvent also contributes to the heat transfer in the system. The constraints between the phenomenological coefficients guarantee that the model does not violate the second law of thermodynamics. This model is able to capture the effect of thermal diffusion in polymer solutions. Polymers typically have a positive thermal diffusion coefficient, and when a temperature gradient develops in a binary polymer solution, polymer chains tend to migrate to the cold side of the temperature gradient [50].

Figure 3.6 shows the concentration and temperature profiles at the onset of evaporation of solvent from a polymer solution. The concentration and

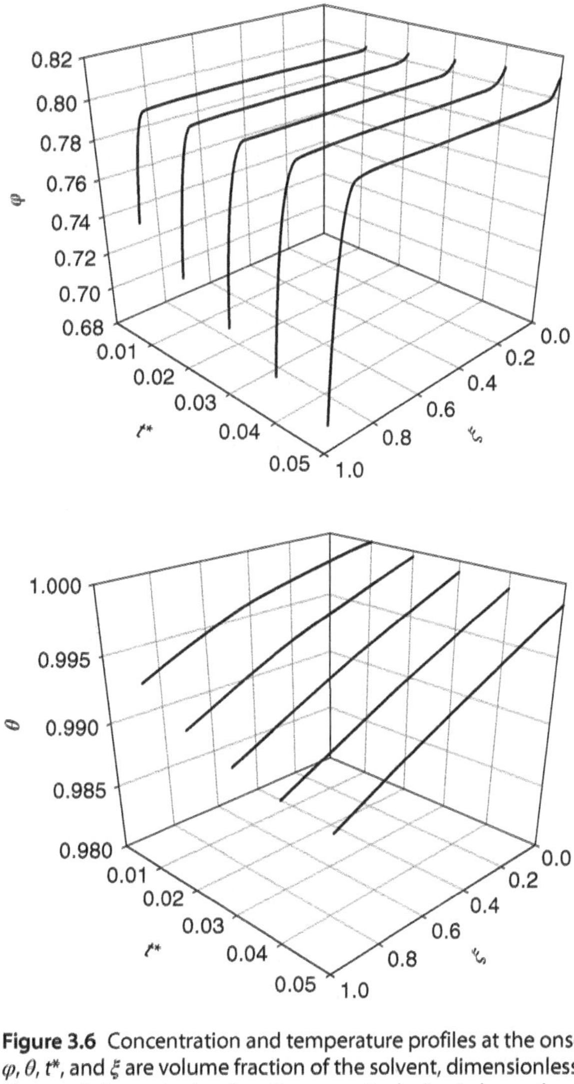


**Figure 3.6** Concentration and temperature profiles at the onset of evaporation of the solvent. $\varphi$, $\theta$, $t^*$, and $\xi$ are volume fraction of the solvent, dimensionless temperature, dimensionless time, and dimensionless length, respectively. $\xi = 0$ shows the substrate and $\xi = 1$ represents the evaporating surface at $t^* = 0$. *Source:* Reproduced by permission from Ref. [49].

temperature profiles are the results of simulation of the model as a 1D mass and heat diffusion problem [47, 49]. Uniform concentration and temperature profiles were considered as the initial conditions. Upon applying the boundary conditions, the solvent evaporates from the surface and temperature decreases, and a temperature gradient develops within the polymer solution film. Because polymer chains migrate to the cold side of the concentration gradient, the local concentration of the solvent increases at the substrate.

So far the mass and heat diffusion in binary polymer solutions were discussed. As mentioned earlier, polymer solutions that are used in solvent casting especially for industrial purposes often contain more than two components. By increasing the number of components in polymer solutions, the transport phenomena become more complicated. For example, considering isothermal mass transport in a ternary system, four diffusion coefficients are necessary to describe the mass transport. The number of mass diffusion coefficients increases rapidly by adding more components to the system. For a quaternary system, nine diffusion coefficients are required to describe mass transport [51]. In the case of a ternary system, mass diffusion coefficients form a $2 \times 2$ matrix whose off-diagonal entries are cross-diffusion coefficients. The process of diffusion becomes more complicated considering a time-varying temperature field in the system because in non-isothermal mass diffusion in a ternary mixture, in addition to four mass diffusion coefficients, two thermal diffusion coefficients are needed to describe the diffusion phenomena accurately [52].

In the following, a brief history of studies on mass and heat diffusion in ternary polymer solutions is presented.

Vrentas et al. [53] derived governing equations for mass diffusion in ternary polymer solutions based on a theory due to Bearman [54]. The mass flux of two solvents derived by Vrentas et al. [53] for 1D diffusion in solvent–solvent–polymer systems is given by

$$j_1 = \frac{1}{N_A^2 |\phi|} \left[ \left( \phi_{22} \frac{\partial \mu_1}{\partial \rho_1} - \phi_{12} \frac{\partial \mu_2}{\partial \rho_1} \right) \frac{\partial \rho_1}{\partial x} + \left( \phi_{22} \frac{\partial \mu_1}{\partial \rho_2} - \phi_{12} \frac{\partial \mu_2}{\partial \rho_2} \right) \frac{\partial \rho_2}{\partial x} \right], \tag{3.24}$$

$$j_2 = \frac{1}{N_A^2 |\phi|} \left[ \left( \phi_{11} \frac{\partial \mu_2}{\partial \rho_1} - \phi_{21} \frac{\partial \mu_1}{\partial \rho_1} \right) \frac{\partial \rho_1}{\partial x} + \left( \phi_{11} \frac{\partial \mu_2}{\partial \rho_2} - \phi_{21} \frac{\partial \mu_1}{\partial \rho_2} \right) \frac{\partial \rho_2}{\partial x} \right], \tag{3.25}$$

where $N_A$, $\rho_i$, $\mu_i$ are Avogadro's number, the mass density of component $i$, and chemical potential of component $i$ and $\phi_{ij}$ and $|\phi|$ are given by

$$\phi_{11} = \frac{\rho_2 \hat{V}_1 \zeta_{12}}{\rho_3 \hat{V}_3 M_2} - \frac{\left(1 - \rho_2 \hat{V}_2\right) RT}{\rho_1 \rho_3 \hat{V}_3 N_A^2 D_{T1}}, \tag{3.26}$$

$$\phi_{12} = \frac{\zeta_{12}\left(1-\rho_1\hat{V}_1\right)}{\rho_3\hat{V}_3M_2} - \frac{\hat{V}_2RT}{\rho_3\hat{V}_3N_A^2D_{T1}}, \tag{3.27}$$

$$\phi_{21} = \frac{\zeta_{12}\left(1-\rho_2\hat{V}_2\right)}{\rho_3\hat{V}_3M_1} - \frac{\hat{V}_1RT}{\rho_3\hat{V}_3N_A^2D_{T2}}, \tag{3.28}$$

$$\phi_{22} = \frac{\rho_1\hat{V}_2\zeta_{12}}{\rho_3\hat{V}_3M_1} - \frac{\left(1-\rho_1\hat{V}_1\right)RT}{\rho_2\rho_3\hat{V}_3N_A^2D_{T2}}, \tag{3.29}$$

$$|\phi| = -\frac{\zeta_{12}^2}{\rho_3\hat{V}_3M_1M_2} + \frac{R^2T^2}{\rho_1\rho_2\rho_3\hat{V}_3N_A^4D_{T1}D_{T2}}, \tag{3.30}$$

where $\zeta_{12}$, $\hat{V}_i$, $M_i$, $R$, $T$ are friction coefficient, partial specific volume of component $i$, molecular weight of the component $i$, gas constant, temperature, respectively, and $D_{T1}$ and $D_{T2}$ are related to the self-diffusion coefficient of the components that are derived using Vrentas–Duda free volume theory [53]. Using the same approach, Shojaie et al. [55] developed a model for the 1D drying process of a ternary polymer solution containing a polymer and two solvents. The governing equation for heat transport in their model was given by

$$\frac{\partial T}{\partial t} = \alpha\frac{\partial^2 T}{\partial x^2}, \tag{3.31}$$

where $\alpha$ is the thermal diffusivity of the polymer solution.

Verros and Malamataris [56] used the governing equations derived by Vrentas et al. [53], and they expressed heat transfer in the system by Eq. (3.31) to study the drying process of formamide–acetone–cellulose acetate system.

Alsoy and Duda [57] studied the drying of ternary polymer solutions. They approximated the diffusion coefficients by self-diffusion coefficients taken from Vrentas–Duda free volume theory, and in their model temperature was only assumed to be a function of time, that is, the gradient of the temperature inside the polymer solution during evaporation was neglected.

Wong et al. [58] used the theory developed by Alsoy and Duda [57] to study the drying process of poly(vinyl alcohol) (PVA)/water/methanol system.

Dabral et al. [59] studied the drying process of ternary polymer solutions with some simplifications. They set the cross-diffusion coefficients to zero and assumed an isothermal mass diffusion process. It is important to note that dropping the cross-diffusion coefficients can result in serious errors in drying prediction [60, 61]. Recently, a simulation study of drying ternary polymer solutions was performed using some of the models cited previously [62].

None of the previously mentioned models considered the effect of thermal diffusion in ternary polymer solutions, that is, the contribution of the temperature gradient to mass diffusion was neglected. Moreover, in cases where non-isothermal mass diffusion was concerned, the heat transfer was modeled using simple heat conduction equation, Eq. (3.31). This formulation does not take into account the effect of mass diffusion on the heat transfer, and the coupling between mass and heat transfer is only due to the temperature dependency of mass diffusion coefficients.

A recent formulation of non-isothermal mass diffusion in ternary systems [52] in the framework of classical irreversible thermodynamics indicates that the details of mass and heat diffusion phenomena are completely expressed by a $3 \times 3$ matrix whose entries are mass diffusion coefficients (four entries), thermal diffusion coefficients (two entries), and three entries that describe the evolution of heat in the system. The entries of the diffusion matrix are related to the elements of Onsager matrix that are bounded by some constraints to satisfy the positive definiteness of entropy production in the system. This formulation proved that setting cross-diffusion coefficients to zero results in a contradiction [52].

## 3.3 Real-Time Tracking of the Drying Process of Polymer Solutions

In order to have a good understanding of the physics underlying the casting process of polymer solutions, accurate experiments must be performed to study the process. Accurate experimental data is also crucial to gain a basic understanding of model development. As mentioned earlier, the most accurate way to study the drying process of polymer solutions is the real-time observation that can track all the changes that happen during the process.

### 3.3.1 Real-Time Measurement System

Recently, a unique real-time measurement system was developed by Cakmak et al. [63, 64] that is capable of tracking thickness, weight, birefringence, and surface temperature during the drying process of polymer solutions. Figure 3.7 shows the schematic design of the real-time measurement apparatus. The main body of the device was designed as a wind tunnel connected to an air blower at position 2. The wind tunnel design provides a uniform airflow for the drying polymer solution that is placed at position 4 on a graphite platform.

The uniformity of the airflow can be tuned using three adjustable vertical baffles located at position 3. The downstream end of the wind tunnel is open to the atmosphere, position 5. The heart of the system is position 4 where the cast polymer solution and sensors for real-time measurement are located. The details of this part are shown in Figure 3.8.

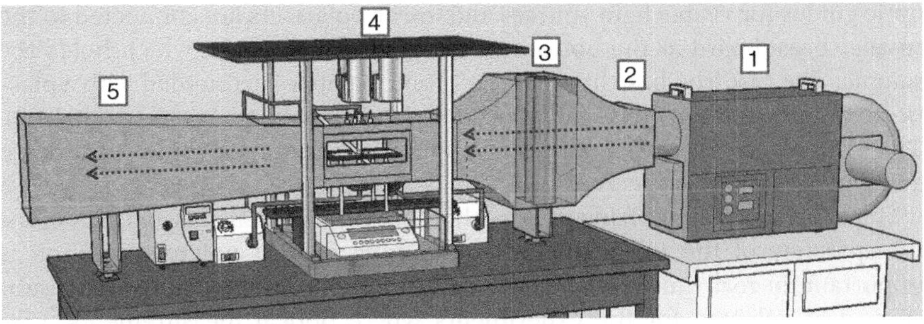


**Figure 3.7** Schematic representation of the real-time measurement system for studying the drying process. From right to left: (1) hot air blower, (2) connection of the blower to the tunnel, (3) vertical baffles, (4) drying sample and sensor location, and (5) open end of the tunnel. *Source:* Reproduced with permission from Ref. [64].

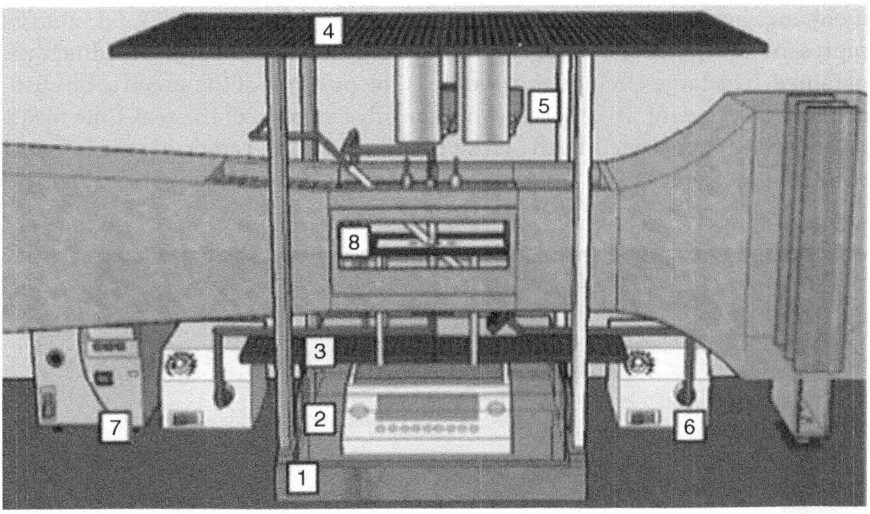


**Figure 3.8** Sample and sensor location: (1) cast iron base, (2) balance and plastic box, (3) small breadboard for optical components, (4) large breadboard for optical components and sensors, (5) laser displacement sensors, (6) visible range light sources, (7) UV light source, and (8) sample platform. *Source:* Reproduced with permission from Ref. [64].

Two breadboards located in position 3 and position 4 of Figure 3.8, one located at the top and one located at the bottom of the sample position, hold all the sensors and optical components. To stabilize the structure, a heavy cast iron base, position 1, is attached to these breadboards. The laser displacement sensors, position 5, for thickness measurement, bifurcated fiber optic cables for in-plane and out-of-plane birefringence measurements, and a universal serial bus (USB) camera are connected to the top larger breadboard. The fiber optic guides for visible light sources and linear polarizers are connected to the smaller breadboard at the bottom of the graphite platform, which holds the sample. The cast iron base holds an electronic balance surrounded with a plastic box to prevent any noise introduced from airflow fluctuations (position 2). The visible and UV wavelength range light sources are shown in positions 6 and 7 in Figure 3.8, respectively.

Since the spectral birefringence method operates in transmission mode, the transparency of the substrate and the coating or cast polymer solution is important for real-time birefringence measurements. Therefore, the glass substrate was selected for the experiments where optical measurements were concerned. In order to improve the signal quality of the laser displacement sensor, the back face of the glass substrate was painted black, while a small portion was left uncoated for optical birefringence measurements. The glass substrate with coated polymer solution is shown in Figure 3.9. The dots shown by arrows and labeled L1, L2, and L3 are the positions where the laser displacement measurements are conducted. The circles labeled P1, P2, P3, and P4 represent the positions where pyrometer surface temperature readings are obtained. The large circle in the middle is the position of the spectral birefringence measurement. In order to avoid the edge effects, all the real-time measurements are performed in the close proximity to the center of cast polymer solutions. After casting polymer solution on the glass substrate using a

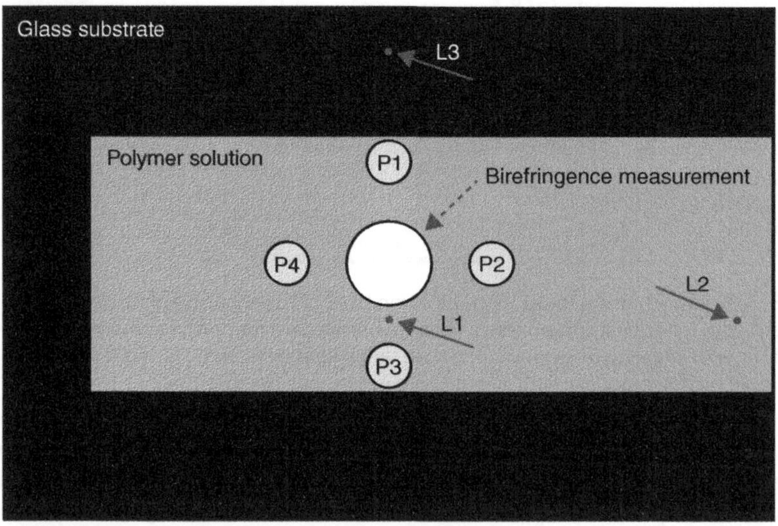


**Figure 3.9** Glass substrate with cast polymer solution. L1, L2, and L3 are the points for laser displacement measurements, and circles labeled P1, P2, P3, and P4 are the locations where the pyrometers read the surface temperature. *Source:* Reproduced by permission from Ref. [64].

Dr. Blade casting system, the coated glass substrate is placed on top of the sample platform inside the wind tunnel, position 8 in Figure 3.8. A magnified view of the sample position with the optical sensors and pyrometers is shown in Figure 3.10.

Real-time optical measurements during evaporation of the solvent are performed by passing a 45° linear polarized light through the sample that was then collected via bifurcated fiber optic cables at 0° and 45° optical paths, as shown in Figure 3.10. For each angle, one of the fiber optic cables is polarized parallel and the other is polarized perpendicular to the initial polarization. The data are transferred to the spectrometers for further calculations. The real-time in-plane, $\Delta n_{12}$, and out-of-plane birefringence, $\Delta n_{23}$, values are calculated by measuring the retardation values at two different angles, 0° and 45°.

Surface temperature is measured by four separate pyrometers, position 2 in Figure 3.10, focused downward on the film at the points where circles labeled P1, P2, P3, and P4 are located. The honeycomb-structured air diffuser is placed inside the wind tunnel to eliminate potential noise due to the airflow.

The thickness of the drying polymer solution is measured by three non-contact laser displacement sensors (position 4 in Figure 3.10) that are attached to the breadboard at the top of the sample platform. The measurements are performed at the points L1, L2, and L3, shown in Figure 3.9. The sensors only measure the position of the top surface of the cast solution through the laser

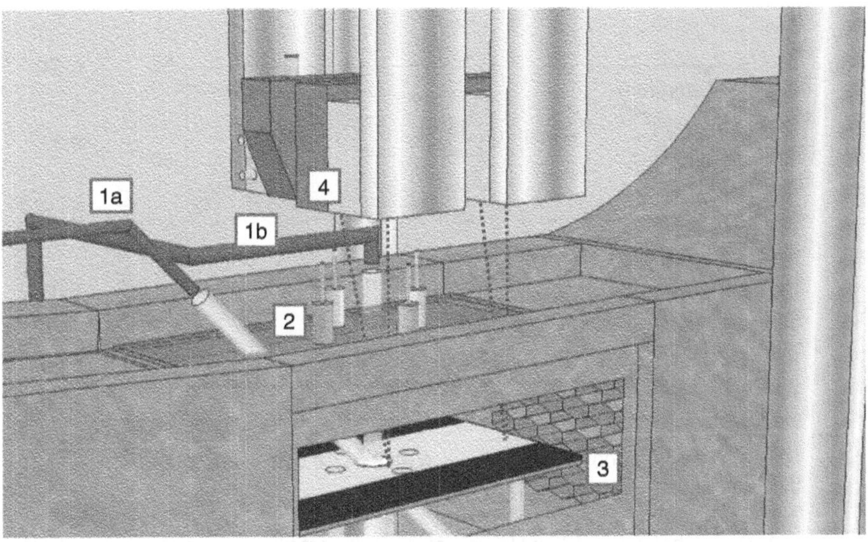


**Figure 3.10** Glass substrate with cast polymer solution located on the sample holder: (1) optical paths and fiber optic cables, (2) pyrometers, (3) honeycomb-structured air diffuser, and (4) position of laser displacement sensors. *Source:* Reproduced with permission from Ref. [64].

reflection. Laser-1 and laser-2 measure the displacement of the surface of the polymer solution at two different points, one in the middle of the film and the other on the upstream toward the heater. Laser-3 measures the displacement of the surface of the glass substrate, and this value is subtracted from the laser-1 and laser-2 values for real-time thickness calculations at those locations.

The apparatus is also capable of monitoring real-time UV curing process of mixtures containing reactive materials. For this, one of the pyrometers can be replaced with a UV light source that is introduced via a fiber optic cable.

In the following, real-time experimental observations made on the drying behavior of different polymer solutions under various processing conditions using the novel measurement system are presented.

### 3.3.2 Drying Process of Polyimide/*N,N*-Dimethylformamide Solutions

Eguchi et al. [65] conducted comprehensive real-time experiments to study the drying process of solutions of a semi-aromatic polyimide in *N,N*-dimethylformamide (DMF) under various processing conditions.

Figure 3.11 displays the results of real-time monitoring of the drying process of a 15 wt % polyimide/DMF solution at 60°C under airflow rate of 0.5 m/s. The drying process can be divided into three stages that are separated by dashed lines in Figure 3.11:

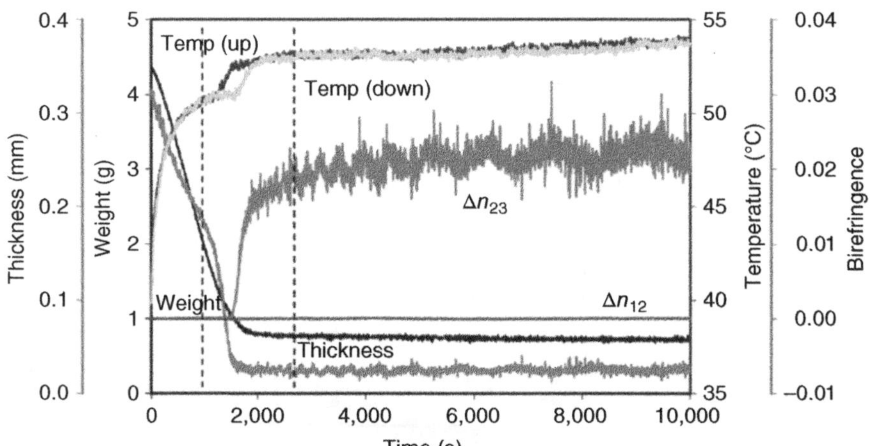


**Figure 3.11** Drying behavior of 15 wt % polyimide/DMF cast film. Drying conditions are as follows: air blower set temperature 60°C, airflow rate 0.5 m/s. Temp (up) and temp (down) are surface temperatures at P2 and P4 points of the substrate, respectively (see Figure 3.9). $\Delta n_{12}$ and $\Delta n_{23}$ are in-plane and out-of-plane birefringence, respectively. *Source:* Reproduced by permission from Ref. [65].

Stage 1: After insertion of the glass substrate coated with polymer solution into the apparatus, the surface temperature increases quickly and then levels off. During the first stage, the weight of the polymer solution decreases rapidly. The thickness that is measured locally also decreases. Since both in-plane, $\Delta n_{12}$, and out-of-plane, $\Delta n_{23}$, birefringences are zero, the sample remains isotropic during this stage. It should be noted that 1 is the airflow direction, 2 is the transverse to airflow direction, and 3 is the thickness direction.

Stage 2: As drying proceeds, a front develops upstream near the air blower and moves in the direction of the airflow. This front is characterized by a rapid local reduction in the thickness of the film. When the front reaches the points where upstream and downstream temperatures are measured, points P2 and P4 in Figure 3.9, an increase in the surface temperature is observed. The increase in the surface temperature indicates that the solvent concentration at the surface of the film becomes very low, and, therefore, the evaporative cooling is no longer effective. Therefore, the surface temperature increases due to the high temperature of the airflow passing by. Thickness nearly levels off at the end of this stage, and a drastic increase is observed in out-of-plane birefringence, while in-plane birefringence remains constant near zero.

Stage 3: At this stage, the drying front moves toward downstream. The trend of variation of temperature, weight, and out-of-plane birefringence becomes very slow so that they can be assumed constant during this stage.

During evaporation of solvent, the most significant changes take place in the second stage of evaporation. The rapid increase in out-of-plane birefringence that is observed during the second stage of the drying process is the direct consequence of development of stress and preferential chain orientation in the plane. The origin of the out-of-plane birefringence in dried polymeric films has been discussed in detail in the literature. Cherkasov et al. [66] observed birefringence in amorphous polymer films of different chemical structures, and they concluded that birefringence is caused by a strong planar ordering of polymer chains in the surface layers. By performing post drying offline birefringence measurement, Prest and Luca [67, 68] concluded that polymer chain orientation responsible for out-of-plane birefringence is the result of both stress supported by the entanglements and the alignments of polymer segments, and it also depends on the difference between glass transition temperature of the coated polymer and the drying temperature. Greener et al. [7, 69, 70] summarized their discussion in the study of birefringence of dried polymer films that optical anisotropy represents a random molecular alignment parallel to the film plane, and the out-of-plane birefringence is closely associated with residual stress produced by the casting process. Coburn et al. [71] studied the optical properties of aromatic polyimide film and concluded that birefringence of these polyimide films is the result of both molecular orientation and stress development in the films. The studies cited previously were conducted using dried polymer films. The evolution of out-of-plane birefringence shown in Figure 3.11 is the first study that tracks birefringence during evaporation. It is important to note that the trend of variation of out-of-plane birefringence during evaporation of the solvent is quite similar to the evolution of stress inside a cast polymer solution during the course of drying. The real-time stress development in polymer coatings during evaporation of solvent reported by Francis et al. [72–75] indicates that before solidification stress development is suppressed due to fast relaxation of polymer chains and at the onset of solidification stress increases rapidly.

The effect of airflow temperature on the kinetics of drying and evolution of out-of-plane birefringence in the drying of 15 wt % polyimide/DMF solution is shown in Figure 3.12.

As expected, increasing drying temperature leads to a faster solvent evaporation at the beginning of the drying process (Figure 3.12a). At the intermediate time scales, drying curves cross each other at the point shown by the arrow in Figure 3.12a. This crossover indicates that increasing drying temperature does not necessarily accelerate long-term drying behavior of the polymer solution.

The development of the out-of-plane birefringence at different drying temperatures is shown in Figure 3.12b. As one can see, increasing drying temperature leads to the development of out-of-plane birefringence at earlier times; however, it decreases the final value of the birefringence when it reaches a plateau. The low value of out-of-plane birefringence at higher drying temperatures is the result of relaxation of polymer chains at higher temperatures due to slower freezing-in process. It is important to note that high drying temperature results in low residual stress in polymer films despite fast solvent evaporation [72–75]. Moreover, by rapid evaporation of solvent from the surface of the polymer solution, the glass transition at the surface increases [76, 77]. This can accelerate the formation of skin that reduces the evaporating solvent flux. In such a situation before the whole sample vitrifies, the polymer chains in the interior have a longer time to relax at higher temperatures.

Figure 3.13 shows the effect of airflow velocity on drying kinetics and development of out-of-plane birefringence in the drying of 15 wt % polyimide/DMF solution. The higher air velocity results in faster drying of the polymer solution. Moreover, at higher air velocities, the onset of development of birefringence takes place earlier, and birefringence exhibits higher value at the plateau region. In the results shown in Figure 3.13, the temperature of the airflow is the same for three different air velocities. Here, faster evaporation of the solvent by airflow results in a fast freezing-in process, and consequently the magnitude of birefringence will be higher. Although both increasing drying temperature and airflow velocity at constant airflow temperature can increase drying kinetics, the latter results in polymer films with more residual stress and out-of-plane birefringence.

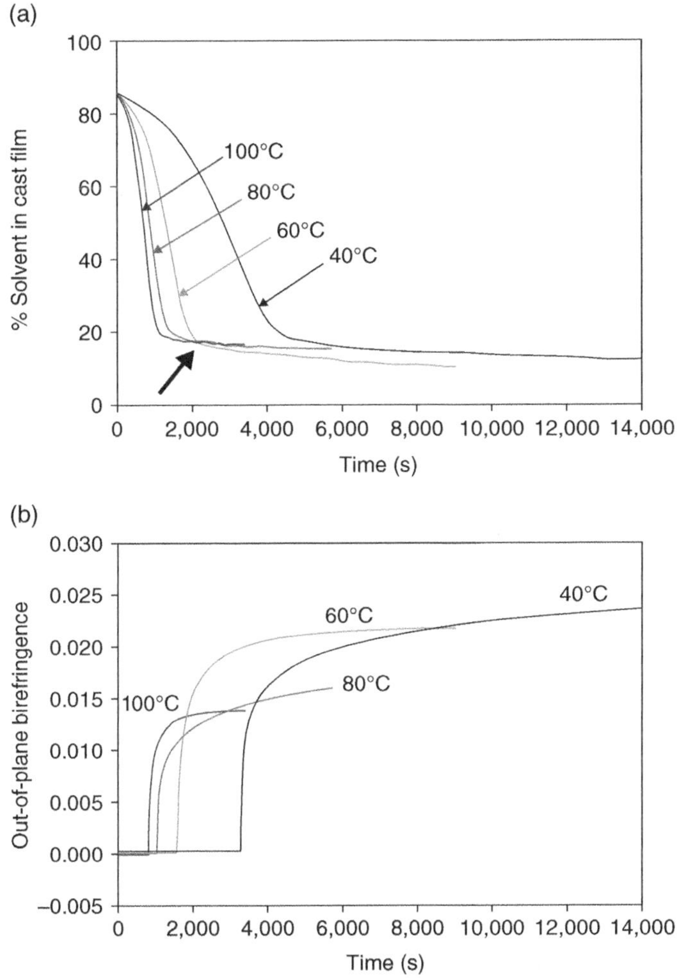


**Figure 3.12** Effect of airflow temperature on the drying process of 15 wt % polyimide/DMF polymer solution with the airflow velocity 0.5 m/s. (a) Temporal development of solvent content and (b) temporal development of out-of-plane birefringence. *Source:* Reproduced by permission from Ref. [65].

The effect of initial thickness of cast polymer solutions on the kinetics of drying and evolution of out-of-plane birefringence in the drying of 15 wt % polyimide/DMF solution is shown in Figure 3.14.

As one can see, the initial thickness of polymer solution plays an important role in drying kinetics and development of optical anisotropy within the film. Thinner polymer solutions dry faster, and out-of-plane birefringence of thinner cast polymer solutions develops much faster than that of the thicker ones.

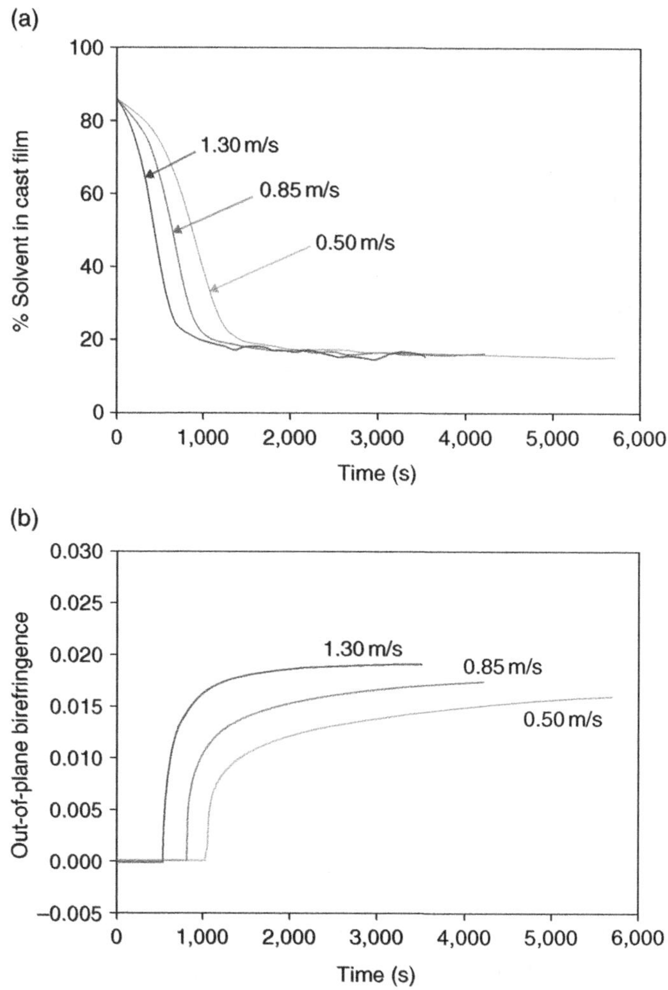


**Figure 3.13** Effect of airflow velocity on the drying process of 15 wt % polyimide/DMF polymer solution with drying temperature 80°C. (a) Temporal development of solvent content and (b) temporal development of out-of-plane birefringence. *Source:* Reproduced by permission from Ref. [65].

Figure 3.15 compares drying kinetics and development of optical anisotropy in five polymer solutions prepared by dissolving 15 wt % polyimide in cyclopentanone (CP), DMF, *N,N*-dimethylacetamide (DMAc), γ-butyrolactone (GBL), and *N*-methyl-2-pyrrolidone (NMP). The drying condition was the same for all the polymer solutions. Drying experiments were performed in the real-time measurement system, where drying temperature and airflow velocity were 80°C and 0.5 m/s, respectively [65].

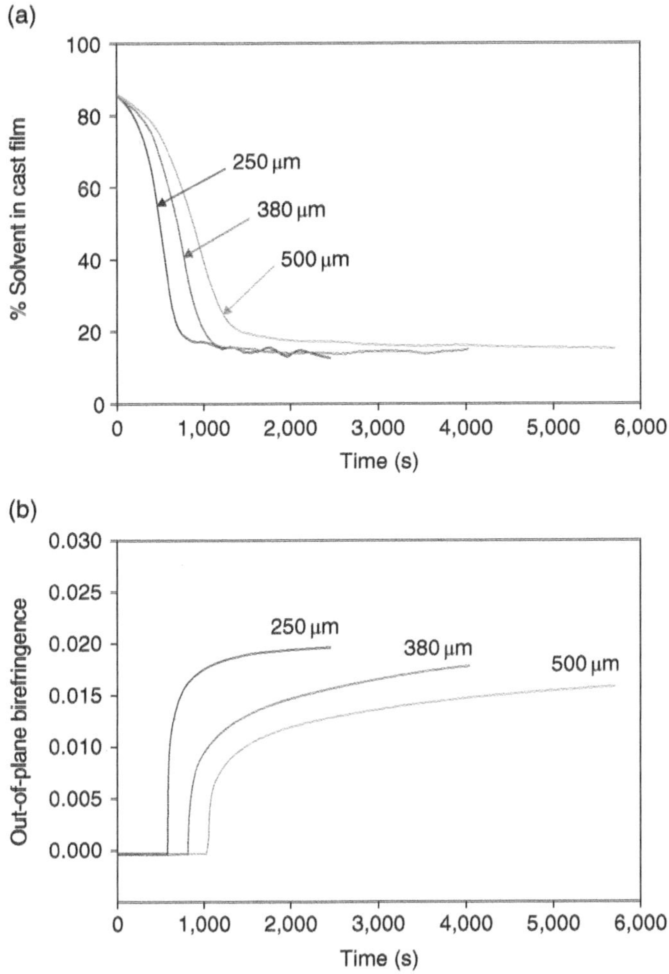


**Figure 3.14** Effect of initial thickness of cast polymer solution on the drying process of 15 wt % polyimide/DMF polymer solution with drying temperature 80°C and the airflow velocity 0.5 m/s. (a) Temporal development of solvent content and (b) temporal development of out-of-plane birefringence. *Source:* Reproduced by permission from Ref. [65].

The most important characteristic of these solvents that affect the drying behavior of polymer solutions is their boiling point. The solvents were mentioned in order from the low boiling point, CP, to the high boiling point, NMP. As Figure 3.15a shows, polymer solutions made from low boiling point solvents such as CP, DMF, and DMAc exhibit a much faster drying kinetics as compared with the polymer solutions made from high boiling point solvents, GBL and NMP. A comparison of solvent contents as a function of time in polymer

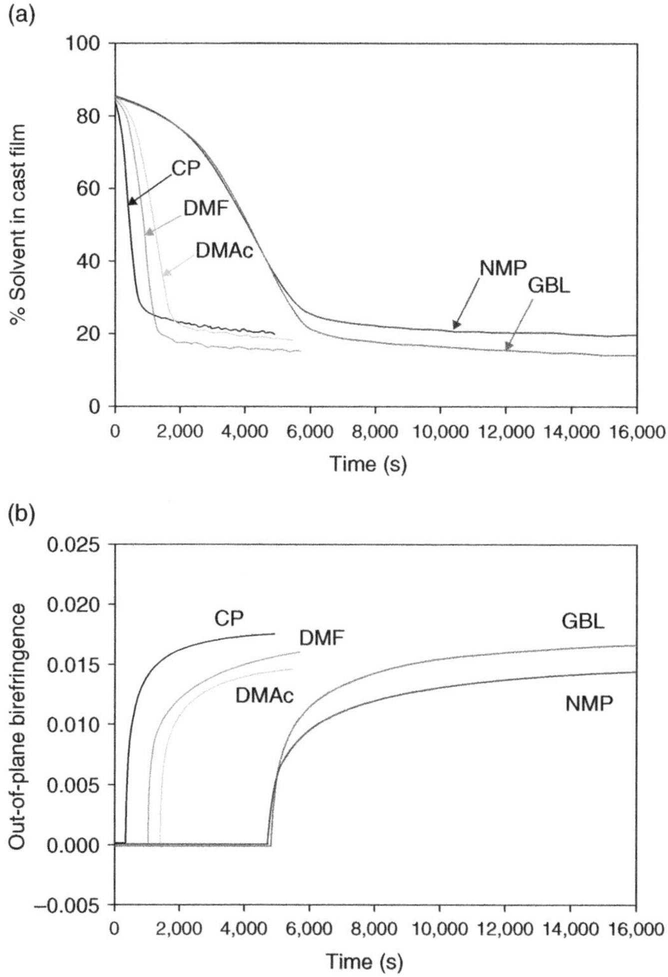


**Figure 3.15** Effect of solvent type on the drying process of 15 wt % polyimide polymer solution with drying temperature 80°C and the airflow velocity 0.5 m/s. (a) Temporal development of solvent content and (b) temporal development of out-of-plane birefringence. *Source:* Reproduced by permission from Ref. [65].

solutions with low boiling point solvents in Figure 3.15a indicates that polyimide/CP solution dries faster, but at the point where drying rate decreases, considerable amount of solvent remains in the solution as compared with polyimide/DMF and polyimide/DMAc solutions. This observation suggests rapid skin formation at the surface of the polymer solution at early stages due to very fast evaporation of CP.

The development of out-of-plane birefringence for different polymer solutions is compared in Figure 3.15b. The polyimide/CP solution exhibits the fastest increase in out-of-plane birefringence. In other solutions, the onset of development of birefringence is the point where the rate of the drying process drastically decreases.

### 3.3.3 Real-Time Study of Drying and Imidization of Polyamic Acid/NMP Solution

Unsal and Cakmak conducted a comprehensive study to investigate the real-time dynamics of imidization reaction of polyamic acid (PMDA–ODA) solution in NMP [78, 79].

A 15 wt % PMDA–ODA/NMP solution was cast on a glass substrate at room temperature, and then the solution-coated glass substrate was placed on the sample holder inside the preheated chamber of the drying apparatus. The sample was first dried for 4 h at low temperature, and then the temperature was increased to the imidization temperature, 200°C for the next 4 h. The airflow speed was kept constant at 0.5 m/s during the experiment. Figure 3.16 shows the whole process of drying and imidization [78].

As shown in Figure 3.16, the whole process is divided into three regimes. The first regime is the drying regime in which the temperature is kept low enough to prevent imidization. The second regime is the transition regime

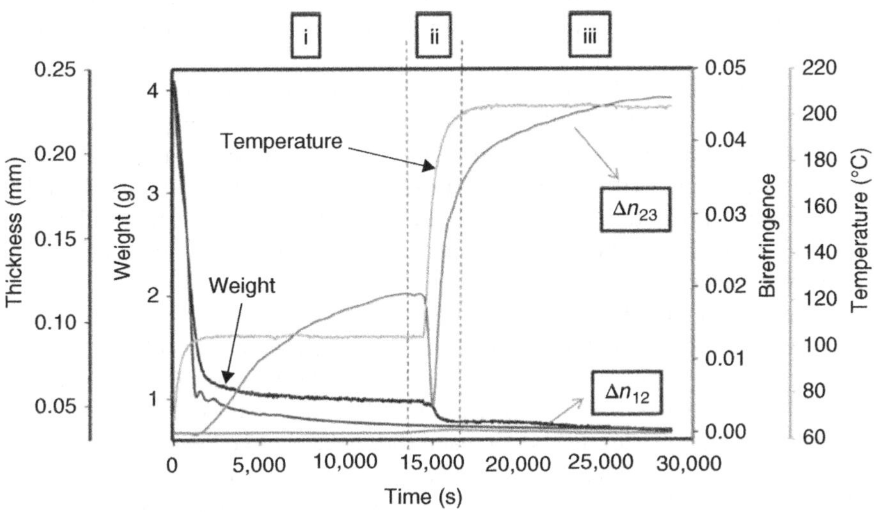


**Figure 3.16** Drying and imidization process. The process was divided into three regimes: (i) drying regime, (ii) transition regime, and (iii) imidization regime. *Source:* Reproduced with permission from Ref. [78].

where the temperature is increased to 200°C to start the imidization. The third regime is the imidization regime. During the first regime, due to evaporation of the solvent, a rapid reduction in weight and thickness is observed. As drying proceeds, the in-plane birefringence remains zero, indicating no in-plane anisotropy; however, at the point where the variation of weight and thickness becomes very low, the out-of-plane birefringence starts increasing. The development of out-of-plane birefringence is the result of a competition between the relaxation and reorientation of polymer chains as the polymer solution undergoes a compression during the drying process [67–70, 80]. After 4 h, at the end of the first regime, the evaporation of solvent substantially decreases. At this stage, the partially dried sample contains about 30–40% solvent by weight. At the transition regime, where the temperature is increased to 200°C, additional weight loss and thickness reduction are observed that is associated with the evaporation of the bound solvent. Moreover, at the transition regime, a temporary decrease is observed in the out-of-plane birefringence that is due to the relaxation at the higher temperature in the presence of decomplexed bound solvent. When the temperature reaches a critical value (~150–160°C), the rate of imidization reaction increases; by conversion of polyamic acid chains into more rigid polyimide chains that have higher intrinsic birefringence and also increasing chain orientation of polyimide chains parallel to the film surface, as the chain mobility decreases due to imidization reaction and consequent increase in glass transition temperature [81, 82], the out-of-plane birefringence develops at a higher rate as compared with the first regime. The weight loss beyond 15,000 s indicates the continuation of the evaporation of the bound solvent and the evaporation of the water that is formed during the reaction.

The effect of initial thickness of cast solution on the drying and imidization process of 15 wt % PMDA–ODA/NMP solution was studied by preparing films with 500, 250, and 100 μm casting thickness [78, 79]. In the first step, the temperature was kept at 100°C at 0.5 m/s air speed and the films were dried for 4 h. In the next 4 h, the temperature was increased to 200°C for the imidization reaction. Figure 3.17 shows the weight change during the whole drying and imidization process.

The effect of initial thickness on drying kinetics is similar to the results shown in Figure 3.14a that compare the drying kinetics at different initial thickness for drying process of 15 wt % polyimide/DMF solution [65]. As initial casting thickness decreases, drying kinetics become faster due to the shorter paths that solvent has to diffuse to reach the surface and evaporates.

The effect of initial thickness of cast solution on the development of in-plane and out-of-plane birefringence during the drying and imidization of 15 wt % PMDA–ODA/NMP solution is shown in Figure 3.18.

During the drying and imidization process, the in-plane birefringence remains zero, and out-of-plane birefringence increases with the decreasing

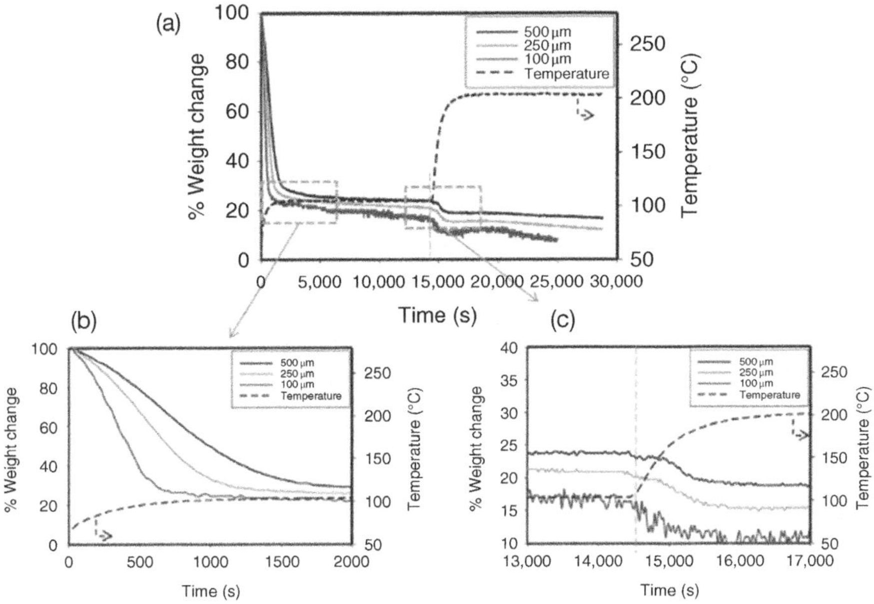


**Figure 3.17** Effect of initial thickness of cast solution on the weight change during the drying and imidization of 15 wt % PMDA–ODA/NMP solution. In the first step the temperature was kept at 100°C at 0.5 m/s air speed, and the films were dried for 4 h. In the next 4 h, the temperature was increased to 200°C to for the imidization reaction. Increasing the initial thickness results in a reduction in the drying kinetics. (b and c) Magnified views of two parts of (a). *Source:* Reproduced by permission from Ref. [78].

initial casting thickness. In the time interval shown in Figure 3.18b, little or no imidization occurs during drying. The development of the out-of-plane birefringence at this stage is mainly due to the shrinkage of the film in the thickness direction. Rapid evaporation in the case of very low initial thickness leads to an increase in overall orientation, and therefore higher value of birefringence is achieved. Figure 3.18c shows the time interval in that the temperature is increased to 200°C for imidization reaction. At the beginning, due to the higher temperature and also plasticization effect of the decomplexed solvent molecules, a reduction is observed in the out-of-plane birefringence. However, beyond 15,000 s where the rate of imidization reaction increases as the temperature goes high, the out-of-plane birefringence increases with a higher rate due to the intrinsic birefringence of polyimide chains and orientation of polymer chains at the surface. This pattern is the same for all three polymer films with different initial thickness except for a slight delay in case of the film with the initial thickness 100 μm. The delay is presumably due to the low content of the bound solvent in this sample. The changes in the behavior for 500 and 250 μm samples occur nearly at the same time. Similar to the

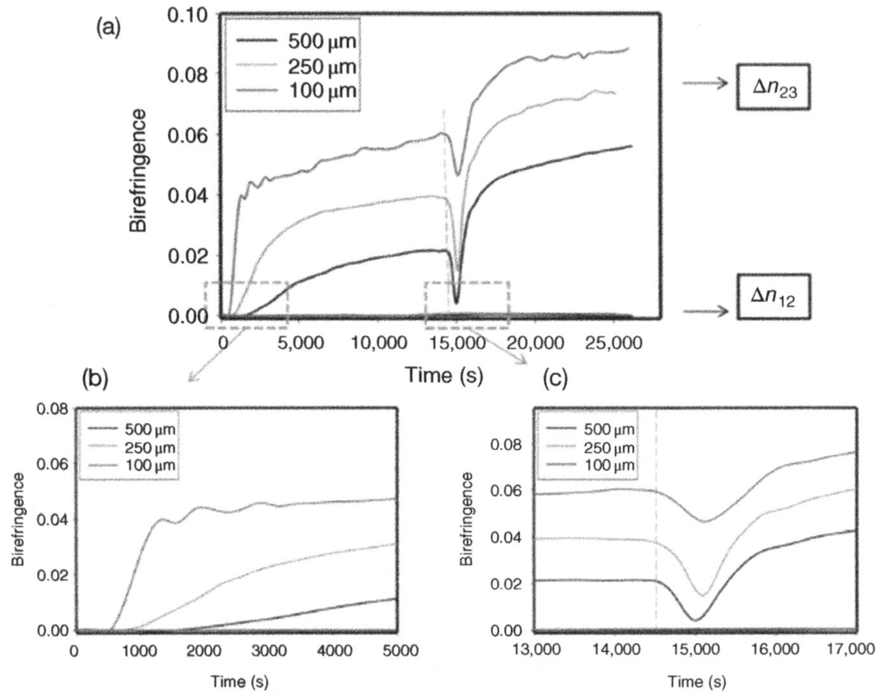


**Figure 3.18** Effect of initial thickness of cast solution on the development of in-plane and out-of-plane birefringence during the drying and imidization of 15 wt % PMDA–ODA/NMP solution. In the first step the temperature was kept at 100°C at 0.5 m/s air speed, and the films were dried for 4 h. In the next 4 h, the temperature was increased to 200°C to for the imidization reaction. The out-of-plane birefringence increases with decreasing initial casting thickness. (b and c) Magnified views of two parts of (a). *Source:* Reproduced by permission from Ref. [78].

drying behavior, the final birefringence value increases with decreasing thickness. The observation is similar to the temporal evolution of the out-of-plane birefringence observed in Figure 3.14b in the drying process of 15 wt % polyimide/DMF solution [65].

To study the effect of drying temperature on drying and imidization process of 15 wt % PMDA–ODA/NMP solution, the first 4 h drying regime was performed at five different temperatures, 100, 120, 140, 160, and 200°C, and then the temperature was increased to 200°C for the next 4 h [78, 79]. Figure 3.19 shows the weight change during the whole drying and imidization process. Sample notation shows the temperature used for the drying and imidization process. For example, PI 140/200 represents the sample that dried at 140°C and then the temperature was increased to 200°C.

The first 2000 s of the evaporation process is shown in Figure 3.19b. The majority of the solvent evaporates at this time interval. As expected, by increasing the drying temperature, the kinetics of drying becomes faster. Moreover, higher drying temperature results in lower weight, indicating that more solvent removed from the cast solutions exposed to higher drying temperatures. Therefore, drying at low temperatures leads to the higher solvent content remaining in the films. The variation of weight in the transition region after 4 h of drying is shown in Figure 3.19c. At this stage, the temperature was increased to 200°C. As one can see in this regime, only samples that are exposed to low drying temperatures in the first regime, 100 and 120°C, exhibited a marked weight loss. The effect of drying temperature on the development of in-plane and out-of-plane birefringences during the drying and imidization of 15 wt % PMDA–ODA/NMP solution is shown in Figure 3.20. During the process, in-plane birefringence remained nearly zero for all drying temperatures, and out-of-plane birefringence increased with increasing drying temperature. Figure 3.20b shows the initial 5000 s of the drying process. According to the kinetics of imidization reaction [78], 100°C is below the imidization

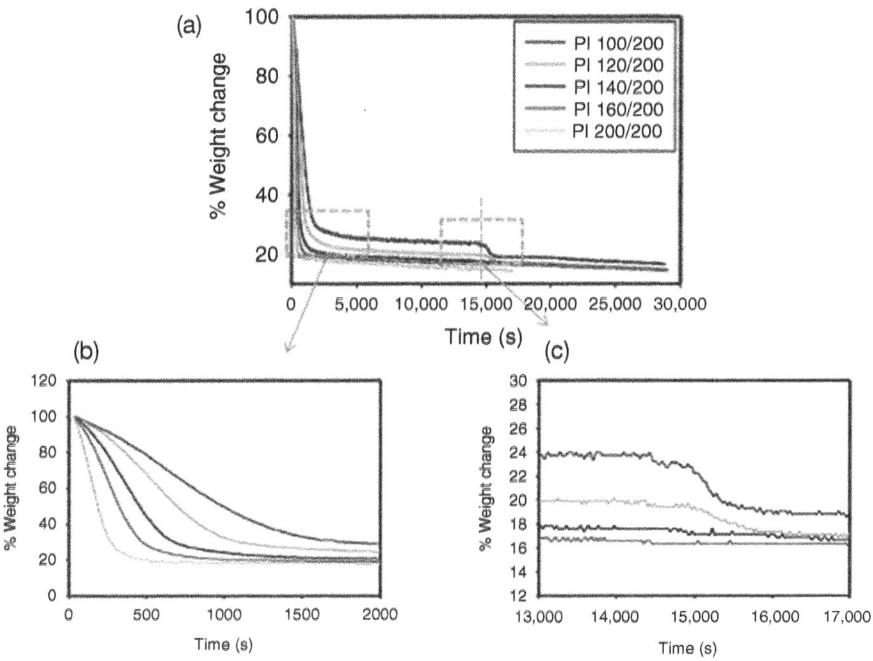


**Figure 3.19** Effect of drying temperature on the weight change during the drying and imidization of 15 wt % PMDA–ODA/NMP solution. The first 4 h drying regime was performed at five different temperatures, 100, 120, 140, 160, and 200°C, and then the temperature was increased to 200°C for the next 4 h. Increasing the drying temperature results in a faster drying kinetics. (b and c) Magnified views of two parts of (a). *Source:* Reproduced by permission from Ref. [78].

temperature, 120°C is the onset temperature for imidization reaction, and the fastest reaction occurs at 150°C. Thus, in the case of samples exposed to airflow with drying temperatures 140, 160, and 200°C, drying and imidization take place simultaneously. In this interval the development of out-of-plane birefringence is the result of evaporation of the solvent, increasing the intrinsic birefringence of the medium, and the liquid crystalline-like order of rigid and semiflexible polymer chains in the concentrated polymer solutions. The transition stage, where temperature was increased to 200°C, is shown in Figure 3.20c. The out-of-plane birefringence was initially decreased for samples dried at 100 and 120°C. This relaxation is the result of the plasticization effect due to the decomplexed solvent molecules. A very small relaxation was observed in the samples dried at 140 and 160°C. At these high temperatures, the evaporation of the majority of the solvent occurs in the first stage (see Figure 3.19), and therefore, the evaporation of the remaining solvent was not enough to enhance the

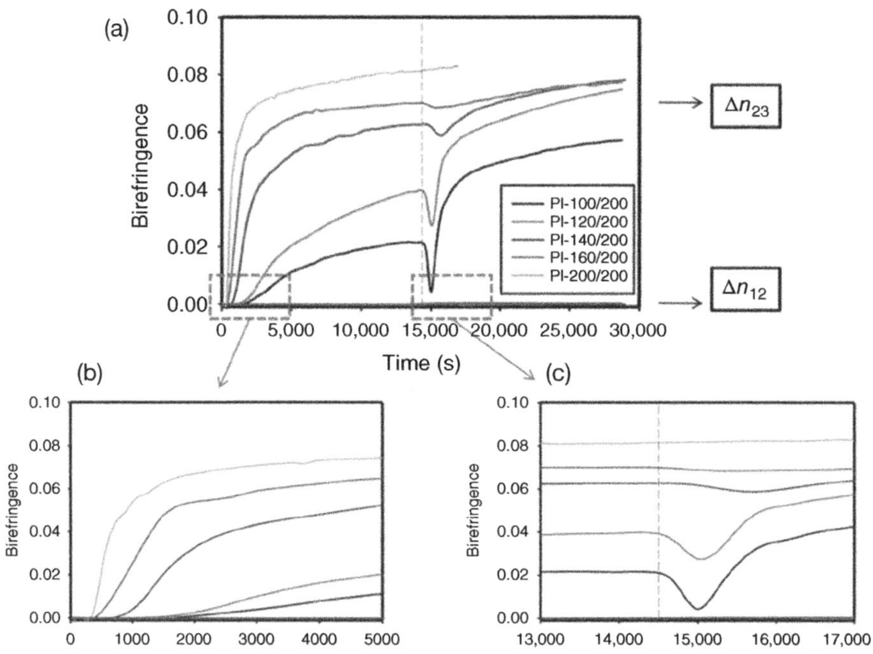


**Figure 3.20** Effect of drying temperature on the development of in-plane and out-of-plane birefringence during the drying and imidization of 15 wt % PMDA–ODA/NMP solution. The first 4 h drying regime was performed at five different temperatures, 100, 120, 140, 160, and 200°C, and then the temperature was increased to 200°C for the next 4 h. The out-of-plane birefringence increases with increasing the drying temperature; PI 100/200 and PI 200/200 exhibited the lowest and the highest birefringence, respectively. (b and c) Magnified views of two parts of (a). *Source:* Reproduced by permission from Ref. [78].

mobility of polymer chains to cause relaxation as observed in the case of samples dried at the lower drying temperatures (100 and 120°C).

These real-time measurements allow one to quantitatively study the competition between chain relaxation due to the evaporation of the decomplexed solvent molecules and orientation development due to decreased chain mobility caused by the imidization reaction and increasing glass transition temperature for the PMDA–ODA/NMP solutions. Moreover, understanding the effect of processing conditions on the drying kinetics and temporal evolution of birefringence will help one to fine-tune the final properties of polyimide films by changing processing conditions.

### 3.3.4 Development of Optical Gradient During Evaporation of Solvent

The development of optical anisotropy gradient induced by evaporation of the solvent in poly(amide–imide) (PAI)/DMAc solution for the first time was reported by Yucel et al. [23, 83]. They discovered the development of optical gradient by performing various experiments using real-time optical measurements with spectral birefringence technique coupled with off-line optical measurements such as Abbe refractometer and optical compensator method.

Figure 3.21 shows the results of optical measurements using an Abbe refractometer during the drying of a 10 wt % PAI/DMAc solution. Optical measurements at the air–liquid and substrate–liquid interfaces indicate that at the early stages of drying, the polymer solution is optically isotropic. As drying proceeds, refractive indices start to increase. This occurs after 4 h at the air–solution interface and after 6 h at the substrate–solution interface. The in-plane (MD) refractive indices become higher than out-of-plane (ND) refractive indices. The difference between refractive indices in MD and ND directions is attributed to planar orientation of highly aromatic and rigid PAI chains induced by evaporation of the solvent. The highly oriented layer development starts at the air–solution interface and continues toward the substrate–solution interface.

In order to capture the development of optical anisotropy through thickness at the early stages of the drying of 10 wt % PAI/DMAc solution, Yucel et al. [23, 83] used an optical compensation technique. Solution casting and sample preparation procedure for the calculation of retardation using the compensation method are shown in Figure 3.22.

Cross-polarized optical micrographs obtained during the course of drying using a Berek compensator are shown in Figure 3.23. The profiles in micrographs represent the retardation gradient from the air to the substrate. In Figure 3.23, the air–solution interface and substrate–solution interface are shown by (a) and (s), respectively, and $\Delta D_{\text{ol}}$ represents the high orientation layer. The quantified gradients at different time intervals are shown in Figure 3.24. At the early stages of the drying process before 140 min, the

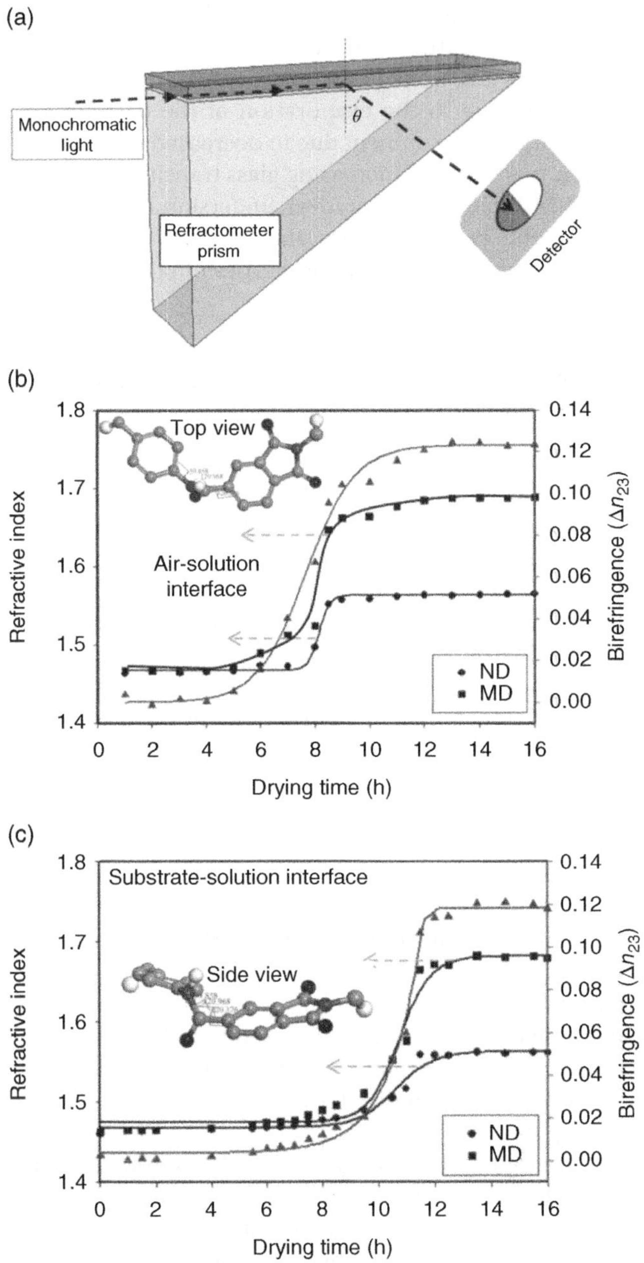


**Figure 3.21** (a) optical measurements using an Abbe refractometer during the drying of 10 wt % PAI/DMAc solution; (b) temporal evolution of principal refractive indices at air–liquid interface and (c) temporal evolution of principal refractive indices at substrate–liquid interface. Inset: molecular drawings depict planar low-energy conformation state for single PAI repeating unit. *Source:* Reproduced with permission from Ref. [83].

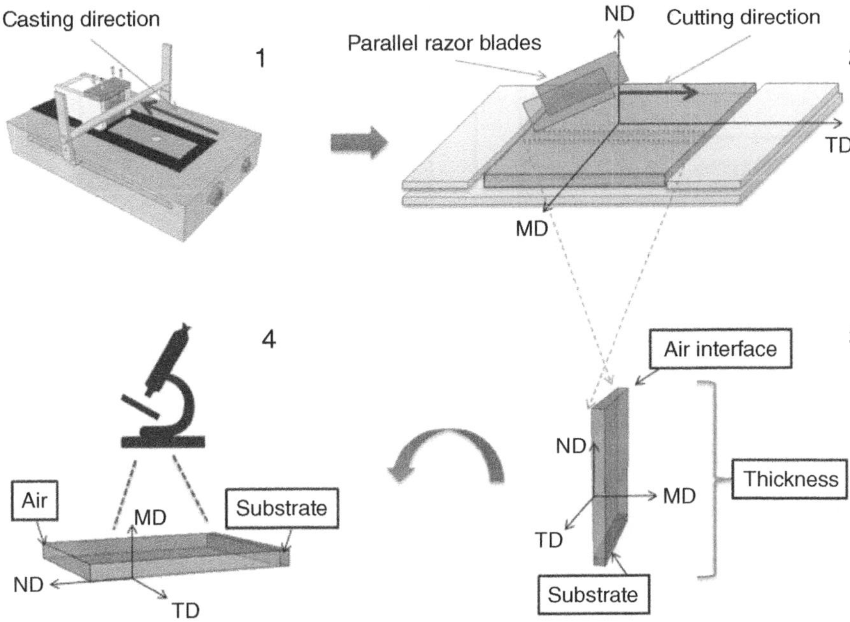


**Figure 3.22** Solution casting and sample preparation procedure for retardation calculation using the compensation technique. *Source:* Reproduced with permission from Ref. [83].

birefringence is slightly higher at the substrate–solution interface that is attributed to the shear-induced planar orientation of PAI chains near the substrate. Starting from 140 min, a rapid increase is observed in the magnitude of birefringence at the air–solution interface. As drying proceeds, at 200 min a highly oriented layer is formed at the air–solution interface, $\Delta D_{ol}$ in Figure 3.23, that progresses through the thickness by the continuation of drying. Due to the high retardation values at the oriented layer, reading of these values was not possible as they went out of the range of the 30-order Berek compensator [23, 83]. The highly oriented layer was shown as the shaded area in Figure 3.24.

### 3.3.5 Effect of Organoclay and Graphene Oxide on the Drying Process of PAI/DMAc Solution

Yucel et al. studied the effect of Cloisite 30B organoclay [23, 84] and graphene oxide (GO) nanosheets [23, 85] on the drying process of PAI/DMAc solution using the real-time measurement apparatus.

Effect of organoclay loading on the kinetics of drying of an 8 wt % PAI/DMAc solution exposed to airflow rate 0.5 m/min at room temperature is shown in Figure 3.25. As one can see, organoclay reduces the evaporation rate, and,

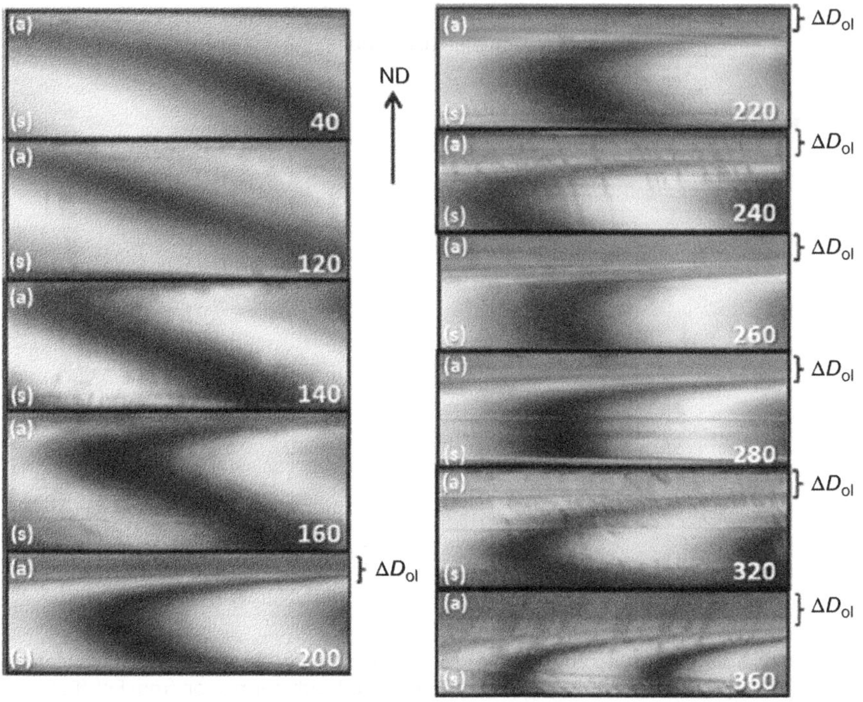


**Figure 3.23** Cross-polarized optical micrographs obtained using 30-order Berek compensator inserted at 45° orientation captured during the drying of PAI/DMAc solution at different drying times. (a) air–solution interface and (s) substrate–solution interface; $\Delta D_{ol}$: high orientation layer. *Source:* Reproduced with permission from Ref. [83].

therefore, it is expected that the total drying time will increase by increasing the organoclay content. In effect, the presence of organoclay inside the solution increases the diffusion paths of the solvent and leads to a slower evaporation.

Figure 3.26 displays the effect of organoclay loading on the development of out-of-plane birefringence during the drying process of 8 wt % PAI/DMAc solution exposed to airflow rate 0.5 m/min at room temperature. The real-time variation of out-of-plane birefringence as a function of solid content in polymer solutions indicates that by adding organoclay to the polymer solution, the critical solid content at which the out-of-plane birefringence develops in the cast film reduced from 40 to 10 wt %. It is interesting to note that further loading of organoclay does not change this critical solid content; however, it leads to higher values of birefringence at the end of the drying process. Yucel et al. [23, 84] found that the relative humidity of the environment may also influence the birefringence of the samples. That effect was attributed to absorption of water since DMAc is a hygroscopic solvent.

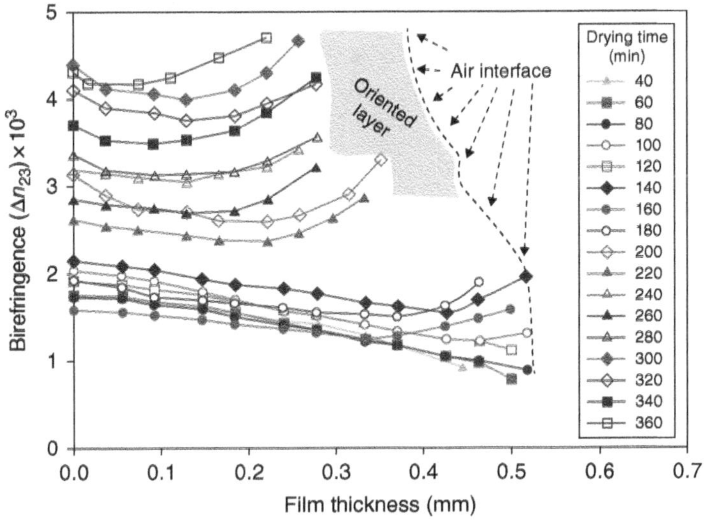


**Figure 3.24** Calculated birefringence values at different drying times (minutes) from optical micrographs using the compensator method. *Source:* Reproduced with permission from Ref. [83].

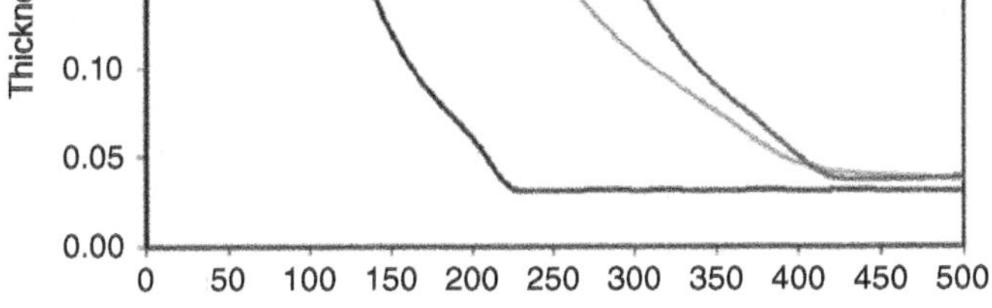


**Figure 3.25** Effect of organoclay loading on real-time thickness of 8 wt % PAI/DMAc solution. The cast polymer solutions were exposed to airflow rate of 0.5 m/s at room temperature. Arrow shows the increase in the clay loading from 0 to 5 wt %. *Source:* Reproduced with permission from Ref. [84].

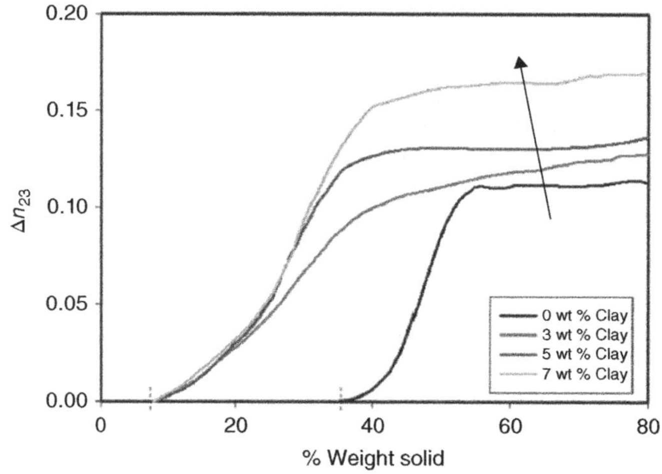


**Figure 3.26** Effect of organoclay loading on the development of out-of-plane birefringence during the drying process of 8 wt % PAI/DMAc solution. The cast polymer solutions were exposed to airflow rate of 0.5 m/s at room temperature. Arrow shows the increase in the clay loading from 0 to 7 wt %. *Source:* Reproduced with permission from Ref. [84].

The effect of GO loading on the drying kinetics and development of out-of-plane birefringence during the drying process of a 10 wt % PAI/DMAc solution exposed to airflow rate 0.2 m/min at room temperature is shown in Figures 3.27 and 3.28, respectively. In contrast to organoclay, by increasing the GO content, the rate of evaporation of solvent increases. Herein, the nanosheets increase the drying surface area and facilitate the diffusion of solvent molecules through the thickness of cast polymer solution. Moreover, by increasing the concentration of GO, the nanocomposite films become hydrophobic [85]. This will reduce the water uptake from the environment that can increase the drying time. By increasing the GO loading, the development of out-of-plane birefringence takes place at earlier times (Figure 3.28).

This is due to the rapid evaporation of the solvent in the presence of high concentration of GO. However, since, the high concentration of GO can disrupt the planar orientation of polymer chains, a high ultimate birefringence cannot be achieved at high GO content.

### 3.3.6 Real-Time Drying Study of Polyetherimide/NMP

Jiang and Cakmak [86] studied the real-time drying behavior of polyetherimide (PEI)/NMP solution under different drying conditions.

The influence of initial polymer concentration on the development of birefringence during the drying process of PEI/NMP solution at 65°C under the airflow rate of 0.5 m/s is shown in Figure 3.29.

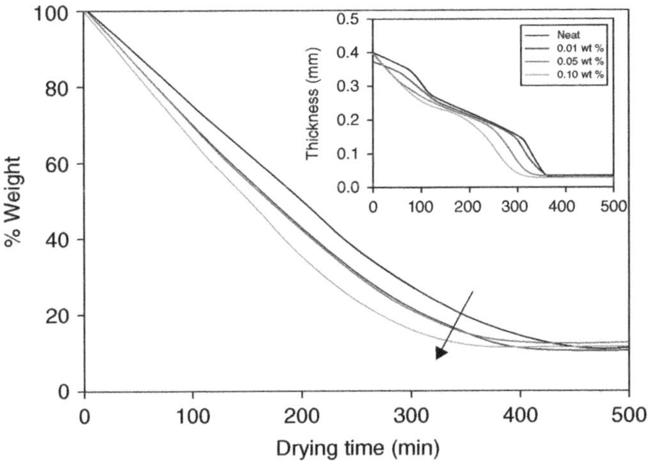


**Figure 3.27** Effect of GO loading on real-time weight and thickness of 10 wt % PAI/DMAc solution. The cast polymer solutions were exposed to airflow rate of 0.2 m/s at room temperature. Arrow shows the increase in the GO loading from 0 to 0.10 wt %. *Source:* Reproduced with permission from Ref. [85].

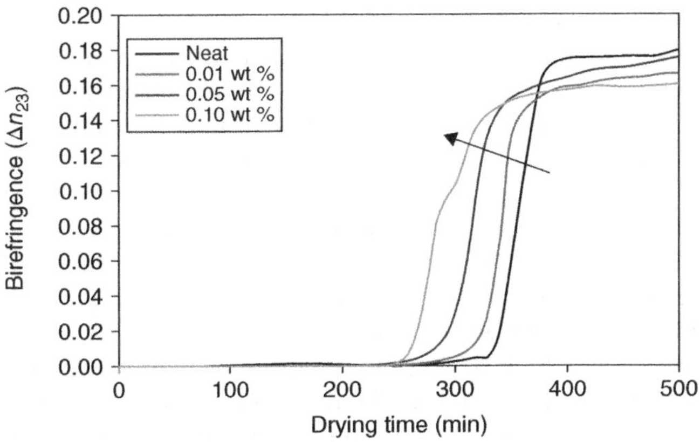


**Figure 3.28** Effect of GO loading on the development of out-of-plane birefringence during the drying process 10 wt % PAI/DMAc solution. The cast polymer solutions were exposed to airflow rate of 0.2 m/s at room temperature. Arrow shows the increase in the GO loading from 0 to 0.10 wt %. *Source:* Reproduced with permission from Ref. [85].

During the evaporation process, in-plane birefringence remains zero, but out-of-plane birefringence develops inside the film after about 100 min of evaporation. The final out-of-plane birefringence is lower in the case of polymer solutions with high initial concentration. That is explained by the better mobility of the polymer chains in more dilute polymer solutions, where more

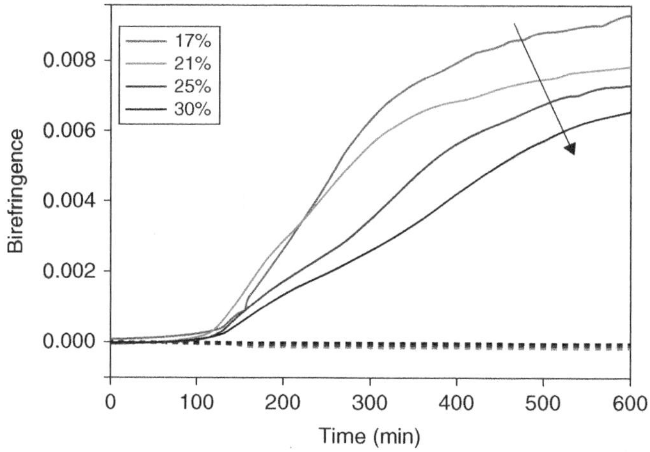


**Figure 3.29** Effect of initial polymer concentration on the development of birefringence during the drying process of PEI/NMP solution at 65°C under the airflow rate of 0.5 m/s. (dashed line is the in-plane birefringence and solid line is the out-of-plane birefringence). Arrow shows the increase in the initial polymer concentration. *Source:* Reproduced with permission from Ref. [86].

efficient planar packing of polymer chains is possible. Furthermore, frustration effects during chain realignment are reduced at low polymer concentrations.

The effect of drying temperature on the temporal evolution of birefringence during the drying process of two 17 wt % PEI/NMP solutions with different initial thickness (300 and 750 μm) under the airflow rate of 0.5 m/s is shown in Figure 3.30.

In-plane birefringence remains zero for all the drying temperatures during the whole drying process. Drying at low temperatures leads to low values of out-of-plane birefringence at the early stages of evaporation; however, the birefringence increases rapidly after a certain time and ultimately reaches a high plateau value. The reason why high drying temperature leads to lower final values for birefringence may be attributed to the rapid skin formation at high temperatures that reduces the evaporating solvent flux. By trapping the solvent under the skin, polymer chains can relax inside the cast film that eventually reduces the ultimate out-of-plane birefringence.

## 3.4 Conclusions

Evaporation of solvent from polymer solutions is an example of a nonequilibrium phenomenon that is important from both scientific and technological perspectives. A solvent-borne polymer coating is the product of transient mass and heat diffusion phenomena. The temporal evolution of concentration and

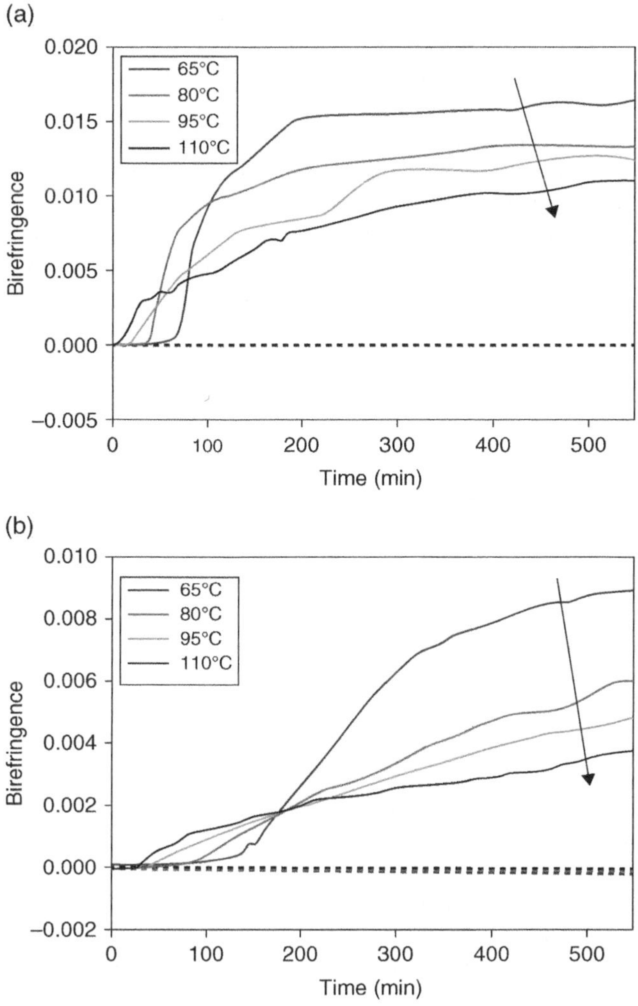


**Figure 3.30** The effect of drying temperature on temporal evolution of birefringence during the drying process of 17 wt % PEI/NMP solution under the airflow rate of 0.5 m/s: (a) 300 μm initial thickness and (b) 750 μm initial thickness (dashed line is the in-plane birefringence and solid line is the out-of-plane birefringence). Arrows show the increase in the drying temperature. *Source:* Reproduced with permission from Ref. [86].

temperature during evaporation of solvent plays a crucial role in determining the final properties of the product. Owing to the fact that diffusion is a dissipative process and contributes to the entropy production of the system, the governing equations for the evolution of mass and heat diffusion must be formulated in the framework of irreversible thermodynamics. The governing equations derived in this framework are thermodynamically consistent and can capture the physical effects such as thermal diffusion during evaporation of solvent. Moreover, transport coefficients can be derived in terms of thermodynamic variables.

The temporal evolution of concentration and temperature during the drying process highly depends on the physicochemical properties of the polymer system in question and also initial and boundary conditions of the diffusion problem. Therefore, the final properties of a solvent-cast polymer film are the result of numerous factors. The complex interaction between different factors contributing to the final state of the polymeric film needs to be understood in order to achieve desired properties in the final polymeric film. A powerful experimental way to investigate this complicated process is real-time tracking of the temporal evolution of all the changes that take place during evaporation of solvent. The results of such real-time monitoring at various processing conditions show how it might be possible to explain and quantify the evolution of changes with respect to the processing conditions. This way of investigating the complex phenomena of evaporation will help one to come up with ideas to fine-tune the final properties of polymeric films based on the knowledge gained by the real-time tracking of the phenomena.

## References


1 Siemann, U. Solvent Cast Technology: A Versatile Tool for Thin Film Production. In *Scattering Methods and the Properties of Polymer Materials*, Stribeck, N. and Smarsly, B. (Eds.); Springer, Berlin, Heidelberg, **2005**; pp 1–14.

2 Baker, R. W. *Membrane Technology and Applications*, 2nd ed.; John Wiley & Sons, Ltd, Chichester, **2004**; Ch.3.

3 Lee, M.; Park, J. K.; Lee, H. S.; Lane, O.; Moore, R. B.; McGrath, J. E.; Baird, D. G. *Polymer* **2009**, 50, 6129–6138.

4 Huang, Y. C.; Wang, T. Z.; Tsai, C. P.; Liu, T. J. *J. Appl. Polym. Sci.* **2003**, 129, 507–516.

5 Zhao, W. Flexible Transparent Electrically Conductive Polymer Films for Future Electronics. Ph.D. Dissertation [Online], The University of Akron, OH, May **2011**. http://rave.ohiolink.edu/etdc/view?acc_num=akron 1297888558 (Accessed on September 23, 2017).

6 Ma, R.; Baldwin, A. F.; Wang, C.; Offenbach, I.; Cakmak, M.; Ramprasad, R.; Sotzing, G. A. *ACS Appl. Mater. Interfaces* **2014**, 6, 10445–10451.

7 Greener, J.; Lei, H.; Elman, J.; Chen, J. *J. Soc. Inf. Disp.* **2005**, 13, 835–839.

8 Liu, J.; Ni, H.; Wang, Z.; Yang, S.; Zhou, W. Colorless and Transparent High-Temperature-Resistant Polymer Optical Films: Current Status and Potential Applications in Optoelectronic Fabrications. In *Optoelectronics: Materials and Devices*, Pyshkin, S.; Ballato J. (Eds.); InTech, Rijeka, **2015**; pp 57–81.

**9** Hungerford, G. P. Manufacture of Polyimide Film by Solvent Casting. U.S. Patent 4,405,550, September 20, **1983**.
**10** Cakmak, M.; Reneker, D. H.; Yalcin, B. Hybrid Manufacturing Platform to Produce Multifunctional Polymeric Films. U.S. Patent 2009/0020921 A1, January 22, **2009**.
**11** Cakmak, M.; Batra, S.; Yalcin, B. *Polym. Eng. Sci.* **2015**, 55, 34–46.
**12** Okazaki, M.; Shioda, K.; Masuda, K.; Toei, R. *J. Chem. Eng. Japan* **1974**, 7, 99–105.
**13** Okazaki, M. Transport Properties in Drying Process and Drying Characteristics. Ph.D. Dissertation [Online], Kyoto University, February **1975**. https://dx.doi.org/10.14989/doctor.r2800 (Accessed on September 23, 2017).
**14** Sosnowski, T. P.; Weber, H. P. *Appl. Phys. Lett.* **1972**, 21, 310–311.
**15** Weber, H. P.; Tomlinson, W. J.; Chandross, E. A. *Opt. Quant. Electron.* **1975**, 7, 465–473.
**16** Yamamura, M.; Uchinomiya, T.; Mawatari, Y.; Kage, H. *AIChE J* **2009**, 55, 1648–1658.
**17** Yamamura, M.; Horiuchi, K.; Kajiwara, T.; Adachi, K. *AIChE J* **2002**, 48, 2711–2714.
**18** Bormashenko, E.; Balter, S.; Pogreb, R.; Bormashenko, Y.; Gendelman, O.; Aurbach, D. *J. Colloid Interface Sci.* **2010**, 343, 602–607.
**19** Cavadini, P.; Krenn, J.; Scharfer, P.; Schabel, W. *Chem. Eng. Process. Process Intensif.* **2013**, 64, 24–30.
**20** Kajiya, T.; Doi, M. *J. Soc. Rheol. Japan* **2011**, 39, 17–28.
**21** Pauchard, L.; Allain, C. *Europhys. Lett.* **2003**, 62, 897–903.
**22** Arai, S.; Doi, M. *Eur. Phys. J. E* **2012**, 35, 1–9.
**23** Yucel, O. Birefringence Gradient Development during Drying of Solution Cast Functional Films and their Mechanical, Optical and Gas Barrier Properties. Ph.D. Dissertation [Online], The University of Akron, OH, December **2013**. http://rave.ohiolink.edu/etdc/view?acc_num=akron1377699282 (Accessed on September 23, 2017).
**24** Powers, G. The Effect of Temperature on the Solvent Removal from Solution-Cast Thin Polymer Films. M.S. Thesis [Online], Ohio University, OH, June **1988**. http://rave.ohiolink.edu/etdc/view?acc_num=ohiou1182868409 (Accessed on September 23, 2017).
**25** Krenn, J.; Scharfer, P.; Kind, M.; Schabel, W. *Eur. Phys. J. Spec. Top.* **2009**, 166, 45–48.
**26** Crank, J. *Free and Moving Boundary Problems*, 1st ed.; Oxford University Press, Oxford, **1984**; Ch.1.
**27** Port, A. B.; Cameron, C. Film Formation. In *The Chemistry and Physics of Coatings*, Marrion, A. (Ed.), 2nd ed.; The Royal Society of Chemistry, Cambridge, **2004**; pp 46–63.
**28** Roehner, R. Mathematical Modeling of Solvent Removal from Thin Polymer Films. M.S. Thesis [Online], Ohio University, OH, June **1982**. http://rave.

ohiolink.edu/etdc/view?acc_num=ohiou1183388452 (Accessed on September 23, 2017).

29 Kokes, R. J.; Long, F. A. *J. Am. Chem. Soc.* **1953**, 75, 6142–6146.

30 Blandin, H. P.; David, J. C.; Vergnaud, J. M.; Illien, J. P.; Malizewicz, M. *Prog. Org. Coat.* **1987**, 15, 163–172.

31 Ion, L.; Vergnaud, J. M. *Polym. Test.* **1995**, 14, 479–487.

32 Tu, Y.-O.; Drake, R. L. *J. Colloid Interface Sci.* **1990**, 135, 562–572.

33 Guerrier, B.; Bouchard, C.; Allain, C.; Benard, C. *AIChE J* **1998**, 44, 791–798.

34 Hwang, D. H.; Cohen, C. *Macromolecules* **1984**, 17, 1679–1685.

35 Price, P. E.; Cairncross, R. A. *J. Appl. Polym. Sci.* **2000**, 78, 149–165.

36 Vrentas, J. S.; Duda, J. L. *J. Polym. Sci. B Polym. Phys.* **1977**, 15, 403–416.

37 Vrentas, J. S.; Duda, J. L. *J. Polym. Sci. B Polym. Phys.* **1977**, 15, 417–439.

38 Masaro, L.; Zhu, X. X. *Prog. Polym. Sci.* **1999**, 24, 731–775.

39 Shams Es-Haghi, S.; Cakmak, M. *Polymer* **2016**, 97, 472–479.

40 Teraoka, I. *Polymer Solutions: An Introduction to Physical Properties*, 1st ed.; John Wiley & Sons, Inc., New York, **2002**; Ch.3.

41 Robinson, D. E.; Higinbotham, A. E.; Wankat, P. C. *Ind. Eng. Chem. Process. Des. Dev.* **1969**, 8, 502–508.

42 Price, P. E.; Wang, S.; Romdhane, I. H. *AIChE J* **1997**, 43, 1925–1934.

43 Alsoy, S.; Duda, J. L. *Dry. Technol.* **1998**, 16, 15–44.

44 Saure, R.; Wagner, G. R.; Schlünder, E. U. *Surf. Coat. Technol.* **1998**, 99, 253–256.

45 Verros, G. D.; Malamataris, N. A. *Comput. Mech.* **2001**, 27, 332–340.

46 Ramesh, N.; Duda, J. L. *AIChE J* **2001**, 47, 972–983.

47 Shams Es-Haghi, S. Modeling and Simulation of Diffusion in Evaporating Polymer Solutions. M.S. Thesis [Online], The University of Akron, OH, May **2015**. http://rave.ohiolink.edu/etdc/view?acc_num=akron1430492857 (Accessed on September 23, 2017).

48 De Groot, S. R.; Mazur, P. *Nonequilibrium Thermodynamics*, 1st ed.; Dover, New York, **1984**; Ch.11.

49 Shams Es-Haghi, S.; Cakmak, M. *Polymer* **2015**, 79, 110–118.

50 de Gans, B. J.; Kita, R.; Wiegand, S.; Luettmer-Strathmann J. *Phys. Rev. Lett.* **2003**, 91, 245501.

51 Vanag, V. K.; Epstein, I. R. *Phys. Chem. Chem. Phys.* **2009**, 11, 897–912.

52 Shams Es-Haghi, S.; Cakmak, M. *Polymer* **2016**, 97, 147–157.

53 Vrentas, J. S.; Duda, J. L.; Ling, H. C. *J. Appl. Polym. Sci.* **1985**, 30, 4499–4516.

54 Bearman, R. J. *J. Phys. Chem.* **1961**, 65, 1961–1968.

55 Shojaie, S. S.; Krantz, W. B.; Greenberg, A. R. *J. Membr. Sci.* **1994**, 94, 255–280.

56 Verros, G. D.; Malamataris, N. A. *Polymer* **2005**, 46, 12626–12636.

57 Alsoy, S.; Duda J. L. *AIChE J* **1999**, 45, 896–905.

58 Wong, S. S.; Altınkaya, S. A.; Mallapragada, S. K. *Polymer* **2004**, 45, 5151–5161.

59 Dabral, M.; Francis, L. F.; Scriven, L. E. *AIChE J* **2002**, 48, 25–37.
60 Price, P. E.; Romdhane, I. H. *AIChE J* **2003**, 49, 309–322.
61 Pourdarvish, R.; Danner, R. P.; Duda, J. L. *J. Appl. Polym. Sci.* **2009**, 111, 417–428.
62 Arya, R. K.; Bhargava, C. K. *Prog. Org. Coat.* **2015**, 78, 155–167.
63 Cakmak, M.; Unsal, E.; Yalcin, B.; Drum, J. E.; Yucel, O.; Nugay, I. Real-Time Measurement System for Monitoring and/or Controlling Properties of a Composition Transitioning from Liquid State to Solid State. U.S. Patent 2014/0230577 A1, August 21, **2014**.
64 Unsal, E.; Drum, J.; Yucel, O.; Nugay, I. I.; Yalcin, B.; Cakmak, M. *Rev. Sci. Instrum.* **2012**, 83, 025114.
65 Eguchi, Y.; Unsal, E.; Cakmak, M. *Macromolecules* **2013**, 46, 7488–7501.
66 Cherkasov, A. N.; Vitovskaya, M. G.; Bushin, S. V. *Polym. Sci. USSR* **1976**, 18, 1865–1873.
67 Prest, W. M.; Luca, D. J. *J. Appl. Phys.* **1979**, 50, 6067–6071.
68 Prest, W. M.; Luca, D. J. *J. Appl. Phys.* **1980**, 51, 5170–5174.
69 Greener, J.; Lei, H.; Rao, Y.; Elman, J. F. Strategies for controlling the optical properties of solvent-cast polymer films as applied to liquid crystal displays. IDMC 2007: International Display Manufacturing Conference and FPD Expo—Proceedings, Taiwan, Province of China, July 3–6, **2007**, 402–405.
70 Machell, J. S.; Greener, J.; Contestable, B. A. *Macromolecules* **1990**, 23, 186–194.
71 Coburn, J. C.; Pottiger, M. T.; Noe, S. C.; Senturia, S. D. *J. Polym. Sci. B Polym. Phys.* **1994**, 32, 1271–1283.
72 Payne, J. A.; McCormick, A. V.; Francis, L. F. *Rev. Sci. Instrum.* **1997**, 68, 4564–4568.
73 Payne, J. A.; McCormick, A. V.; Francis, L. F. *J. Appl. Polym. Sci.* **1999**, 73, 553–561.
74 Francis, L. F.; McCormick, A. V.; Vaessen, D. M.; Payne, J. A. *J. Mater. Sci.* **2002**, 37, 4717–4731.
75 Vaessen, D. M.; McCormick, A. V.; Francis, L. F. *Polymer* **2002**, 43, 2267–2277.
76 Croll, S. G. *J. Coatings Technol.* **1978**, 50, 33–38.
77 Croll, S. G. *J. Appl. Polym. Sci.* **1979**, 23, 847–858.
78 Unsal, E.; Cakmak, M. *Macromolecules* **2013**, 46, 8616–8627.
79 Unsal, E. Integrated Real Time Studies to Track all Physical and Chemical Changes in Polyimide Film Processing from Casting to Imidization. Ph.D. Dissertation [Online], The University of Akron, OH, December **2013**. http://rave.ohiolink.edu/etdc/view?acc_num=akron1384129535 (Accessed on September 23, 2017).
80 Chen, P. J.; Liu, T. J.; Wu, P. Y.; Tseng, C. F.; Leu, C. M. *AIChE J* **2010**, 56, 790–800.
81 Kotera, M.; Nishino, T.; Nakamae, K. *Polymer* **2000**, 41, 3615–3619.

**82** Pottiger, M. T.; Coburn, J. C.; Edman, J. R. *J. Polym. Sci. B Polym. Phys.* **1994**, 32, 825–837.
**83** Yucel, O.; Unsal, E.; Cakmak, M. *Macromolecules* **2013**, 46, 7112–7117.
**84** Yucel, O.; Unsal, E.; Harvey, J.; Graham, M.; Jones, D. H.; Cakmak, M. *Polymer* **2014**, 55, 4091–4101.
**85** Yucel, O.; Unsal, E.; Harvey, J.; Graham, M.; Jones, D. H.; Cakmak, M. *J. Membr. Sci.* **2015**, 495, 65–71.
**86** Jiang, F.; Cakmak, M. *Polymer* **2015**, 68, 168–175.

# 4

# In-Line Vacuum Deposition

*C. A. Bishop*

*C. A. Bishop Consulting Ltd., Loughborough, Leicestershire, UK*

## 4.1 Introduction

Vacuum deposition is often only one step within the full roll-to-roll (R2R) manufacturing line, with other process steps proceeding and/or following the vacuum coating step. Most commonly these different process steps are separated by the substrate being conveyed from one process step to another, with each step being run independently of the other steps. In an ideal R2R operation, all the different process steps are combined into a single process line with the raw substrate being unwound at one end and the final product web being wound up or sheeted at the other end of the line. This ideal flow is not often achieved as the different process steps may have to run at different line speeds and may need different cleaning and maintenance routines. Currently the most common type of vacuum deposition process is still a batch process where a roll of material is loaded into a vacuum system, pumped down to the desired pressure with the substrate conveyed through a process zone and then wound up. The process zone may be relatively simple involving a single layer of coating deposited, but it is usually more complex with the options of plasma pretreatment followed by multiple coatings deposited in a single pass before rewinding. In some systems it may be necessary for the web to be wound back and forth repeatedly either to build up the coating thickness or to increase the number of layers. Coatings can be organic or inorganic or combinations of the two.

A common feature to all R2R operations is that the substrates are flexible enough to allow conveying through the system. Substrates can be thin metal foils, polymer films, thin flexible glass, ceramics, papers, fabrics, nonwoven materials, and foams. Whatever the flexible substrate materials, there are some common problems related to the quality of the bulk materials and the surface

*Roll-to-Roll Manufacturing: Process Elements and Recent Advances*, First Edition.
Edited by Jehuda Greener, Glen Pearson, and Miko Cakmak.

on which the coating is to be deposited. Often many coatings are deposited with some intrinsic stress as each material may have a different coefficient of thermal expansion, which may lead to distortion and stress at the interface due to mismatch of thermal expansion. Dimensional stability issues can be critical especially in many electronic devices as different layers need to be scribed or etched in register with other patterned layers. As the device size is reduced, the precision is increased and so the dimensional stability of the substrate during all of the different processing steps has to be well understood and properly managed.

In order to achieve good quality coatings, it is important to start with high-quality substrates with a defect-free, pristine surface and good adhesion between coating and substrate. As we shall see, this is not easy to achieve. In vacuum deposition the substrate is a critical component and so we will comment on substrate differences and then discuss two critical process factors: (i) the management of defects and (ii) the management of heat load. We will then discuss different deposition processes and systems followed by highlighting a number of potentially important products that require vacuum-deposited coatings.

## 4.2 Substrates

Just about any material that can be made flexible enough to be wound into a roll can be vacuum coated. Polymer films represent the materials that have the largest area vacuum coated per year but are mostly used in low value-added products such as packaging or decorative metalized products. In higher value-added applications, the substrate materials include also flexible glass, flexible ceramics, metal foils, paper, and even fabrics, fibers, and nonwoven materials. Metalized film can be micro-slit and twisted to make a metalized yarn for sewing. At the other end of the value-added scale, polymer or metal fiber can be coated by a variety of different methods, including vacuum deposition, to make electroluminescent, piezoelectric, photovoltaic, and supercapacitor fibers that can be woven into fabrics. In some cases the coatings may be applied to the already woven fabric. In other applications, the use of nonwoven materials is possible, and these too may be vacuum coated and used in applications such as medical dressings or as part of touch switch fabrics [1–7].

Choosing the right substrate for a given application can be challenging partly because there are so many options to choose from. Polymer substrates provide a wide range of options. These include various types of homopolymers, copolymers, co-extruded structures, and filled polymer composites, fabricated into film. We will look at some of the different types of substrate materials and highlight some of the benefits and limitations of each. In terms of relative cost, polyethylene naphthalate (PEN) is approximately 3× more expensive than polyethylene terephthalate (PET) special treatments or coatings notwithstanding, polyimide is approximately 4×, stainless steel is approximately 2.5×, and flexible glass is of the order 25× more expensive than PET. Flexible glass is expensive, but, considering that heat stabilized and planarized PEN can be 10× the cost of PET, the glass is then only 2.5× the cost of the specialized PEN. The difference between these two materials is still significant but glass is in addition a superior barrier material, whereas PEN would need to be laminated or coated with a specialized barrier layer, which can add significantly to the cost. If a conducting coating is required, flexible glass can be coated at a much higher deposition temperature, which is an advantage when coating indium tin oxide (ITO) to prepare a highly transmissive and conductive substrate. With these two important features, the more expensive glass substrate becomes cost competitive with the other flexible substrates.

### 4.2.1 Polymer Substrates

Within the vacuum coating industry, the most commonly used substrate materials are PET and polypropylene (PP). The corresponding films are also known as biaxially oriented PET (BOPET) or polypropylene (BOPP). These two materials dominate the packaging industry, and a large segment of this industry is coated using a simple aluminum metalizing process.

As we move into higher value-added products, the coating method often changes to meet tighter coating specifications and higher quality requirements. This may also dictate a change in substrate.

PET is still a good substrate material for many electronic applications, and many suppliers have produced specific grades that are designed to meet the specialized market needs of the various products. BOPET can be manufactured with an added heat stabilization process that allows the substrate to be processed at a higher temperature with minimal shrinkage. When electronic circuits are being deposited sequentially, it is important that the circuits are in register with each other and any laser scribing is also in register with the coatings. This means that the substrate has to be highly uniform and dimensionally stable. With a BOPET used on a packaging metalizer, the shrinkage may be greater than 1% at 100°C. After heat stabilization this can be reduced to a fraction of a percent at 140°C. It is also possible to further extend the temperature range by using PEN, whereby the usable temperature range can be increased to over 200°C. Another critical issue with polymer substrates is the migration of oligomers to the surface. This low molecular weight species can cause adhesion problems and produce pinholes in the deposited inorganic coating (see Section 4.3 on defects). This issue has been well recognized and addressed by many film suppliers. The film, as produced, may also have a particulate contamination problem in addition to oligomers. Some of the particles may be cleaned off, but it is generally hard to remove particles of approximately

300 nm and smaller. This problem is usually addressed by adding an overcoat that covers these contaminants and produces a new clean surface to deposit onto. This surface may be very difficult to keep clean after winding, and so it is common to insert an inert material (interleaf) to separate the top and bottom surfaces during rewinding. This process is known as interleaving. Another option is to co-extrude a protective layer together with the substrate layer with relatively poor adhesion between them to allow peeling off from the clean substrate material just prior to entering the deposition zone. In this way the substrate surface is only exposed for a very short time within the vacuum system before it is coated and low molecular weight and particulate contamination is minimized [8–11]. This method is referred to as "clean-on-demand" or "peelable-clean-surface" film.

Various polyimide films have been used for high temperature processing with some processes going as high as 500°C for short coating times. When using polymer films at high temperature, there is often a problem of a mismatch in the coefficient of thermal expansion between the substrate and coated layers, which can result in curling and deviation from planarity once the film is cooled to ambient temperature. A common solution to this problem is to coat the back surface of the polymer substrate with a layer that counteracts the thermal expansion of the coating on the top surface [12].

### 4.2.2 Flexible Glass

The defensive strategy taken by the glass industry to protect their sales in the display market has been to offer ever thinner glass. Thinner glass not only offers a reduction in weight but also is more flexible and it enables curved displays. As the available flexible glass has reduced in thickness from 200 to 50 μm, with 25 μm glass under development, the flexibility has increased proportionately. This flexible glass can be wound into rolls and it could offer the cost advantages of R2R processing. The main functional advantages of flexible glass are a higher transparency, a smoother surface with lower haze, a better thermal stability with the option of using higher processing temperatures, a harder surface that permits more robust cleaning to minimize surface contamination, and a superior barrier performance [13–16]. Flexible glass also presents some challenges. It is difficult to cut without shattering if any defects are present. The glass can be cut using scribing or by carbon dioxide laser [17] wherein the edge of the glass is melted and has to be rapidly quenched to prevent formation of differential thermal stresses causing cracking to occur. The glass may still be susceptible to cracking from the edges, and so the edges must be protected from contact using, for example, a simple polymer bead applied along the edges. This not only protects the edges but also provides interleaving for the glass surface during winding.

### 4.2.3 Metal Foils

Metal foils, like flexible glass, are much stiffer than polymer substrates, leading to difficulties in roll winding. Specifically, profile variations can lead to breakage due to tension differences and difficulty in preparing tightly wound rolls because of high winding pressures. Foils are manufactured by taking a large metal billet and sequentially passing it between successive nip rolls to squeeze it down to the desired thickness [18]. The metals are work hardened and need to be periodically annealed. With the thinner foils the crystal size may be of the same order as the foil thickness, which makes it easy to tear and increases surface roughness that is related to the crystal size. The foil may require a planarizing layer to provide a suitably smooth surface [19]. The metal foils are often lubricated to make it easier for the metal to pass between the nip rolls, and thus a final washing step is needed. This may not remove all the contaminants that may have migrated into the grain boundaries and that are slow to migrate to the surface. It may require heating, for example, by a scanned electron beam, either to volatilize or to carbonize the contaminant. Alternatively high power plasma cleaning or scrubbing may be used on the foil to provide a similar effect [20–25].

The foil may have variations in oxide content and roughness across the web so a planarizing layer may be required to smooth out its surface. The planarizing layer can also be used to prevent any contaminant from migrating from the foil into the applied coatings.

### 4.2.4 Fibers, Fabrics, Nonwovens, and Foams

We are entering a new age of flexible materials. In the past, polymers were considered to be very flexible compared with the alternatives such as paper, foil, or glass. Now we have a new range of materials that are being coated such as nonwoven and woven fabrics or the fibers that are used to make the fabrics and foams. All these materials can be processed by R2R, and hence vacuum coating is possible although they present some challenges. A number of these materials will not survive the high tension needed in typical winding operations and so the winding system needs to be relatively gentle [26] or, if the material is too fragile, it can be wound up supported by a carrier web.

A number of these materials have a sufficiently open structure that any depositing material will pass through the gaps and, if the substrate is placed on a cooled deposition drum, the drum would become coated. To prevent the deposition drum from being coated, it is possible to wrap a sacrificial flexible material around the drum, which can be removed and replaced as required. But this approach can have a deleterious effect on the heat transfer coefficient and could impact the coating process. If a system is being designed specifically for coating these materials, then it may be necessary to replace a deposition drum with a free span design [27–29]. In this case the atoms passing through will deposit onto a shield that can be cleaned periodically. When the deposition source is a magnetron sputtering source, it is possible to have two sources opposite each other, one on either side of the web. This will coat both sides of the web simultaneously, assuming the substrate can tolerate the corresponding heat load.

The free span system also has an advantage when using a plasma treatment by allowing either treating or depositing onto the fabric or fibers. The fabric can be fully immersed in the plasma, which provides the best opportunity for the plasma to penetrate the structure and treat the individual fibers uniformly [30, 31]. A plasma treatment process may be changed into a deposition process by using a precursor gas with the plasma and so making it a chemical vapor deposition (CVD) process. If the deposition pressure is increased toward atmospheric pressure, the mean free path is reduced down to very small dimensions and this too helps the plasma penetrate the fabric or fiber bundle and deposit the coating. A special version of the plasma-enhanced CVD (PECVD) process is the atomic layer deposition (ALD) process. In this process there are two precursor gases that are used sequentially to build up the coating monolayer by monolayer, and this too can be carried out at pressures close to atmospheric pressure, making it useful for directly coating fabrics. For more details of the ALD process, see Section 4.6.

### 4.2.5 Paper

Paper has been vacuum coated for a long time primarily for decorative and packaging products. Paper has a rough surface and so when metalized the metallic coating will appear matt, which is not regarded as being aesthetically appealing. To address this issue, paper is precoated to smooth the surface to be coated and improve the specular reflectivity of the vacuum-deposited metal. The other main problem with vacuum deposition onto paper is the moisture content of the paper. This can vary widely, depending on the storage conditions and the ambient humidity. In a high humidity climate, the water content of paper may be as high as 20%. If the relative humidity for storing the paper is kept around 50%, then the water content of the paper will be approximately 7%. This is high for materials to be processed inside a vacuum chamber as the paper will outgas and release the water into the vacuum system. Vacuum systems that are regularly used to process rolls of paper typically have extra cryopumps installed in the system [32–36] to help reduce the moisture content in the chamber. Reducing the paper water content before it is placed in a vacuum system can improve the productivity by reducing the pumpdown time and improving the coating quality by reducing the partial pressure of water during deposition. Drying the paper out completely is not an option as the moisture content affects the tensile strength of the paper. Too little moisture content causes the paper to lose strength and become fragile and easy to tear, making it very difficult to wind and convey. After the paper is coated in the vacuum system, it needs to be rehydrated once it is back at atmospheric pressure. This will happen slowly in the coated paper roll, but this can be a problem as the uptake of water will cause the paper to swell, leading to increase in the hardness of the roll as it swells and ultimately damaging the paper product. Often the roll will be rewound in atmosphere and may have a fine mist of water sprayed onto the back surface to speed up the rehydration process.

Paper is currently being used for various disposable, limited lifetime electronics applications. The limited lifetime of the product makes it possible to use a low-cost substrate such as paper. Examples of limited lifetime electronics include light-emitting diodes for Christmas decorations and disposable batteries.

## 4.3 Managing Defects

Two types of defects on vacuum-deposited coatings are particulates that sit on the surface and classified as gross contaminants and material that migrates from the bulk to the surface and classified as molecular contaminants. These two different types of surface contaminants need to be treated differently, and cleaning the surface of each type requires different techniques. The gross particulate contamination will result in pinholes in the coating as the particles are coated and may at a later stage in the processing be removed, leaving behind an uncoated area known as a pinhole or pin window, as shown in Figure 4.1. Particulates may be cleaned down to a diameter of approximately 300 nm below, with the van der Waals forces holding the particulates attached to the surface large, making them very difficult to remove. Particulates of 300 nm and smaller are still a significant problem as the multilayer coating total thickness of an organic light-emitting device is also of the order of 300 nm, and, for barrier coatings for food packaging, the coating thickness is often less than 0.1 of this particulate size at only 30 nm.

The surface contamination by the low molecular weight species affects particularly the coating adhesion. But this can also lead to pinholes in the coating due to the low molecular weight material evaporating during exposure to high temperature while in the deposition zone. This causes the material to vaporize and thus prevents the coated atoms from depositing or creating pinholes through localized pick-off of the coating where the coating is deposited onto the low molecular weight material. The latter may fail cohesively, allowing the coating to be lifted off and become attached to the back surface of the film during winding. Because the low molecular weight material reduces adhesion, it will also cause the coating to be more susceptible to stress cracking, which can lead to a reduction in performance in applications where film flexibility is critical.

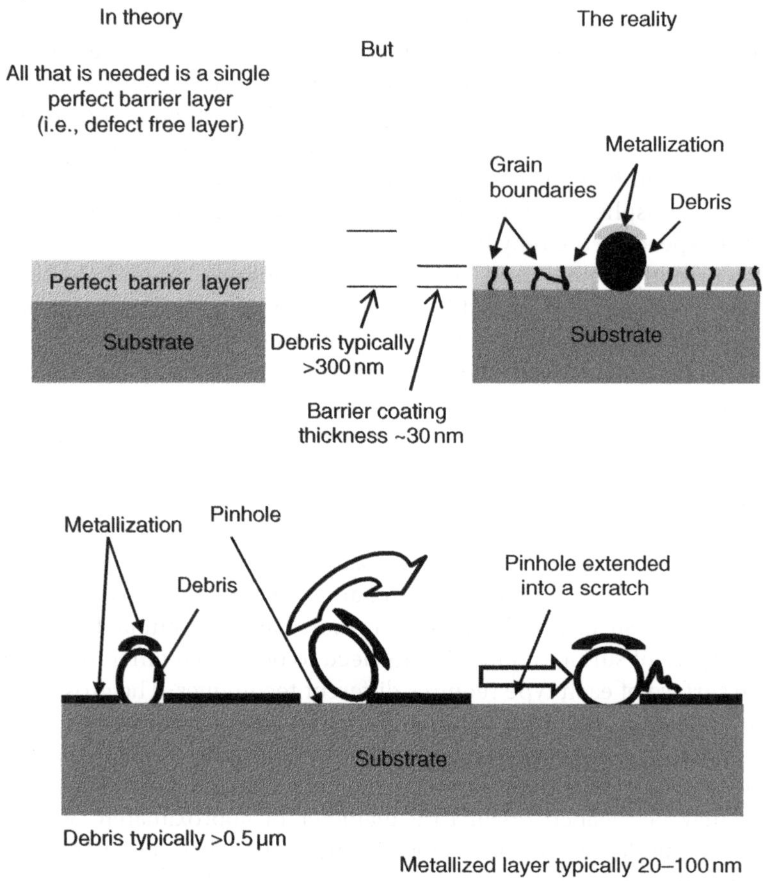


**Figure 4.1** The formation of pinholes or scratches by debris on a substrate surface.

The vacuum deposition industry uses primarily polymer substrates. Hence the focus of this discussion will be on defect management for polymer substrates. Hard substrates such as metal foils or flexible glass have the additional advantage of being less susceptible to scratches during contact with hard surfaces. Thus some of the contact cleaning options such as scrubbing or wiping can be used without detriment to the surface. With polymers, some scratching and damage to the surface can happen during cleaning with abrasive materials. The main source of particle contamination is the atmosphere. Airborne particles can settle onto the web surface during manufacture and slitting operations, and this would be exacerbated by triboelectric charges that attract particles to the surface, thus increasing the extent of contamination.

With polymer film manufacturing, many airborne particles originate from either white powder from the fumes from the polymer melt or slitting dust and so will be of similar hardness to the film substrate. Fillers may also be released during slitting, and these could be harder than the film surface and may lead to damage as they are moved around on the surface by wiping or scrubbing. In general, it is preferred to try to remove any particles from the surface without any lateral movement. Most webs are cleaned at atmospheric pressure. When a web is conveyed in air, there is a boundary layer of air that is dragged along with the web. The thickness of this boundary layer depends on the winding speed and may be 10–100 μm thick. Most of the particulates on the surface are confined to this boundary layer. The use of an air knife or a vacuum extract will do little to clean the surface as they are unable to break through the boundary layer well enough to disturb the particulates. It is possible to use a combination of techniques to remove particulates. If an ultrasonically pulsed jet of neutralized and filtered air is directed at the web, the web will flutter due to the ultrasonic pulsing, causing the particles to separate from the web surface. If there is also a vacuum extract applied in close proximity to the gas jet on the surface, the particulates may be removed [37], as shown in Figure 4.2. This same principle has been refined using finite element analysis to minimize the dimensions of the system, reduce the volume of air required, and lessen the corresponding noise. This design also enables both sides of the web to be cleaned simultaneously using a compact cleaning head [38], as shown in Figure 4.3.

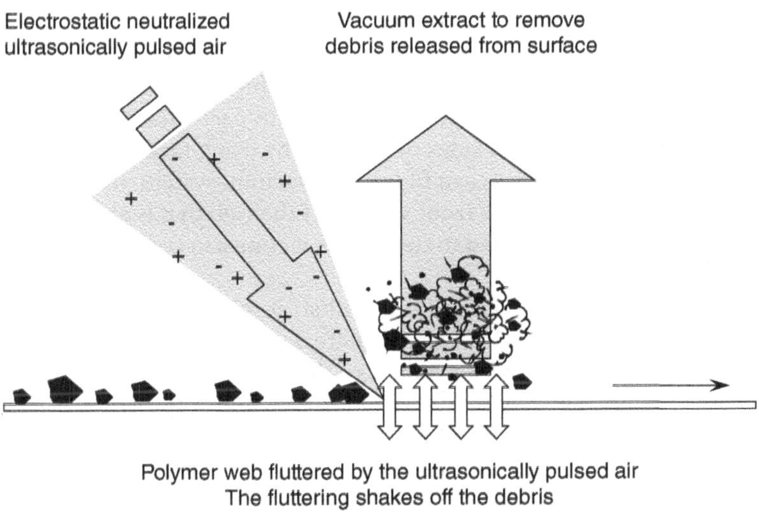


**Figure 4.2** A schematic of a gas cleaning system. Ultrasonically neutralized air is used to flutter the web enabling the release of debris from the surface that is then removed by vacuum extraction.

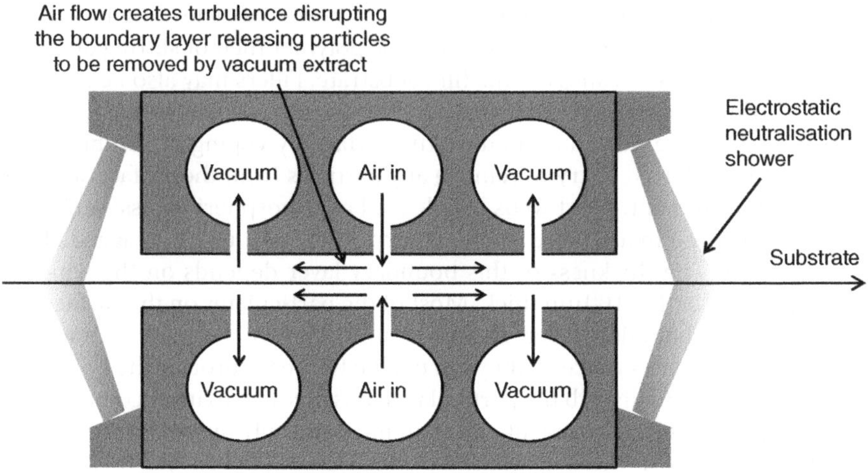


**Figure 4.3** A schematic of a modern compact air cleaning head for web materials. The turbulent air disturbs the debris, which are then removed by vacuum. The air also acts as air bearing, preventing the web from making contact with the head surfaces and thus preventing surface scratching.

The other, possibly more widely used, web cleaning technique involves passing a tacky roll in contact with the web surface and causing the particulates to be removed from the web by sticking to the tacky roll. The tacky roll could quickly become saturated with particulates and so, in order to keep the tacky roll effective, a high adhesive roll is run in contact with the tacky roll, thus transferring the accumulated particulates to the high adhesive roll (Figure 4.4). The high adhesive roll is made up of multiple layers of high adhesive film, enabling each layer to be removed as it becomes filled with particulates. This system can also be used to clean both sides of the film, by using tacky rolls as nip rolls [39–41]. This cleaning process can fit into the vacuum system. In this case it is generally used to clean only the front surface of the web that is about to be coated where the tacky roll is applied to the web as it comes in contact with the deposition drum [42].

Another significant problem with polymer substrates is the contamination of the surface with low molecular weight material species. This material is primarily an oligomer that is a by-product of the polymerization and film-making process [43–49]. The oligomer content is typically of the order 1–3%, depending on the processing conditions and the polymer. It is possible to minimize the oligomer content, but this usually requires a slower processing time with a concomitant increase in film cost. PET film has been produced with an oligomer content of less than 0.5%, but even at these low levels, there will be some oligomer that migrates to the surface. In fact, any low molecular weight material contained within the bulk polymer may migrate

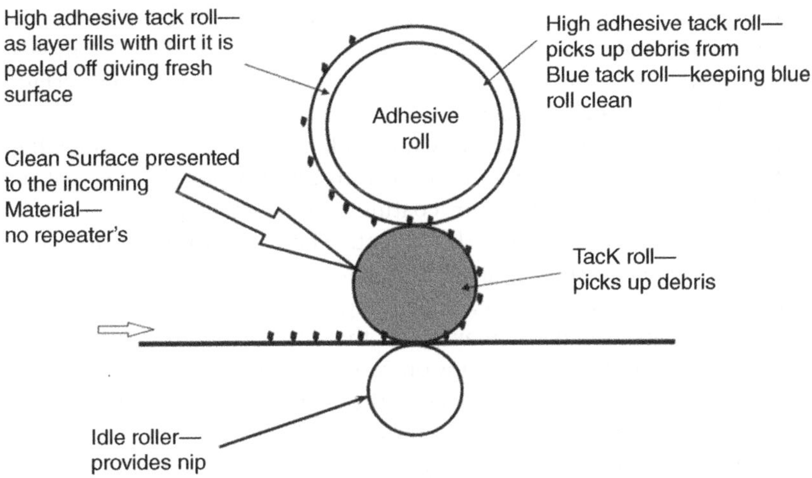


**Figure 4.4** A schematic of a tacky roll cleaning process with the tacky roll transferring the debris from the web to the high adhesive accumulation roll.

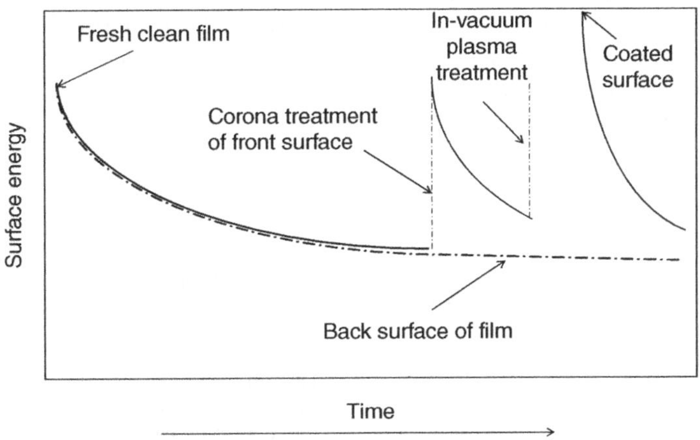


**Figure 4.5** A schematic showing the change in surface energy of a polymer substrate due to migration of low molecular weight contaminant, such as oligomer, to the surface.

to the surface. Polymers often contain various additives designed to aid the substrate manufacturing process or to modify the downstream processing such as the addition of slip agents used to make the handling of the film easier by reducing the coefficient of friction. These slip agents are designed to migrate to the surface and to be mobile and hence they too will affect adhesion. The migration process is time and temperature dependent, and so films can be considered to age, as shown in Figure 4.5. The migration rate will generally increase with temperature. Thus it is important to know the temperature history of the film since substrate manufacture. The low molecular weight material will change the chemical composition of the surface, leading to a change in surface energy.

If a coating is deposited onto the low molecular weight material, the adhesion of the coating to the substrate is compromised. It is possible to treat the surface before it reaches the vacuum coating system either by corona or by atmospheric plasma to minimize this effect. These treatments can either remove the low molecular weight contaminant from the surface or cross-link this material into the surface, depending on the energy level and gas composition. Any pretreatment may clean the surface but the surface will begin to degrade again from that point onward. If the cleaned front surface is brought in contact with the back surface of the film during rewinding some of the low molecular weight contaminant on the back, the unclean surface can transfer to the clean surface (Figure 4.6). Thus, it is important to perform the cleaning immediately prior to coating in the vacuum chamber to minimize recontamination of the front surface.

The plasma treatment can be applied just before film reaches the vacuum deposition system. Oxygen in the plasma may react with any hydrocarbons and produce volatile compounds that can be pumped away. Or if the energy is sufficient, the polymer or oligomer chains can be fragmented and some may be cross-linked into the surface. In this way, the low molecular weight material may either be removed or may be immobilized on the surface.

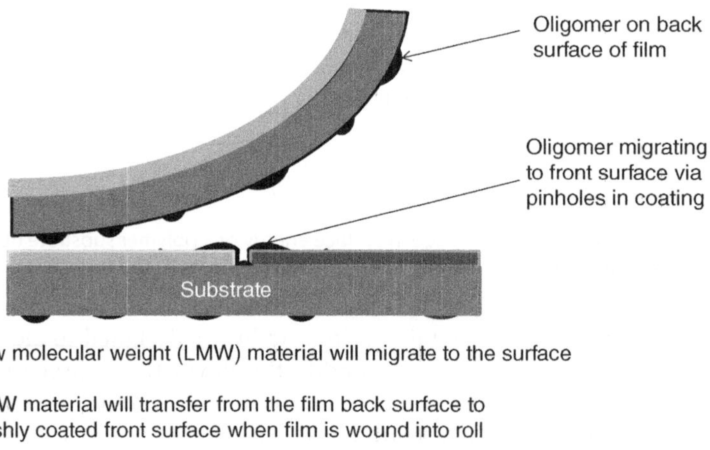


**Figure 4.6** A schematic showing how oligomer, or debris, can be transferred from the back surface to the cleaned, or freshly coated, front surface.

## 4.4 Managing Heat Load

Generally, by managing heat load we mean web cooling after the web exits the deposition zone for processes where the substrate must be heated prior to entering the vacuum chamber.

When deposition onto hot substrates is required, the substrate choice is limited to those materials that can withstand the desired deposition temperature. In this case the most commonly used substrates are metal foils, flexible glass, and high temperature polymers such as polyimide. It is relatively easy to heat the substrate by radiant heaters, but it is essential to apply the heating uniformly across the whole substrate width. If the web is not uniformly heated, there can be differential thermal expansion, which can result in the web deforming and wrinkling. The applied tension in the web along the machine direction generates a transverse compressive force. This force coupled with the thermal expansion needs to be managed to prevent wrinkling in the web, which may be partially controlled by using a spreader roll.

The high temperature processes need efficient web cooling after the deposition zone to prevent development of large compressive stresses in the web as it cools after winding. In processes such as coating photovoltaics, the deposition zone temperature of the web can be as high as 500°C, which needs to be reduced down to ambient temperature before winding. This is very challenging and may require considerably longer web path lengths than for the heating step.

The addition or removal of heat is generally controlled by radiation, convection, and conduction mechanisms. As the process operates in a vacuum, natural convection is relatively unimportant, which leaves radiation and conduction as the dominant heat transfer mechanisms. There is some heat lost by conduction to each roller that the web comes in contact with, and there is some heat lost by radiation to all the line-of-sight surfaces that the web passes through. The rate of heat lost by conduction depends on the contact area between the web and cooled surface, which is increased to some extent by web tension, as shown in Figure 4.7. Instead of a solid surface, using multiple deformable fibers that can conform to the web surface can increase the contact area and improve the cooling efficiency [50, 51]. However, using a roll with a porous surface, which allows cooled gas to fill the gap between the web and the roll, is the best method of increasing the heat transfer coefficient and cooling efficiency [52–55]. It is possible to cool the web between rolls by radiating heat to cooled blocks with black surfaces that absorb heat well. This works well initially but becomes less effective over time as the surface quality degrades gradually.

When the web needs to be cooled through the deposition zone, the use of gas between the web and deposition drum can increase the heat transfer coefficient considerably. For relatively wide webs or high winding speeds, a gas bar directing the gas to the point where the web first touches the deposition drum can be

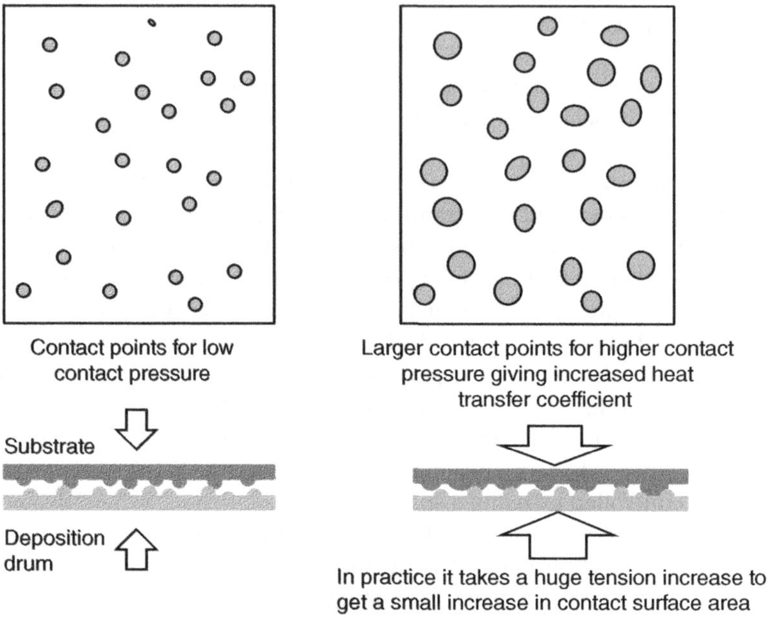


**Figure 4.7** A schematic showing the marginal change in contact surface area with increasing web tension.

an effective cooling option. The gas becomes trapped between the web and drum for long enough for the heat transfer coefficient to remain high throughout the time the web is in contact with the drum. When the web is narrow or the speed is low, the gas may leak out at the edges before the web reaches the deposition zone and the heat transfer coefficient does not change significantly [56]. This is shown schematically in Figure 4.8 with the continuous feed of gas through the drum surface to reduce the gas loss problem shown in Figure 4.9.

A heat transfer model of the cooling of a web in a vacuum deposition process can provide useful insights on the effects of various system variables such as substrate thickness, line speed, and so on, on the cooling efficiency of a particular system [57–60]. This is especially useful when wishing to change the process or materials and system design without undertaking costly experimental trials.

## 4.5 Vacuum Deposition Systems

Vacuum deposition systems for R2R manufacturing operations come in all shapes and sizes, and selection of a particular system depends on product specifications, production volume, and expected manufacturing outlays. Some typical systems are briefly described in the following text.

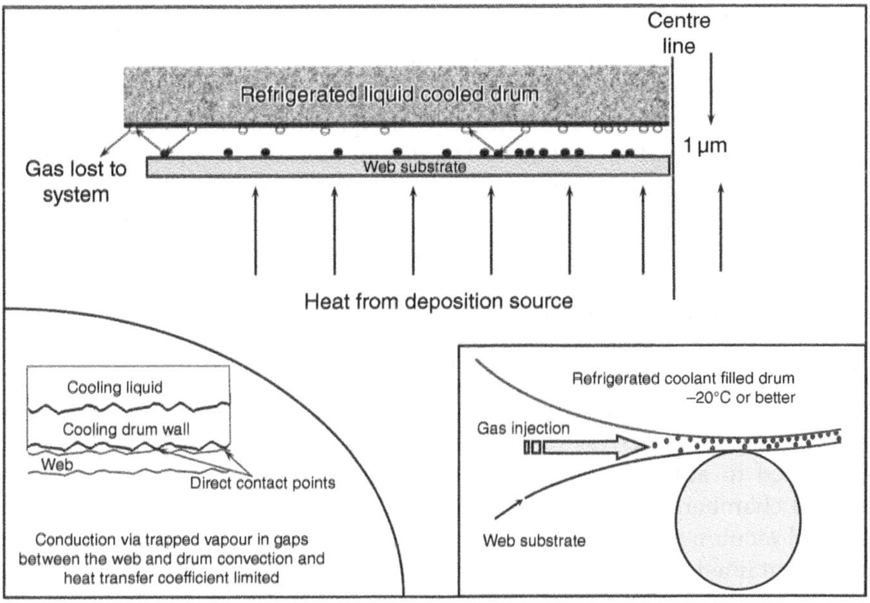


**Figure 4.8** A schematic showing how gas injection can improve the heat transfer coefficient and how gas can be lost from the edges.

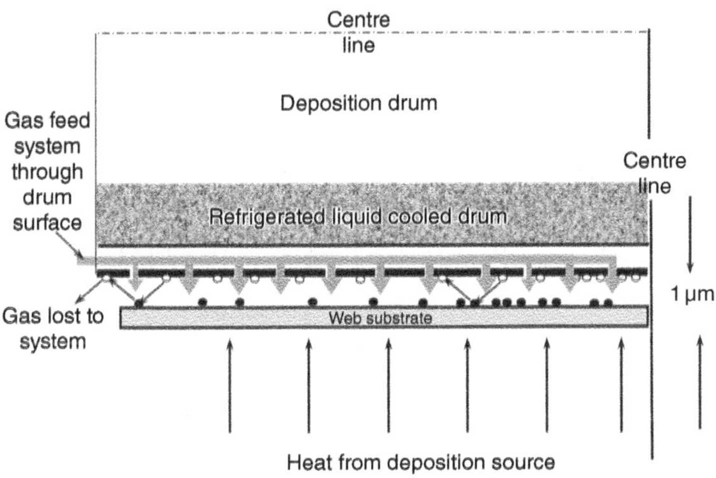


**Figure 4.9** A schematic showing how gas can be fed through the drum surface to maintain a gas cushion, with gas lost from the edges being continuously replenished.

### 4.5.1 Batch Systems

When a substrate is wound into a roll and loaded into a vacuum chamber, the process is in effect a batch rather than a continuous process. When the process speed is low and the roll sufficiently long, the process time may be very long, of the order of several days. The design of batch R2R coaters can vary considerably [61, 62], depending on the deposition source type, number of sources, number of coatings, number of deposition drums or free span, single or double side coating, and whether the rolls need to have an interleaving layer removed at unwinding and reapplied during rewinding. Bringing the system to atmospheric pressure and pumping the system down can be a slow part of the process, and there can be advantages to using separate but linked chambers for the unwind, deposition process zone, and rewind zone. The chambers can be isolated from each other so that the main deposition chamber can be kept under vacuum, while the unwind and rewind chambers can be vented to atmospheric pressure. This enables the smaller unwind and rewind chambers to be vented, replenished, and pumped back down to the desired vacuum more quickly than if the whole winding and deposition zones were contained in a single chamber. Periodically the main chamber has to be vented to enable the sources to be replenished and the deposition shields to be cleaned. A schematic illustrating these two different batch systems is shown in Figure 4.10.

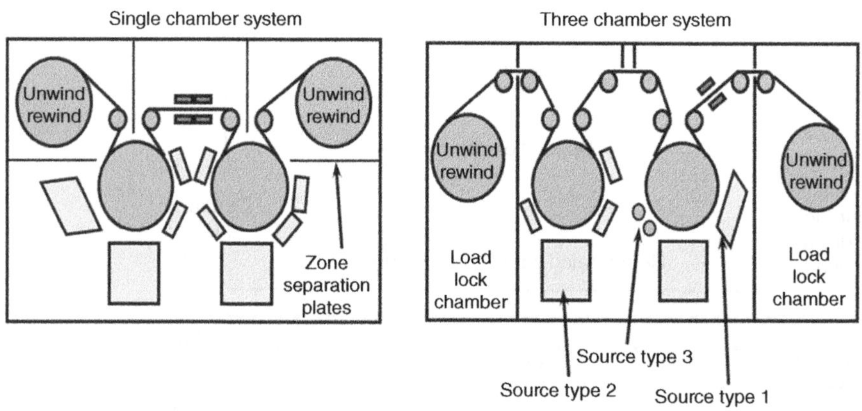


**Figure 4.10** Left: A schematic of a system where the winding and deposition systems are combined in a single chamber with zone separation plates to allow for a pressure difference between the unwind, rewind, and deposition zones. Right: A system with load locks for unwind and rewind chambers. A sealing mechanism between the load locks and main deposition chamber allows the main chamber to remain under vacuum while the load locks are vented.

### 4.5.2 Air-to-Air Systems

The concept of an air-to-air R2R vacuum system is often considered in relation to photovoltaic and display products in which vacuum deposition is only one process step along an R2R production line with one unwind and one wind-up roll at each end [63]. Between these two rolls, the web may be coated by an atmospheric wet coating step followed by laser scribing, a vacuum deposition step, lamination, and so on, before the final product web is wound up at the end of the line. This is the ultimate R2R process envisioned to minimize costs and maximize productivity. In reality the different process steps often operate at different optimal speeds, which are difficult to reconcile. Designing systems with matched processing speeds can be expensive, and often it is cheaper, at least initially, to have a series of batch operations rather than a fully integrated single R2R process line.

To work toward the goal of a single integrated manufacturing line requires the web to be able to pass from the air into a vacuum chamber to be coated and then brought back out of vacuum to atmospheric pressure. There have been a number of attempts to do this with varying degrees of success [64–69]. In the early attempts the web was passed through a narrow conductance gap into a series of chambers where the pressure in each chamber was gradually lower than in the previous one. Similarly, after exiting the deposition chamber, the web passes through a series of chambers with gradually increasing pressure. This arrangement involves a large footprint and a significant amount of pumping. This also requires a large number of contact rolls to guide the web through the many chambers, increasing the likelihood of damage to the substrate.

More recently a new design with a miniaturized small aperture was proposed to replace the series of large chambers. The new design utilizes a zone with a series of channels cut into an extended inlet, or exit, slot. The channels either have some gas fed in, to provide an air bearing, or are pumped out in sequence to increasingly lower pressure. This design reduces the size of the system and the amount of required pumping and significantly reduces the system cost. A schematic comparing the two designs is shown in Figure 4.11.

Where air-to-air systems have proved to be most successful is in the coating of strip steels with protective coatings such as Al or ZnMg and copper or aluminum foils with light-absorbing coatings to make solar absorbers [70–75]. Key to the success of this process is the development of the high material efficiency deposition sources that deposit of the order of 98–99% of the source material, which reduces the need to open the system for cleaning.

Metalizers and sputtering systems fall well short of this level of material efficiency, and the deposition shields that collect the stray deposition need to be cleaned regularly. With the high efficiency sources, some of the strip coaters run for many days before opening for cleaning. However, this has some drawbacks.

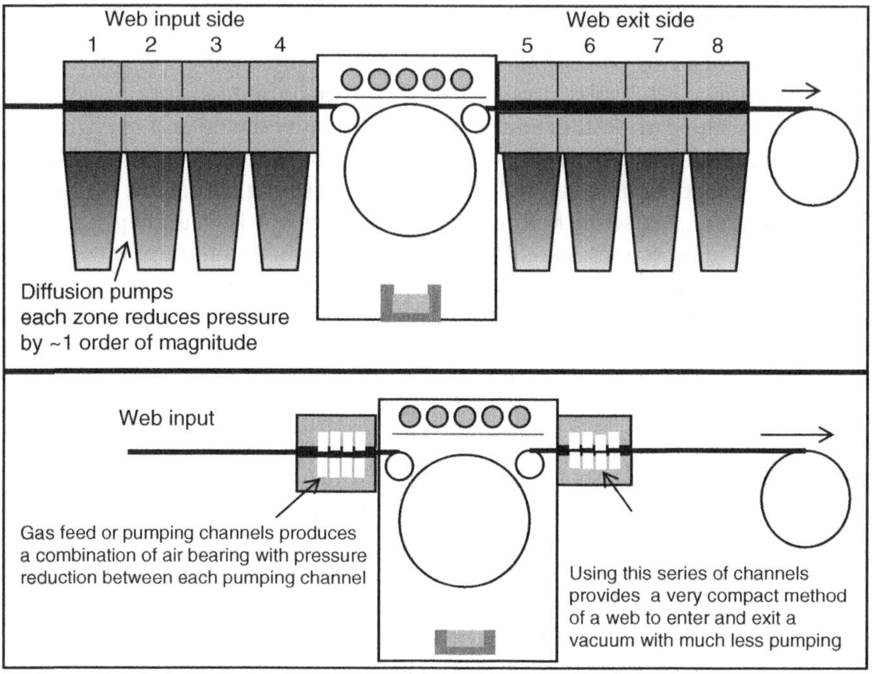


**Figure 4.11** Top: A schematic showing a large system with a series of chambers with large pumps. Bottom: A schematic showing a much smaller system with precision machined channel plates requiring considerably less space and pumping.

The source of depositing material needs either a continuous or periodic method of replenishment. The purity of the supply material becomes an issue over the latter stages of the long deposition process as the small levels of impurities will accumulate and may eventually reach high enough level to form an alloy to affect the deposition rate as well as the depositing coating composition [76]. These issues suggest that there are still improvements to be made in the web coating industry with respect to material efficiency.

## 4.6 Vacuum Deposition Processes

Three common vacuum deposition processes used for a wide range of coatings and product categories are described in the following text.

### 4.6.1 Physical Vapor Deposition (PVD)

PVD covers a wide range of different deposition processes that fall into two basic techniques: evaporation or sputtering [77–82]. With evaporation the source material is heated above its melting point, and the vapor pressure increases with the substrate passing through the flux so that the evaporated atoms condense on the substrate and a coating forms. This also includes materials that sublime rather than melt. The method of getting the source material to melt varies from resistance heated evaporation sources as used in aluminum metalizers, induction-heated sources, radiant-heated sources, and electron beam-heated sources. Evaporation is a high rate deposition process, but is often not as stable as the sputtering sources, which are generally preferred for high precision deposition such as for coatings that require a precise stoichiometry as for transparent conducting coatings. In the case of sputtering, the target source material is the cathode that is being impinged by plasma ions, causing the ejection of material that condenses on the substrate thus forming a coating. This process can be enhanced by using magnets to densify the plasma, but even with this enhancement the deposition rate is approximately 1000× slower than in evaporation. The magnetron sputtering sources can have problems of long-term stability particularly when using a reactive deposition process as well as with arcing associated with particulates or target defects. Some of these problems have been addressed by using dual magnetrons and dual rotatable magnetrons where the targets are kept clean and alternate as the cathode and anode. In this way cathodes can operate continuously throughout the target lifetime.

Magnetron sputtering is a more energetic deposition process than the evaporation processes, and the coating deposited can be denser and the adhesion better than for evaporated coatings. This can be addressed by the use of an additional plasma situated between the evaporation source and substrate. This will help to both densify the coating and sputter off any loosely bonded material and thus improve adhesion. This bombardment results in slower deposition rate, but this is still considerably faster rate than could be obtained by magnetron sputtering.

The choice of deposition source depends primarily on the specified properties of the coating. Evaporation will produce a more porous coating than magnetron sputtering, which can be an advantage for some products. The deposition of pyrotechnics for use in airbags is done by evaporating magnesium onto a polymer substrate. In this case the way the material is ignited and propagated requires that the hot gas percolates ahead of the ignition front preheating the magnesium. If the coating is too dense, the material would be difficult to ignite and it would extinguish. With solar cells the thick coatings are electron beam deposited with a dopant added to the coating by sputtering. Here the difference in deposition rates helps achieve low levels of dopant throughout the depth of the coatings. When very dense, hard and thick coatings are required, and electron beam deposition with an additional plasma is used by feeding the additional plasma with a reactive gas to produce a compound such as alumina or titania.

### 4.6.2 Chemical Vapor Deposition (CVD)

Rather than starting with a solid source material as in PVD, the CVD process starts with a precursor gas or vapor being fed into the vacuum system. This gas is either thermally decomposed or more often fed into a plasma where the precursor gas is fragmented, and some fragments either combine with other atoms or produce the material that will form the desired coating. This process too can be enhanced using magnetic confinement to densify the plasma and increase the deposition rate [83, 84]. This process was originally developed for the deposition of silica for barrier applications. The silica is a dielectric, which over time covers the anode with an insulating coating and prevents it from working, as with the disappearing anode problem, also experienced in some reactive magnetron sputtering processes. Strategies for keeping anodes active include running them very hot to prevent atoms from sticking to the surface and feeding gas through the anode also to prevent the atoms from reaching the anode.

### 4.6.3 Atomic Layer Deposition (ALD)

ALD can be regarded as a special case of CVD and it also uses gas or vapor precursor. In this case, a first precursor gas is fed into the system to form a monolayer on all surfaces. This gas is then pumped away and the system flushed with an inert gas, and then a second gas is introduced that deposits another monolayer on all surfaces. These two monolayers interact with each other and form a monolayer of the desired coating material. The sequence of introducing precursor gas "A," flushing and introducing precursor gas "B," is repeated multiple times until the coating builds up to the required thickness.

To be able to coat a moving web with enough monolayers to achieve a useful coating thickness requires multiple sequences to be completed. This can be achieved using a special deposition head having a series of slots across the width with alternate slots being pumped or fed with a gas [85–90]. As the web passes over the head, it goes through a pumped vacuum slot followed by a gas feed slot for the first precursor, followed by another vacuum slot, a buffer gas feed, a vacuum slot, the second gas precursor slot, and then a final vacuum pumped slot. Once the web has passed over all of these slots, a single monolayer is deposited. If this series of slots is repeated, another layer will be deposited. If each of these slots is small, many slots can be compressed into a short space and so many multiples can be used sequentially to build up the layers in a single pass.

As shown in Figure 4.12, if the gas feed and pumping slots are machined into the winding drum, it is possible to have multiple layers deposited during the time the web passes around the drum circumference, assuming the drum is stationary. If the drum rotates so that the surface is moving in the opposite direction to the web, it is possible to build up many more layers during the time

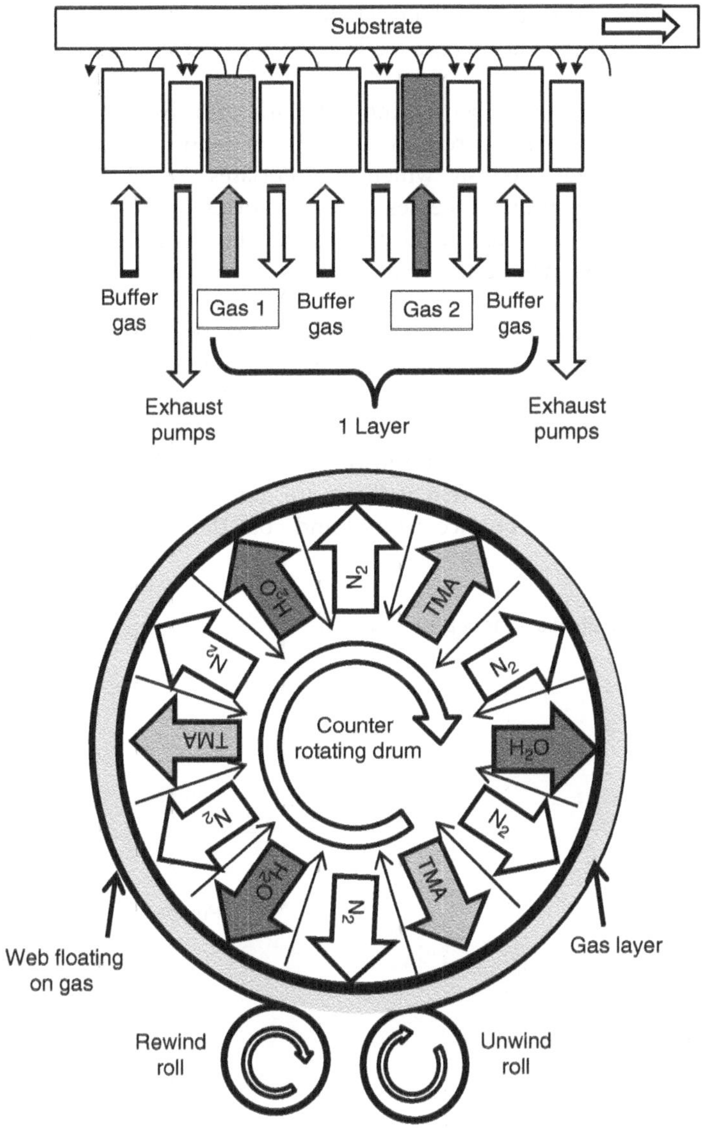


**Figure 4.12** Top: A schematic showing the sequence of slots in an ALD deposition head for depositing one layer. Bottom: A schematic showing a design where the deposition head is machined into a rotating drum that rotates in the opposite direction to the web direction in a way, allowing more layers to be deposited in a single pass [91]. © American Chemical Society.

the web passes over the drum surface. The web does not come in contact with the drum surface because the various gas feeds act as a gas bearing that separates the web from the drum.

An alternative design is to use a serpentine web path with the web moving between zones [87], as shown in Figure 4.13. This system deposits a coating on both sides of the web simultaneously, which differentiates the ALD process from other precision deposition processes. Where fabrics or fiber bundles need to be coated, this process has the advantage that the gas may penetrate the fibers from both sides, making the treatment or deposition on the fibers more easily achieved.

In ALD, since all that is required is to drive the gas to the surfaces to be coated and then remove the excess gas, there is no need to maintain low vacuum. In fact, in some cases it is possible to run the ALD process at atmospheric pressure. As the pressure is relatively close to atmospheric, the mean free path for the gas molecules is short and there is considerable molecular scattering. This has the advantage that if there are any particulates on the surface, the gas molecules will bounce around and coat the substrate under

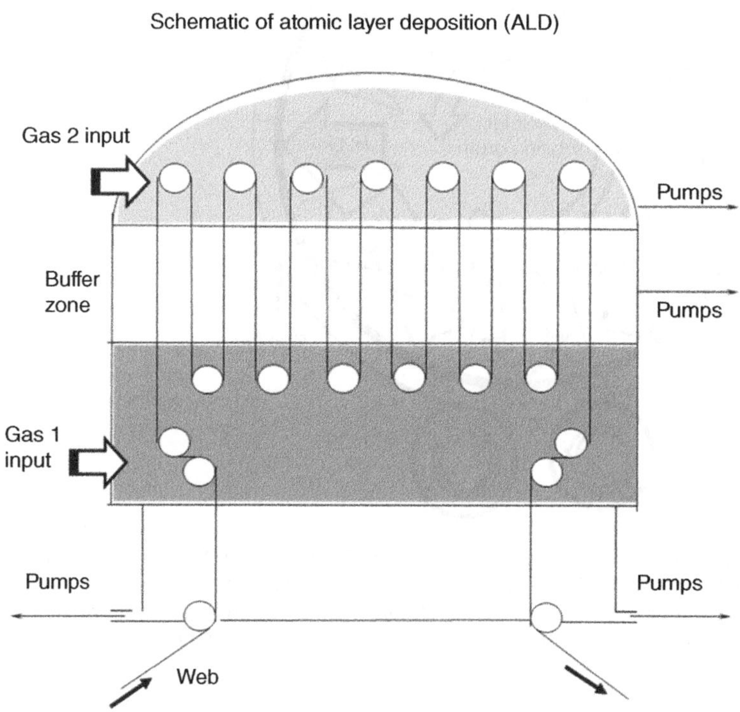


**Figure 4.13** A schematic of a serpentine web path for ALD deposition.

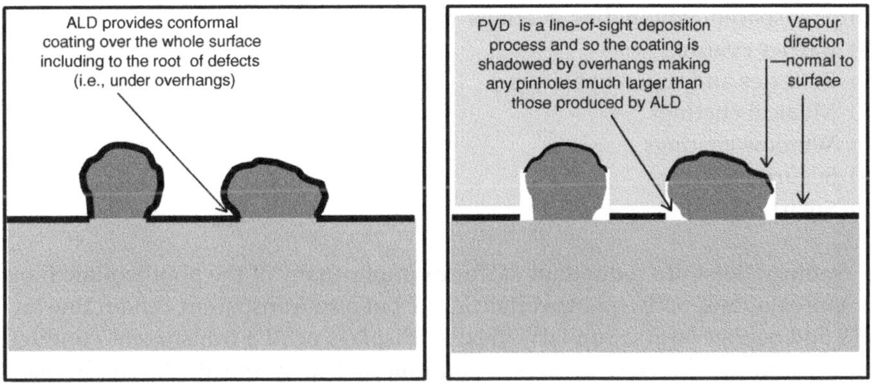


**Figure 4.14** A schematic comparing ALD coating (left) with PVD coating (right) in their ability to deposit around particles present on the coating surface.

the particle overhangs. This compares favorably with PVD processes where the deposition is more in line of sight and particles tend to produce a "shadow" on the substrate, as shown in Figure 4.14. If the particle is removed at a later stage of the process, the pinhole will be much bigger in PVD coating compared to ALD coating. For this reason ALD is the preferred deposition method for barrier coatings.

Although the ALD process has some obvious advantages, it has some limitations that restrict its widespread use. Generally, the precursor chemicals can be prohibitively expensive for many applications. Aluminum oxide used in some ALD processes is relatively inexpensive in high volume operations. However, large production-scale ALD R2R coaters are still at an early stage of development. The required precision for the deposition source and drum is at least an order of magnitude finer in tolerance than required for evaporation or magnetron sputtering coaters. This is easily addressed for narrow web coaters but for high volume, wide coaters, this is much harder and costlier to achieve. Until these cost issues are fully resolved, ALD will have only limited use in high volume R2R applications.

## 4.7 Vacuum-Deposited Coatings for Growth Markets

There are many areas for vacuum deposition coatings where the market is expected to grow substantially in the coming decade. The following are key markets with significant growth opportunities, listed in no particular order:

1) Flexible photovoltaics
2) Flexible displays

3) Transparent conducting coatings
4) Barrier coatings
5) Batteries and supercapacitors
6) Medical coatings
7) Window coatings
8) Security coatings
9) Smart fabrics

Some of these are redundant as, for example, many of the photovoltaic films require not only active photovoltaic layers but also transparent conducting layers and barrier layers. Similarly, flexible displays need a transparent conducting coating and barrier coatings in addition to display-specific layers. In many of these areas, several competing technologies are vying for the same market. For example, there are several technology options of producing transparent conductive coatings (TCCs) such as nanowire mesh, silver nanowires, carbon nanotubes, conducting polymers, and graphene [92]. The benchmark for all the new TCCs is ITO [93], which is currently the established technology so that any new TCC must have a compelling advantage to make the transition to the new technology viable. One key issue of ITO, which limited its utility in flexible electronics, is brittleness. This coupled with relatively high cost is one of the main incentives for searching for an alternative TCC.

The need for ultra-barrier coatings in flexible photovoltaics and display (especially OLED) is another major challenge of the vacuum coating industry. The required levels of water vapor and oxygen transmission rates are orders of magnitude lower than those currently achieved in food packaging [94]. The difficulty in producing these ultra-barrier coatings is primarily due to coating defects [95–97], which are directly related to the quality and cleanliness of the substrates and formation of pinholes in the barrier layer.

Organic planarizing undercoat can improve the quality of the coating surface on which the barrier layer is deposited and thus minimize the number of defects. Similarly, by adding an organic protective overcoat on top of the inorganic vacuum-deposited coating, the number of defects can be reduced, thus improving barrier performance. A combination of undercoat and overcoat is shown in Figure 4.15. Improvements in barrier performance can be achieved also by adding nanoplatelets to the polymer overcoat [98] or by creating a multilayer organic/inorganic structure, as shown in Figure 4.16, that produces a tortuous path for the diffusing species, thus reducing the gas transmission rate [99, 100].

The energy industry utilizes a variety of vacuum deposited products. Batteries have been vacuum deposited for many years [101–103], and new supercapacitors [104–107] are currently under development. In addition to power storage, power generation by materials such as ZnO piezoelectric coatings deposited onto PEN films are also under development [108, 109]. This is another area

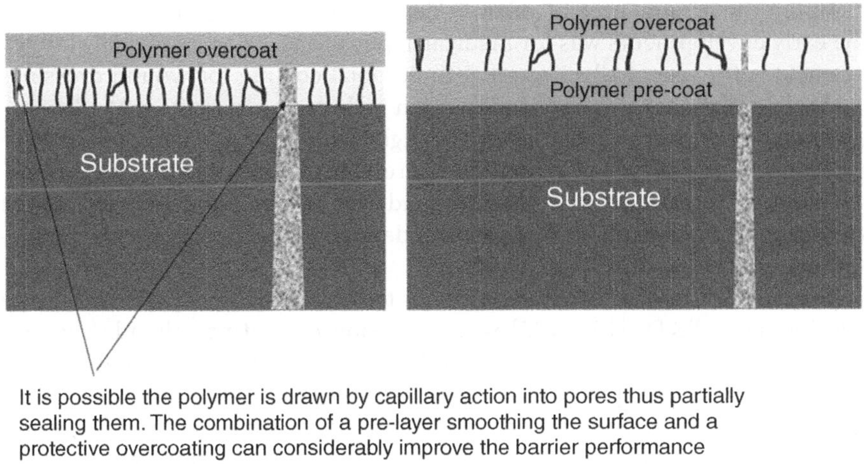


**Figure 4.15** A schematic showing how a polymer overcoat and planarizing layer can improve the barrier performance of an inorganic vacuum deposited coating.

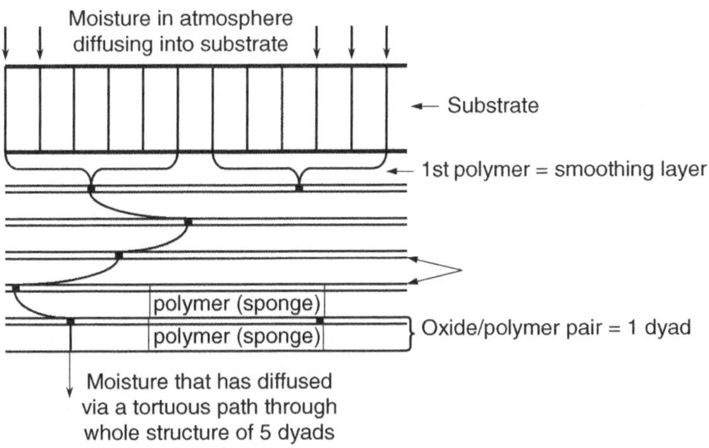


**Figure 4.16** A tortuous path of diffusing moisture in a multilayer organic/inorganic film structure.

where many different deposition options are utilized as well as a variety of different flexible materials used as substrates, including fabrics and paper [110, 111]. Photovoltaics are also being deposited onto fabrics. The challenge here is to be able to deposit the coatings to generate power, conduct the power to energy storage devices, and produce switches to enable some smart functionality, such as lighting within the fabric such as used in safety jackets [112].

Within the general area of vacuum-deposited coatings onto fabrics, one of the early developments was for medical applications. The most widely known example is the use of silver as an antimicrobial coating for wound dressings [113]. The fabric for this application often includes nonwovens that allow the moisture from the wound to pass through while having sufficient silver to provide a supply of ions to attack the microbial enzymes. There is an increasing interest in antibacterial clothing, and this involves the growing use of vacuum plasma treatments and vacuum deposition onto textiles and nonwoven materials [114–121].

Flexible photovoltaic devices continues to be an active area of research and development (R&D) [122–125] with extension of coatings already successfully deposited on glass substrates and with new materials such as perovskites [126–130] and quantum dots [131–134]. As with the other electronic applications, the new materials vary from fully vacuum-deposited coatings for the perovskites through mixed coating techniques to totally atmospheric wet-coated processes [135]. The substrates for the various photovoltaic devices include flexible glass [136], titanium foil, PET, paper [137], fibers, and fabrics [138–140].

## 4.8 Conclusions

There is a considerable R&D activity underway into R2R vacuum deposition of coatings for a wide range of applications. Some key problems that run through all the R&D activities are associated with substrate quality and defect management. Defects affect not only the coatings for electronic devices but also barrier coatings used for device encapsulation.

Coupled to managing defects and cleaning the coating surface is the need, in most cases, to improve the adhesion of the coating to the substrate. This usually requires a plasma treatment within the vacuum system immediately prior to the deposition zone. As most substrate surfaces change with age, it is often important to know the substrate history to ensure optimized pretreatment of the substrate.

Management of heat load is another critical element of the vacuum deposition process. This applies to processes where the substrate is preheated and to processes where the temperature must be kept below some critical temperature. It is generally easier to add heat to the substrate than it is to remove heat after processing, and several methods described in this chapter provide hardware and process options for efficient heating and cooling of webs.

ALD appears to be the coating method of choice for many new applications. ALD produces a continuous coating at a smaller thickness compared with most currently used R2R vacuum coating methods. It also has the advantage of improved flexibility and in producing coatings with fewer pinholes and less "shadow" due to foreign particles, thus improving barrier performance. The ALD process may also be operated at higher pressures, in some cases close to atmospheric, which can be advantageous when the substrate surface is particularly rough such as in uncoated paper or with fibers or textiles. One main drawback of ALD is the cost of some special precursor materials, which are not readily available.

As the drive for low-cost flexible devices increases, there is a more intense interest in R2R vacuum processing as well as significant R&D effort underway. However, R2R production lines with fully integrated vacuum coating process steps are not yet fully realized because of challenges in operating R2R lines under vacuum and mismatch in line speeds among the various process steps. It will be interesting to follow the R&D progress and see how quickly R2R vacuum technology becomes fully operational for the wide range of applications currently under development.

## References


1 Stoppa M. and Chiolerio A. 'Wearable electronics and smart textiles: a critical review'. *Sensors.* **2014**;14: 11957–11992.

2 Harrison D., et al. 'A coaxial single fibre supercapacitor for energy storage'. *Phys. Chem. Chem. Phys.* **2013**;15: 12215–12219.

3 NMP.2011.4.0-3: Advanced textiles for the energy and environmental protection markets. EU FP7 project Coordinator: Ian Jones, TWI Ltd, UK. Available from: www.powerweave.eu (Accessed on September 23, 2017).

4 Vatanseveret D., et al. 'Hybrid photovoltaic-piezoelectric flexible device for energy harvesting from nature'. *Adv. Sci. Technol.* **2013**;77: 297–301.

5 Weigel M., et al. 'iSkin: Flexible, stretchable and visually customisable on-body touch sensors for mobile computing'. In: CHI '13 Extended Abstracts on Human Factors in Computing Systems (CHI EA '13), Seoul, Republic of Korea, April 18–23, **2015**.

6 Xu R., et al. 'Fabric-based stretchable electronics with mechanically optimized designs and prestrained composite substrates'. *Extreme Mech. Lett.* **2014**;1: 120–126.

7 Rio L., et al. 'Comparison of methods for evaluation of the bactericidal activity of copper-sputtered surfaces against methicillin-resistant *Staphylococcus aureus*'. *Appl. Environ. Microbiol.* **2012**;78(23): 8176–8182.

8 MacDonald W.A., et al. 'Engineered films for display technologies'. *J. Mater. Chem.* **2004**;14: 4–10.

9 MacDonald W.A., et al. 'Latest advances in substrates for flexible electronics'. *J. Soc. Inf. Disp.* **2007**;15: 1075–1083.

10 MacDonald W.A., et al. 'Polyester film substrates for the flexible electronics industry: an overview and where next'. In: Proceedings of the AIMCAL

Annual Technical Conference and International Vaccum Web Coating Conferences, Charleston, SC, October 27–30, **2013**.
11 Gordon S., et al. 'Polyester films for the next generation of flexible electronics'. In: 2016 Flexible and Printed Electronics Conference and Exhibition, Monterey, CA, February 29 to March 3, **2016**.
12 Woods L.M., et al. 'Multilayer thin-film back surface contact for flexible photovoltaic devices on polymer substrates'. US 2013/0061927 A1, March 14, **2013**.
13 Murata T., et al. 'Ultra thin glass roll for flexible AMOLED display'. In: Proceedings of the18th International Display Workshops, Nagoya, Japan, December 7–9, **2011**. pp. 1735–1737.
14 Garner S., et al. 'Flexible glass substrates for roll-to-roll manufacturing'. In: Proceedings of the AIMCAL European Technical Conference, Cascais, Portugal, June 8–11, **2012**.
15 Tamagaki H., et al. 'Roll-to-roll sputter deposition on flexible glass substrates'. In: Proceedings of the 56th SVC Annual Technical Conference, Providence, RI, April 20–25, **2013**. pp. 1–5.
16 Garner S., et al. 'Roll-to-roll flexible electronic device fabrication on glass web'. In: Proceedings of the AIMCAL Fall Annual Technical Conference, Myrtle Beach, SC, October 19–22, **2014**.
17 Xinghau L. and Garner S. 'Laser cutting of flexible glass'. In: Conference on Lasers and Electro-Optics (CLEO): 2014, OSA Technical Digest (Online) (Optical Society of America, San Jose, CA, June 8–13, **2014**, Paper ATu3L.1.
18 Eue M. 'Hydro paper on foil'. In: Proceedings of the AIMCAL European Technical Conference, Myrtle Beach, SC, October 19–22, **2014**.
19 Sheehan C. and Matias V. 'Solution deposition planarisation of long-length flexible substrates'. In: Proceedings of the AIMCAL Annual Technical Conference and International Vaccum Web Coating Conferences, Myrtle Beach, SC, October 17–20, **2010**.
20 Holland L. 'Some characteristics and uses of low-pressure plasmas in materials science'. *J. Vac. Sci. Technol.* **1977**;14(1): 5–15.
21 Baker M.A. 'Plasma cleaning and the removal of carbon from metal surfaces'. *Thin Solid Films* **1980**;69(3): 359–368.
22 Vossen J.L. *Contemporary preparation techniques*; Academic Press: London; **1989**. p. 202.
23 Clow H. and Belkind A. 'Cleaning oil from metal substrates using a hollow cathode discharge'. In: Proceedings of the 8th Vacuum Web Coating Conference, Las Vegas, NV, November 6–8, **1994**. pp. 191–198.
24 Belkind A., et al. 'Linear plasma source for reactive etching and surface modification'. In: Proceedings of the 38th Annual Technical Conference SVC, Chicago, IL, April 2–7, **1995**. pp. 432–436.
25 Barankova H. and Bardos L. 'Hollow cathode plasma cleaning and surface activation'. In: Proceedings of the 45th Annual Technical Conference SVC, Lake Bueno Vista, FL, April 13–18, **2002**. pp. 109–112.

**26** Cohen B. 'Metallization in elastic non-woven webs'. In: Proceedings of the 37th Annual Technical Conference SVC, Boston, MA, May 8–13, **1994**. pp. 227–229.

**27** Casey F., et al. 'Properties of metallized film in a free span web metallizer'. In: Proceedings of the 42nd Annual Technical Conference SVC, Chicago, IL, April 17–22, **1999**. pp. 480–483.

**28** Yadin E., et al. 'Vacuum coating of materials having an open structure'. In: Proceedings of the 10th Vacuum Web Coating Conference, Fort Lauderdale, FL, November 10–12, **1996**. pp. 245–251.

**29** Lipin Y. and Yadin E. 'Specialities of non-woven materials coating technology and equipment'. In: Proceedings of the 13th Vacuum Web Coating Conference, Englewood, NJ, **1999**. pp. 188–193.

**30** Palmers J. and Ruys L. 'Roll-to-roll treatment of textiles in industrial plasma batch machines'. In: Proceedings of the Techtextil. Symposium Session 5.1; Messe Frankfurt, Germany, May 9–12, **1997** Plasma Finishing.

**31** Hollander A. and Behnisch J. 'Semi-continuous plasma treatment of textiles'. In: Proceedings of the Techtextil. Symposium Session 5.1; Messe Frankfurt, Germany, May 9–12, **1997** Plasma Finishing.

**32** Schwarz W. 'Thermal limitations of the pumping speed of refrigerator cooled surfaces'. *J. Vac. Sci. Technol. A* **1987**;5(4): 2568–2571.

**33** Casey F., et al. 'Design considerations of a high efficiency cryogenic pumping system for paper metallization'. In: Proceedings of the 36th Annual Technical Conference SVC, Dallas, TX, April 25–30, **1993**. pp. 213–219.

**34** Sack N., et al. 'Properties of vacuum deposited water/ice films'. In: Proceedings of the National Institute of Standards and Technology/AVS Workshop 'Water: Its Measurement and Control in Vacuum', Gaithersberg, MD, May 23–25, **1994**.

**35** Mahoney P. 'Design and placement of cryogenically cooled surfaces in web coating chambers: theory and practice'. In: Proceedings of the 9th Vacuum Web Coating Conference, Tucson, AZ, November 12–14, **1995**. pp. 62–74.

**36** Stenhouse J. and Spencer A.G. 'Coating improvements through better control of water vapour in vacuum deposition'. In: AIMCAL 2003 Proceedings of the 17th Vacuum Web Coating Conference Session 2B, Section 1, Paper 3, Tamaya, Santa Ana Pueblo, NM, October 26–29, **2003**.

**37** Schneider R.N. 'Removing of dust particles from a relatively moving material web'. Patent No US5,916,373 A, September 6, **1994**.

**38** PFFC. 'Meech System Cleans Web with No Contact'. Available from: http://www.pffc-online.com/surface-prep/web-cleaning/10630-meech-system-cleans-web-with-no-contact (Accessed on September 23, 2017).

**39** 'Web cleaning efficiency using contact cleaning roll (CCRs)'. Available from: http://www.polymagtek.com/pdf/dirt_rpt.pdf (Accessed on February 28, 2011).

**40** Hamilton S. 'Yield improvement in metallizing lines by using contact cleaning technology'. In: Proceedings of the 51st Annual Technical Conference SVC, Chicago, IL, April 19–24, **2008**. pp. 778–780.

41 Hamilton S. 'Advances in contact cleaning for yield improvement'. In: Proceedings of the AIMCAL Fall Annual Technical Conference, **2008**.

42 Hamilton S. 'Substrate cleaning in a high vacuum environment'. In: Proceedings of the AIMCAL Fall Annual Technical Conference, Charleston, SC, October 27–30, **2013**.

43 Peebles L.H., et al. 'Isolation and identification of the linear and cyclic oligomers of poly(ethylene terephthalate) and the mechanism of cyclic oligomer formation'. *J. Polym. Sci. A-1* **1969**;7(2): 479–496.

44 Goodman I. and Nesbitt B.F. 'The structures and reversible polymerization of cyclic oligomers from poly(ethylene terephthalate)'. *Polymer* **1960**;1: 384–396. or J. Polym. Sci. **1960**;48(150): 423–433.

45 Giuffria R. 'Microscopic studies of Mylar film and its low molecular weight extracts'. *J. Polym. Sci.* **1961**;49(152): 427–436.

46 De Freire M.T.A., et al. 'Thermal stability of polyethylene terephthalate (PET): oligomer distribution and formation of volatiles'. *Packag. Technol. Sci.* **1999**;12(1): 29–36.

47 Shukla S.R. and Kulkarni K.S. Estimation and characterization of polyester oligomers. *J. Appl. Polym. Sci.* **1999**;74: 1987–1991.

48 Ross S.D., et al. 'Isolation of a cycle trimer from polyethylene terephthalate film'. *J. Polym. Sci.* **1954**;13(70): 406–407.

49 Perovic A. and Sundararajan P.R. 'Crystallization of cyclic oligomers in commercial poly (ethyleneterephthalate) films'. *Poly. Bull.* **1982**;6: 277–283.

50 Metzner C., et al. 'Device and method for cooling strip substrates'. Patent WO 2006/042575 A1, April 27, **2006**.

51 Heinß J.-P. and Lang P. 'Ensuring of high deposition rates by increasing of cooling efficiency'. In: AIMCAL Annual Technical Conference Proceedings 27th Vacuum Web Coating Conference, Charleston, SC, October 27–30, **2013**.

52 Clow H., et al. 'Coating apparatus for thin plastic webs'. Application serial no 07/254,088 filed October 6, **1988** now abandoned. Continued as application no 626,320—Patent No. US 5,076,203.

53 Krug T. 'Apparatus and method for cooling films coated in a vacuum'. US patent 5395647, March **1995**.

54 Casey F., et al. 'Vacuum metallising using a gas cushion and an attractive force'. UK Patent GB 2326647 (A), December 30, **1998**.

55 Yadin E. and Pipkevich G. 'Improvement of web heat condition by the deposition drum design'. In: Proceedings of the 50th Annual Technical Conference Society of Vacuum Coaters, Louisville, KY, April 28 to May 3, **2007**. pp. 749–753.

56 Bishop C.A. and McCann M.J. 'An investigation of the limitations of the heat transfer mechanism when using a gas wedge between the web and drum'. In: Proceedings of the 48th Annual Technical Conference Society of Vacuum Coaters, Denver, CO, April 23–28, **2005**. pp. 626–630.

57 McCann, M.J. and Jones D.P. 'Web coating dynamic thermal and wrinkling model'. In: Proceedings of the Society Vacuum Coaters 41st Annual Technical Conference, Boston, MA, April 18–23, **1998**. pp. 412–417.
58 McCann M.J., et al. 'The power of simple models used on complex processes'. In: Proceedings of the Society Vacuum Coaters 54th Annual Technical Conference, Chicago, IL, April 16–21, **2011**. pp. 629–633.
59 McCann M.J., et al. 'Heat load variations on web substrates as seen by deposition modelling'. In: Proceedings of the AIMCAL 1st European Web Coating Conference, Prague, Czech, June 11–15, **2012**.
60 'TopWeb: Heat load model'. Rheologic Ltd. Leeds, UK. Software Available from: www.rheologic.co.uk (Accessed on September 23, 2017).
61 Kukla R., et al. 'R2R Production Solutions for Touch Panels, Flex PV and other Flex Electronic Applications'. In: ProFlex 2010, Dresden, Germany, September 22, **2010**.
62 Lehmann C., et al. 'Roll-to-roll processing of flexible OLEDs'. In: ProFlex 2013, Dresden, Germany, September 9, **2013**.
63 Brémaud D. 'Manufacturing and applications of Cu(In,Ga)Se2 solar cells on flexible substrates'. In ProFlex 2010, Dresden, Germany, September 22, **2010**.
64 Hartwig E.K., et al. 'Air-to-air metallizer: suitability and profitability'. In: Proceedings of the 34th Annual Technical Conference Society of Vacuum Coaters, Philadelphia, PA, March 17–22, **1991**. pp. 152–161.
65 Taguchi T., et al. 'Air-to-air metallizer: design and operational data'. In: Proceedings of the 35th Annual Technical Conference Society of Vacuum Coaters, Baltimore, MD, March 22–27, **1992**. pp. 135–140.
66 Taguchi T., et al. 'Air-to-air metallizer'. In: Proceeding of the 35th Annual Technical Conference Society of Vacuum Coaters, Baltimore, MD, March 22–27, **1992**. pp. 424–426.
67 Visintin S. and Golfetto E. 'An air-to-air feedthrough for vacuum coating of films'. In: Proceeding of the 53rd Annual Technical Conference Proceedings on Society of Vacuum Coaters, Orlando, FL, April 17–22, **2010**. pp. 613–615.
68 Furukawa H., et al. 'Method of vacuum deposition plating steel strip'. US patent US 4,676,999, June 30, **1987**.
69 Crucet R., et al. 'Atmospheric pressure-to-vacuum transition modules for web processing'. In SVC 54th Annual Technical Conference Proceedings, Chicago, IL, April 16–21, **2011**. pp 604–607.
70 Yadin E. 'Deposition of coatings or free foils of sublimating metals'. In: Proceedings of the 40th Annual Technical Conference Society of Vacuum Coaters, Boston, MA, April 27 to May 1, **1987**. pp. 390–393.
71 Yadin E. and Andreev Y. 'Zinc and magnesium vapor generators in a steel strip coating system'. In: Proceedings of the 42nd Annual Technical Conference SVC, St Louis, MO, April 24–28, **1989**. pp. 39–42.
72 Baptiste L., et al. 'Electromagnetic levitation: a new technology for high rate physical vapour deposition of coatings onto metallic strip'. *Surf. Coat. Technol.*

**2007**;202(4–7): 1189–1193. In: Proc. 34th International Conference and Metallurgical Coatings and Thin Films; 2007.
**73** Van Westrum S., et al. 'Method and device for coating a substrate'. US Patent 7,323,229, January 29, **2008**.
**74** Choquet P. and Chaleix D. 'Developpements des technologies sous vide pour le traitement au défilé de produits plats'. In: 'Traitements plasma et industrie automobile' Saint-Etienne, December 6, **2007**.
**75** Pelerin J. 'ARCEO: acier et ruptures technologiques'. In: Liège Creative Colonster, October 9, **2012**.
**76** Zoestbergen E., et al. 'Influence of contamination on the thermal evaporation of a zinc melt'. *Surf. Coat. Technol.* **2013**;218(15): 108–113.
**77** Mattox D.M. *Handbook of physical vapor deposition (PVD) processing* 2nd Edn.; Elsevier/William Andrew: Oxford and Burlington, MA, **2010**.
**78** Martin P.M. *Handbook of deposition technologies for films and coatings* 3rd Edn.; Elsevier/William Andrew: Oxford and Burlington, MA, **2010**.
**79** Bishop C.A. *Vacuum deposition onto webs, films and foils* 3rd Edn.; Elsevier/ William Andrew: Oxford and Burlington, MA, **2015**.
**80** Glocker D.A. and Ismat Shah S. (Eds.) *Handbook of thin film process technology*, Vol. 1 & 2; Institute of Physics Publishing: Philadelphia, PA, **1995**.
**81** Vossen J.L. and Kern W. (Eds.) *Thin film processes*; Academic Press: New York, NY, **1978**.
**82** Wasa K., et al. *Thin film materials technology: sputtering of compound materials*; William Andrew: Norwich, NY, **2004**.
**83** Park J.-H. (Ed.) *Chemical vapor deposition*; ASM International: Materials Park, OH, **2001**.
**84** Reece R.J. *Industrial plasma engineering*, Vol. 1 & 2; Institute of Physics: Bristol and Philadelphia, PA, **2001**.
**85** Maydannik P., et al. 'A continuous ALD process for deposition on flexible substrates'. In: Proceedings of the Society of Vacuum Coaters 53rd Annual Technical Conference, Orlando, FL, April 17–22, **2010**. pp. 138–141.
**86** Alasaarela T., et al. 'Recent developments in roll-to-roll ALD technology'. In: Proceedings of the Society of Vacuum Coaters 55rd Annual Technical Conference, Santa Clara, CA, April 28 to May 3, **2012**. pp. 675–678.
**87** Poodt P., et al. 'Spatial atomic layer deposition: a route towards further industrialization of atomic layer deposition'. *J. Vac. Sci. Technol. A* **2012**;30(1): 010802-1/11.
**88** Dickey E.R. 'Advances in roll to roll atomic layer deposition'. In: Proceedings of the 55th Annual Technical Conference SVC, Santa Clara, CA, April 28 to May 3, **2012**. pp. 43–80.
**89** Soderlund M. and Soininen P. 'Piloting roll-to-roll ALD: status and results with 500mm wide web system'. In: Proceedings of the PROFLEX 'Vacuum roll-to-roll processing of flexible materials', September 24–25, **2013**.

**90** Knaapen R. 'Equipment for atmospheric, spatial atomic layer deposition in roll-to-roll processes'. In: Proceedings of the AIMCAL European Technical Conference, Myrtle Beach, SC, October 19–22, **2014**.

**91** Roozeboom F., et al. 'Spatial atmospheric atomic layer deposition for next-generation flexible devices'. In: Proceedings of the PROFLEX meeting. Vacuum roll-to-roll processing of flexible materials, Dresden, Germany, September 24–25, **2013**.

**92** Virkar A. 'New materials for transparent conductors'. In: AIMCAL Annual Technical Conference Proceedings of the 29th Vacuum Web Coating Conference, Naples, FL, USA, October 25–28, **2015**.

**93** Hartnagel H.L., et al. *Semiconducting transparent thin films*; Institute of Physics: Bristol and Philadelphia, MA, **1995**.

**94** Bishop C.A. *Roll-to-roll vacuum deposition of barrier coatings* 2nd Edn.; Scrivener Publishing: Beverly, MA, **2015**.

**95** Jamison E.H.H. and Windle A.H. 'Structure and oxygen barrier properties of metallized polymer films'. *J. Mater. Sci.* **1983**;18: 64–80.

**96** Weiss J. 'Parameters that influence the barrier properties of metallized polyester and polypropylene films'. *Thin Solid Films* **1991**;204: 203–216.

**97** Decker W. and Henry B.M. 'Basic principles of thin film barrier coatings'. In: Proceedings of the 45th Annual Technical Conference SVC, Lake Buena Vista, FL, April 13–18, **2002**. pp. 492–502.

**98** Ramadas S., et al. 'Nanoparticulate barrier films and gas permeation measurement techniques for thin film solar and display applications'. In: Proceedings of the AIMCAL Annual Technical Conference and 22nd International Vacuum Web Coating Conference, Myrtle Beach, SC, October 19–22, **2008**.

**99** Erlat A.G., et al 'Ultra-high barrier coatings for flexible organic electronics'. In: Proceedings of the 48th Annual Technical Conference SVC, Denver, CO, April 23–28, **2005**, pp 116–120.

**100** Fayet P., et al. 'Polysiloxane super-toughening of silica gas barrier coatings on polymer substrates'. In: Proceedings of the AIMCAL Technical Fall Conference and 23rd International Vacuum Web Coating Conference, Amelia Island Plantation, FL, October 18–21, **2009**.

**101** Malaspina S.P. 'Lithium-ion rechargeable batteries and the portable communications user's expectations'. In: Proceedings of the 39th Annual Technical Conference Society of Vacuum Coaters, Philadelphia, PA, May 5–10, **1996**. pp. 341–343.

**102** Gross M.E., et al. 'Li3PO4: N/LiCoO2 coatings for thin film batteries'. In: Proceedings of the 45th Annual Technical Conference Society of Vacuum Coaters, Lake Buena Vista, FL, April 13–18, **2002**. pp. 119–124.

**103** Swisher R., et al. 'Web coating with lithium technical solution for metal anode structures in Li batteries'. In: Proceedings of the 45th Annual

Technical Conference Society of Vacuum Coaters, Lake Buena Vista, FL, April 13–18, **2002**. pp. 535–538.
104 Lim J.H., et al. 'Thin film supercapacitors using a sputtered RuO2 electrode'. *J. Electrochem. Soc.* **2001**;148(3): 275–278.
105 Kim H.-K., et al. 'Electrochemical and structural properties of radio frequency sputtered cobalt oxide electrodes for thin-film supercapacitors'. *J. Power Sources* **2001**;102(1–2): 167–171.
106 Jayalakshmi M. and Balasubramanian K. 'Simple capacitors to supercapacitors: an overview'. *Int. J. Electrochem. Sci.* **2008**;3: 1196–1217.
107 Iglesias L., et al. 'Development of electrostatic supercapacitors by atomic layer deposition on nanoporous anodic aluminium oxides for energy harvesting applications'. *Front. Phys.* 25 March **2015**. http://dx.doi.org/10.3389/fphy.2015.00012.
108 Lee E., et al. 'Characteristics of piezoelectric ZnO/AlN2stacked flexible nanogenerators for energy harvesting applications'. *Appl. Phys. Lett.* **2015**;106: 023901.
109 https://www.youtube.com/watch?v5P9uXQZReLzw (Accessed on September 25, 2017).
110 Yoon G.C., et al. 'High-performance hybrid cell based on an organic photovoltaic device and a direct current piezoelectric nanogenerator'. *Nano Energy* **2015**;12: 547–555.
111 Kumara B. and Kim S.-W. 'Energy harvesting based on semiconducting piezoelectric ZnO nanostructures'. *Nano Energy* **2012**;1: 342–355.
112 Stoppa M. and Chiolerio A. 'Wearable electronics and smart textiles: a critical review'. *Sensors* **2014**;14: 11957–11992.
113 Matson J.C. 'Antimicrobial wound dressings'. US Patent No. 4,728,323, March 1, **1988**.
114 Brunon C., et al. 'Characterization of plasma enhanced chemical vapor seposition_physical vapor deposition transparent deposits on textiles to trigger arious antimicrobial properties to food industry textiles'. *Thin Solid Films* **2011**;519: 5838–5845.
115 Ghoranneviss M., et al. 'Influence of plasma sputtering treatment on natural dyeing and antibacterial activity of wool fabrics'. *Prog. Org. Coat.* **2011**;70: 388–393.
116 Hegemann D., et al. 'Nanostructured plasma coatings to obtain multifunctional textile surfaces'. *Prog. Org. Coat.* **2007**;58: 237–240.
117 Scholz J., et al. 'Investigations on fabrics coated with precious metals using the magnetron sputter technique with regard to their antimicrobial properties'. *Surf. Coat. Technol.* **2005**;192: 252–256.
118 Shahidi S., et al. 'Investigation of antibacterial activity on cotton fabrics with cold plasma in the presence of a magnetic field. *Plasma Process. Polym.* **2007**;4: S1098–S1103.

**119** Shahidi S., et al. 'Investigation of metal absorption and antibacterial activity on cotton fabric modified by low temperature plasma'. *Cellulose* **2010**;17: 627–634.

**120** Wei Q., et al. 'Evaluation of the interfacial bonding between fibrous substrate and sputter coated copper'. *Surf. Coat. Technol.* **2008**;202: 4673–4680.

**121** Yuranova T., et al. 'Antibacterial textiles prepared by RF-plasma and vacuum-UV mediated deposition of silver'. *J. Photochem. Photobiol. A Chem.* **2003**; 161: 27–34.

**122** Caironi M. and Noh Y.-Y. *Large area and flexible electronics*; Wiley-VCH: Weinheim, **2015**.

**123** Peng H. *Fiber-shaped energy harvesting and storage devices*; Springer: Berlin, Heidelberg, **2015**.

**124** Shauddin S.M. 'Comparison among various emerging PV cells with history, current status and future challenges'. *Energy Power* **2013**;3(6): 91–105.

**125** Jean J., et al. 'In situ vapor-deposited parylene substrates for ultra-thin, lightweight organic solar cells'. *Org. Electron.* **2016**;31: 120–126.

**126** Docampo P., et al. 'Efficient organometal trihalide perovskite planar-heterojunction solar cells on flexible polymer substrates'. *Nat. Commun.* **2013**;4: 2761.

**127** Li Y., et al. 'High-efficiency robust perovskite solar cells on ultrathin flexible substrates'. *Nat. Commun.* **2016**;7: 10214.

**128** Ahmed M.I., et al. 'Perovskite solar cells: potentials, challenges, and opportunities'. *Int. J. Photoenergy* **2015**; 592308.

**129** Liu M., et al. 'Efficient planar heterojunction perovskite solar cells by vapour deposition'. *Nature* **2013**;501: 395–398.

**130** Ono L.K., et al. 'Organometal halide perovskite thin films and solar cells by vapor deposition'. *J. Mater. Chem. A* **2016**;4: 6693–6713.

**131** Zhao L., et al. 'Development of new polymer systems and quantum dots: polymer nanocomposites for flexible OLED display applications'. Available from: http://www.hpl.hp.com/techreports/2011/HPL-2011-68.pdf (Accessed on September 25, 2017).

**132** Kramer I.J., et al. 'Colloidal quantum dot solar cells on curved and flexible substrates'. *Appl. Phys. Lett.* **2014**;105: 163902.

**133** Alexander H.I., et al. 'Efficient, air-stable colloidal quantum dot solar cells encapsulated using atomic layer deposition of a nanolaminate barrier'. *Appl. Phys. Lett.* **2013**;103: 263905.

**134** Chen L., et al. 'Colloidal quantum dot absorption enhancement in flexible Fano filters'. *Appl. Phys. Lett.* **2010**;96: 083111.

**135** Troughton J., et al. 'Highly efficient, flexible, indium-free perovskite solar cells employing metallic substrates'. *J. Mater. Chem. A* **2015**;3: 9141–9145.

**136** Tavakoli M.M., et al. 'Highly efficient flexible perovskite solar cells with antireflection and self-cleaning nanostructures'. *ACS Nano* **2015**;9(10): 10287–10295.

**137** Wu J. and Lih L.Y. 'Inkjet printable flexible thin-film NCQD photodetectors on unmodified transparency films'. *IEEE Photon. Technol. Lett.* **2014**;26(7): 737–740.

**138** Bedeloglu A.C., et al. 'A photovoltaic fiber design for smart textiles'. *Text. Res. J.* **2010**;80(11): 1065–1074.

**139** Bedeloglu A.C., et al. 'Development of energy generating photovoltaic textile structures for smart application'. *Fibers Polym.* **2010**;11(3): 378–383.

**140** Singh M.K. Flexible photovoltaic textiles for smart applications. In: *Solar cells: new aspects and solutions,* Kosyachenko L.A. (Ed.); INTECH. Available from: http://www.intechopen.com/books/solar-cells-new-aspects-and-solutions/flexible-photovoltaic-textiles-forsmart-applications (Accessed on September 25, 2017).

# 5

# Web Handling and Winding

*David R. Roisum*[1], *Gustavo Guzman*[2,3], *and S. Shams Es-haghi*[2,3]

[1] *Finishing Technologies, Inc., Neenah, WI, USA*
[2] *Birck Nanotechnology Center, Purdue University, West Lafayette, IN, USA*
[3] *School of Materials Engineering, Purdue University, West Lafayette, IN, USA*

## 5.1 Web Handling

A web is anything that is long, thin, and flexible. Traditional web industry sectors include textiles (2000 years old), paper (200 years), tissue (100 years), film (120 years), foil, nonwovens, and many others. New product areas such as microelectronics, solar cells, medical devices, and many more are joining the traditional sectors as they move from low volume sheets to high volume webs. Making a product as a web gives us a much-improved economy of scale over manufacturing as a sheet. Some webs require additional processing before they are ready for the end use customer. Many industries rely on web *converting* processes to achieve some of those properties. Common converting processes include calendering, coating, laminating, printing, slitting, and winding. Converting also enjoys the economy of scale when carried out as a roll-to-roll (*R2R*) process instead of as a sheet-to-sheet. Here, the web is unwound from a long supply roll and run through a calender, coater, etc. and then wound up into a roll again. A few products are converted in a roll-to-sheet process where a crosscut and stacker are on the end of the line. Familiar sheeted products range from office copy paper to personal care products to sheets of drywall. All are manufactured as a web, converted in one or more R2R line and then converted again in a roll-to-sheet process.

Web handling is a century-old art and now mostly science of getting a web through a machine as fast as possible and with minimum waste. Since more traditional web manufacturing and converting operations are routinely done at extremely high speeds, easily exceeding 2000 m/min, we will just note that any

*Roll-to-Roll Manufacturing: Process Elements and Recent Advances*, First Edition.
Edited by Jehuda Greener, Glen Pearson, and Miko Cakmak.

speed concerns for R2R should be strictly chemistry and process related, not web handling related. The same is *mostly* true of web width where webs 3 m (film) and even 10 m (paper) wide are routine. Waste, on the other hand, cannot be easily dismissed as speed and width related. Web handling issues are usually among the top three causes of waste, delay, and customer complaints in traditional web industries. We should expect that web handling would be no less troublesome for newer R2R processes. Table 5.1 lists some of the most common web handling-related issues.

Web handling is documented in more than 4000 articles, books, columns, conference papers, short courses, and other written documents. Even so, the assumption that web handling can be learned as you go is one of the main reasons for failure to successfully scale chemistries from bench top to step-and-repeat and finally to the economies of R2R operations. Indeed, some products that can be made on a bench top are not possible to be made as a web, regardless how desirable the economies are. Information provided in this chapter may help avoid some of pitfalls in the practice of web handling. Table 5.2 lists the main topics associated with web handling.

**Table 5.1** Examples of waste, delay, and customer complaints associated with web handling.

| |
|---|
| Baggy webs (lack of flatness as sheet) |
| Curl (about two different axes) |
| Length and width (dimension control) |
| Registration (location, position control) |
| Web breaks and other tensile damage |
| Winding defects |
| Wrinkling |

**Table 5.2** Web handling main topics.

| |
|---|
| Rollers |
| Tension control |
| Nip control |
| Temperature/moisture effects and control |
| Guiding (path control) |
| Wrinkling |
| Slitting |
| Winding |

## 5.2 Design for Manufacturability (DFM) for Web Handling

While manufacturing *produces* a web, and converting *changes* a web, the objective of web handling is to pass the web down through the machine *without* change. This cannot be done solely by web handling steps. Product design must consider design for manufacturability (DFM). Some products cannot be handled well, and some cannot be handled at all as *webs*. Often the limiting step is winding, at the end of the R2R process. Some of the many web challenges include products that are too tacky or too slippery or too weak or too brittle and those that are too thick. One of the winding concerns may be the ability of upstream manufacturing and converting to make a web with uniform cross-web thickness. Variations in thickness of less than 1–10% may be challenging in many R2R operations.

Unintentional variations in thickness are one level of concern, but many products are designed with an *intentional* thickness variation. Thickness variations will always limit the size that a roll can be successfully wound to. This, in turn, affects productivity as well as the ability to meet customer expectations. Thickness variations are also one of the primary reasons for bagginess, a lack of web flatness, that is a major cause of wrinkling and many other manufacturing issues. Very often the needs of the winder are web profiles better than can be read on many scanners or test labs and occasionally better than can be made at all [1]. Most thickness variations are unintentional. Intentional or designed thickness variations are sometimes the result of some other property the customer wants. Because these variations tend to be bigger, the results can be exceedingly problematic, especially when the design is asymmetric.

Once a suitable *web* is designed, we next consider *machine* design. Here web properties such as width, thickness, strength, modulus, and coefficient of friction (COF) are just a few that will guide machine design. Machine speed and wound roll size and wound roll handling are additional specifications that the machine designer must have before beginning. These web properties and machine specifications are also the inputs to the science of web handling where rollers are sized and selected, a topic we will take up next. Following that we will discuss the machine *controls* of tension, nip, temperature, and speed. In addition to being machine design considerations, these are also machine operating parameters that can have a profound effect on productivity and quality.

## 5.3 Rollers

Rollers are the building blocks of web machines. Even if there is a single manufacturing line in the plant, it will undoubtedly have scores of rollers. A single defective roller can change the entire waste and timeline profile of a plant [2, 3]. For example, roller diameter must be sufficient to keep deflection due to the combined effects of gravity, tension, and nip pressure at a minimum. Another example is the need for precision of roller alignment [4]. These and many other design calculations are given both in Ref. [2] and in AbbottApps that runs on any smart device connected to the Internet [5]. As a final example, we consider the required precision for roller cylindricity. Here, the tolerances may be on the order of a few micrometers for web handling, such as for transport rollers, but tend to be much tighter than for web manufacturing and converting elements including die lips, calendering rollers, and coating heads and rollers.

There are two types of rollers: *Transport* rollers, the most common of which are called “idlers,” which merely convey the web without intentional permanent change. However, even *driven* rollers can be transport rollers if all they change is tension. In this regard changing tensions and cross-direction (*CD*) web path positions down through the line are elastic, acting as if the web had no memory. If the web moves modestly to the left on one roller and modestly back on the next, its properties would be unchanged. This is what is meant by “no memory.” *Process* rollers, on the other hand, change the web permanently. These include the value-added processes of calendering, coating, laminating, and the like that are central to most R2R machines.

There are distinctly different design philosophies related to the transport and process rollers. In the case of process rollers, the number of the rollers is dictated by the process. For example, in a laminating process, one can consider a single coating roller (to apply adhesive) and a single laminating roller. However, one may not use a printing roller. The number of transport rollers, however, is to be minimized for two reasons. First, by minimizing the number of transport rollers, one can reduce the initial cost and subsequent maintenance costs during the life of the machine. Second, by minimizing the number of transport rollers, the probability of facing the problems relating to the malfunction of such rollers is reduced.

Different styles may be adopted for mounting the rollers. Figure 5.1 demonstrates three mounting styles in order of precision. The *live* shaft tends to be the most precise and is used for almost all process rollers and many transport rollers. The *dead* shaft tends to be less expensive but is seldom suited for most process rollers. Instead, it may be used on less demanding idler rollers. The *cantilever* mounting is used exclusively on narrow web ($\ll$1 m wide) machines. The cantilever style may be difficult to align and have low resistance to deflection. Thus, cantilevers may force one into choosing to lower standards or to incorporate unusual dual wall frame designs that are even more expensive than dead or even live shaft designs. Lowered standards may lead to wrinkling, path control, and many other issues, particularly on thinner webs. This does not mean that the cantilevers do not have their place. When L/Ws (web R2R span

Figure 5.1 Live shaft, dead shaft, and cantilevered roller mounting styles.

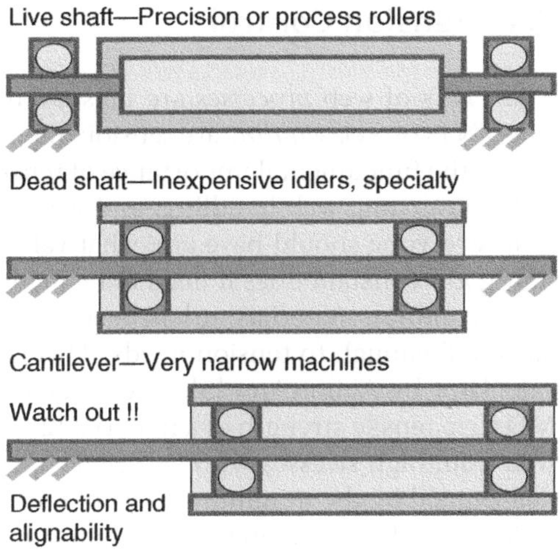


length-to-width ratio) become very high and ribbonlike, *some* of the required alignment and deflection precisions are relaxed.

*Framework* design depends greatly on the details of the application and is beyond the scope of this chapter. Although one can anticipate competing demands, web handlers would like frames that support rollers with precisions at least as good as what they expect from the rollers themselves. At higher speeds, it is desired to avoid vibration. However, maintenance and operators would like visibility during operation and ease of access when the machine is stopped and locked out. These needs can be met with a good framework design. A good framework design cannot be done just by placing expensive metal parts together. This can be achieved by strategic placement of joints and structures to design a bridge capable of carrying maximum load with minimum metal parts.

The foundations of the machines may vary from table-sized machines that might safely be set on an ordinary floor (but only with three legs and extremely stiff frames) to slightly bigger machines, the size of a small car, that could be safely bolted to the standard 6″ (15–20 cm) thick industrial slab floors. Larger and faster and more demanding machines, such as in the paper industry, may require a meter-thick foundation. An important factor that is easily overlooked is that the easiest task of a foundation suitable for web machines is to support weight. The foundations must be strong enough to avoid vibration and to avoid distorting machine alignments due to inevitable movements over time.

There are two operational mechanical concerns [6]: the first is maintenance to repair what is broken, and the second when product quality degrades. This is mostly in the area of web manufacturing, such as cleaning of dies.

## 5.4 Tension Control

The TNTs of web *processes* are tension, nip, temperature, and speed (time). The TNTs of web *winding* are tension, nip, torque differential, and speed. Here we see the fundamental importance of tension and nips in web manufacturing, web converting, web handling, and web winding. The two immediate questions everyone should have are: what value of tension (or nip) is the best and how good/constant does it need to be maintained to?

For common materials, it has been found that thickness alone often correlates well enough to tension needs. This is summarized in the simple tension calculator by AbbottApp [5]. The guideline is that tension be 10–25% of the MD strip tensile strength of a material [2]. The higher limit is set to protect the web from high stresses with sufficient safety margin. The lower limit is often determined by the minimum value of tension required to maintain flatness as well as to stabilize it along the machine direction. The guidelines like this allow designers a place to begin to size the equipment such as load cells, dancers, drives, rollers, and so on, sometimes even if the product is unique to them or even to the world. Of course, there are exceptions to the simplistic rules mentioned earlier, but these simple rules provide a good starting point when specific experience is lacking.

One design challenge that can quite easily arise is the desire to have a single machine that is suitable for a wide range of products. If one considers the need to size quality/precision at 10% of the weakest product and size strength at 25% of the strongest product, the tension may quickly exceed the 10:1 range of tensions that a drive can cover "well." The tension should be typically maintained to 5% of the set point during steady-state conditions and 10% during speed changes and perhaps open to 50% on violent short events such as roll change [2, 3, 6, 7].

Herein, we want to define the machine quality, a concept analogous to web quality but one that is far less familiar. It should not be surprising to note that if one wants to make a uniform product, the key aspects of the web machine and control parameters need to be kept adequately uniform. The drive system design and configuration expertise are essential in this regard.

A more sophisticated view of tension (nips and other controls) goes beyond mere guidelines and gets into economics [8]. The objective is to find the best compromise between tight and loose defects in web handling. The machine mechanical and control qualities fit into this bigger picture because poor quality can result in tight and loose defects, *at the same time*. We will use an identical concept in winding and its TNT controls.

Tension control schemes fall into four different types in order of increasing precisions: draw/speed control, open-loop torque control, dancer control, and load cell control. The draw and speed control are required for manufacturing

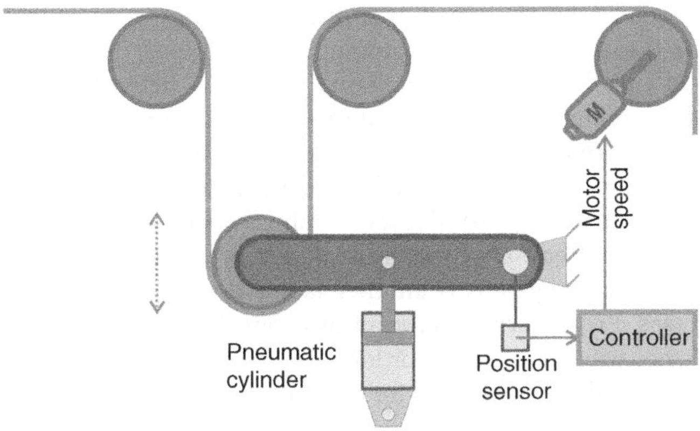


**Figure 5.2** A dancer for controlling tension.

where the products may have near-zero tensile strength, such as the wet end of a paper machine as well as film and rubber extrusion. The torque control is used for modest motor sizes where its effect on tension is determined primarily by a combination of calibration and tuning. Because there is no web tension sensor, both draw and torque controls are open loop on the web. The dancer and load cells, on the other hand, are used in closed-loop tension control methods.

The dancer control is where the position of a pivoting roller, shown in Figure 5.2, increases or decreases the speed of an adjacent drive motor ever so slightly. The dancer control is a form of tension control because in steady state the combined weight of the dancer and force from a controllable cylinder determine the tension in the two web strands. Two major design problems are common with dancers. The first is not calibrating dancers to proper tension units of PLI (lb/in in US system) or kN/m in metric system as a function of cylinder pressure. The second is to not check for excessive friction that would cause tension to vary with time [9]. Often just the presence of a pneumatic cylinder is enough to upset the dancer tension control due to seal friction.

A roller mounted on load cells facilitates the most direct web tension reading. However, it is also in many ways the most demanding for drive motor control. Part of this is because the spring rate of a load cell is so high that tiny errors in the motor speed cause tension to spike. This is in contrast to a dancer that can absorb many centimeters of drive position error. One might consider a dancer to be a small airplane that one can learn to fly in a year. A load cell, however, can be imagined as a jet aircraft that might have much better performance but only with the best pilots. Dancers also have a necessary accumulation function for step-and-repeat operations.

In either dancers or load cells, the speed or torque of the motor is changed slightly by comparing the dancer/load cell output to a set point. This slight changing of speeds is called *trimming* of nominal speed as determined by machine speed and roller diameter. Trimming of motor torque would be based on models of tension difference, roller diameter, and parasitic drag torque. Depending on the drive model, the authority of the sensor is small (tension into torque) or very small (tension into speed). These trimming authorities are around 2 and 10% of motor rating, respectively, for a full-scale sensor reading. The inertia of the winding and unwinding rolls is another factor that complicates the drivers on R2R processes. Proper drive *tuning* always considers inertial effects. That means one does not wait for a dancer or load cell to show the effects of a speed change. They are already compensated for by a calculated value and added to the drive's nominal output as a feed-forward. The winders and unwinders complicate this so that the calculated value changes with roll radius so that the control gain must change accordingly. The web drive language such as the aforementioned probably sounds quite foreign to the reader. Yet, absolute fluency is needed to safely connect any motor to a roller. This control fluency goes way beyond knowing the programming language peculiar to a model and brand of drive or PLC. It is important to know the customer's needs and to be concerned about the web and its requirements that can lead us to a good web handling.

## 5.5 Nip Control

A nip is the linear (force per width) line stress between a roller and roller in processing or a *roll* and roller in winding. Of the TNTs, nips are only slightly less common than tension zones. However, nip control is no less important in terms of web manufacturing, web processing, web handling, and web winding. An uneven nip almost always results in an uneven, streaky product in processes. In web handling and winding, an uneven nip risks wrinkling and other problems. For these reasons, nips need to be treated with the same care as tension and other controls [10]. Unfortunately, the nips are not treated in such a way in most cases. For example, most nips are not calibrated in engineering units of PLI (lb/in) or kN/m. Instead, a machine specific PSI (lb/in$^2$) or Pascal or bar of air pressure on a loading cylinder is most common. This would mean that an air or hydraulic pressure that works well on one machine section may not work well on a different part of the same machine and will certainly be different on a machine made by another builder. The lack of an independent calibration procedure in engineering units also means that there would be no easy way to detect variability of nip (due mostly to cylinder friction) or the drift of the nip load setting for any other reason.

The first step to achieve a good nipped roller design is to choose the design range for the average nip based on the application or process. The average nip load may vary from a fingertip light 0.1 PLI (0.017 kN/m) on some winder lay-on rollers to a solid 100 PLI (17 kN/m) on a laminator nip to well more than a crushing 1000 PLI (170 kN/m) on calenders, metal rolling lines, and even currency printers. Even despite this large application range found on R2R processes, the needs for calibration and uniformity across the width and uniformity with time remain unchanged. The second design step is to size rollers to be sufficiently large in diameter to resist excessive deflection under the maximum nip load as was described earlier. The reason why nip rollers tend to be noticeably larger in diameter than transport rollers is that they must resist much larger loads than idler rollers. Of course, the loading system and supporting framework and other elements must also be suitably sized. The third design step is to make sure both rollers in the nipped pair are sufficiently cylindrical. This is important in case of fussy processes that the rollers might not be allowed to vary more than 10 μm (plus or minus an order of magnitude) across their width nor around their circumference. This involves precision grinding and sometimes frequent regrinding as the rollers wear, especially those with rubber covers. The subject of rubber covers is beyond the scope of this chapter but is covered in the literature [11].

In addition to the variability of nip pressure along the width and around the circumference, we also consider temporally randomly varying mechanical friction like those discussed on dancer roller earlier. We can use nip impressions as a quality check for all these and many other sources of variability. The nip impressions involve any number of different thin pressure products or sensors, the most common being pressure sensitive papers (or films). The nip impression quality control techniques include the static (most common) or dynamic methods [12, 13]. In either case the variation of pressure across the width results in either a variation of nip pressure "footprint" width (static) or nip pressure reading (dynamic) of paperlike products as shown in Figure 5.3. Electronic sensors measure width or pressure directly but are more expensive and may be more time consuming to use because the sensor must be moved several times to establish a profile across the width.

## 5.6 Temperature, Speed, and Gravity

No web handling treatment would be complete without addressing some of the many common fallacies associated with certain parameters. Let us begin with temperature (in the case of most materials) and moisture (in the case of paper and some other materials such as certain coatings, nylon, and textiles). In general, there are far fewer temperature/moisture-controlled zones in web machines. When these parameters are not in equilibrium with the environment, that is, room conditions, they tend to dominate many problems. In effect, temperature and moisture tend to be the strongest of the TNT knobs in

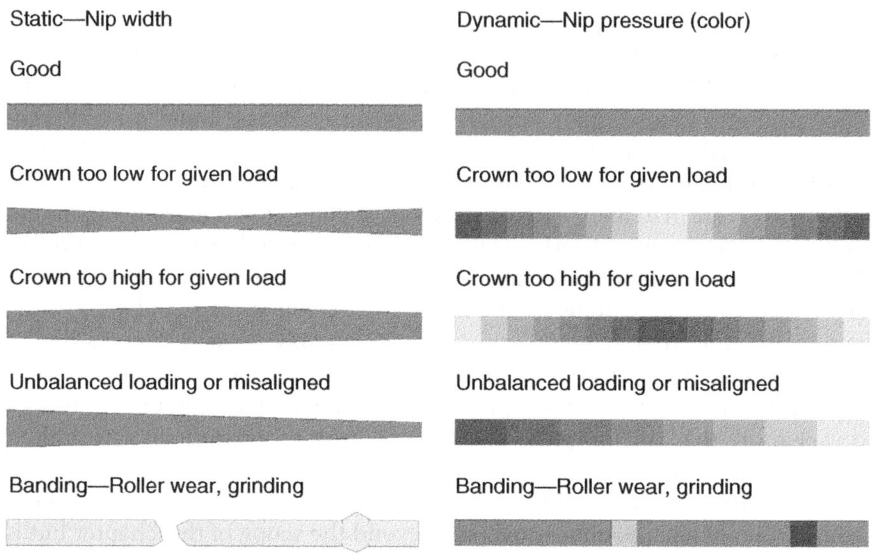


**Figure 5.3** Static and dynamic nip impression techniques.

terms of processing. This means, for example, that temperature may need to be controlled to perhaps within 2–5°C on heated rollers and ovens. This control must be stable with respect to time and with respect to CD (cross-direction position, a.k.a. "profile") to maintain product consistency. Moisture, the paper counterpoint to temperature, may need to be held to better than 1% to maintain product and process consistency. Both temperature and moisture may also have notable effects on web handling. Just one example would be the profound influence of paper moisture on both ductility and strength and, as a consequence, web breaks. One great troubleshooting tool for both temperature and moisture is the infrared (IR) camera.

Every web machine has a **speed** knob. Some *processes*, such as laminating and printing, are mostly insensitive to speed. However, since, in some cases, the speed may have profound effects on the process and the product, a recipe may specify the speed in most manufacturing and converting processes. Some processes, such as reverse roller coating, can only be run without defect in a relatively narrow range of speeds. The oven is similarly constrained though its range might be made wider by reducing temperature as speeds increase and vice versa.

These sensitivities have led many to believe that web *handling* is also speed sensitive. It is, but in only two ways that are well understood. This understanding has allowed paper manufacturing to reach speeds exceeding 2000 m/min and printing on a variety of materials to exceed 1000 m/min as just two examples.

Indeed, paper winding machine design speeds are now up to 4000 m/min. Thus, we know how to go fast in web handling if the process and economics allow. The first limitation is **air entrainment** between webs and rollers that can cause a loss of traction when speeds exceed 100 m/min on smooth webs. The limits and treatments for this type of air entrainment are well known and covered in detail in a chapter in a book [2] and even apps that can be run on a smart device [5]. An only slightly more complicated version of air entrainment is between layers of a wound roll. The limits and treatments for this are well covered in the literature [3, 14]. The most common and effective treatment is to wind with a nipping roller, provided that the web caliper control is quite level across the width and the product is not marked by a nip.

The last of the major fallacies covered here is **gravity**. Essentially, gravity affects *almost* nothing in web handling. The major exception here is that gravity is one of three forces including tension and nip loading that act on rollers. As mentioned earlier, this is easily dealt with by making sure that the rollers have sufficiently large diameter to resist deflection. Therefore, since gravity does not affect the web handling, in most cases, mounting the machine right side up, upside down, and even sideways may be possible with only a little design detail challenge and essentially no web handling challenge.

## 5.7 Web Path Control, Guiding, and Oscillators

For economic reasons, it is important to control the path of the web through fixed width elements such as coating heads, laminators, printers, and slitters. For example, if one lets the web wander sideways 1 cm, then *another* 1 cm of trim will be lost beyond what might have already been done because the edges might be "bad." As we learn in web handling, there are three common mechanisms that can upset the path of the web [3]. The first is simple in-plane roller misalignment, such as calculated by the Wrinkle Predictor by AbbottApp [5] when coupled with a change in tension/traction. Focused attention to the right mechanical and control design and maintenance areas will eliminate this issue. The second is similar but involves a variation of diameter of nip pressure on rollers across the width. This was covered earlier in the section on nip control. However, it should be noted that a perfect machine does not mean perfect path control. A **baggy web** will determine its own path through a perfect machine and it will *not* be straight. It is important to note that, by far, the largest root cause of baggy webs is the winding of webs that have thickness profile variation across the width [15]. Finally, these mechanisms are made worse as the number of rollers and the total path length of the web increase.

Because the machine and the web are not perfect, a guide may be needed to bring the web back to a given position, whether it is centered (two sensors) or

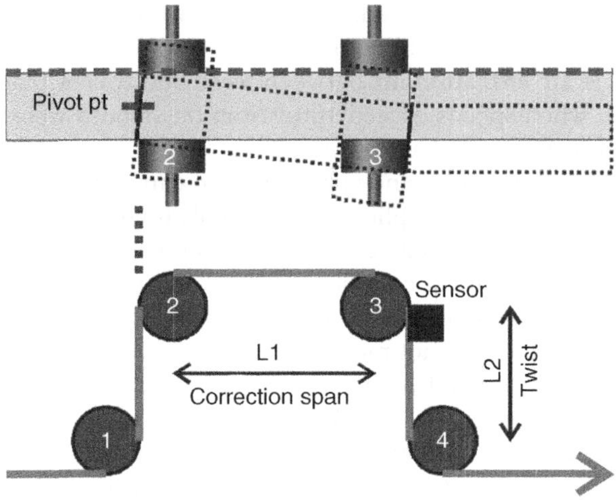


**Figure 5.4** A displacement guide.

guided off one edge (one sensor). Web guiding is the oldest of the web handling sciences, so that the solutions for a given application will be almost precisely the same because all builders should be using the same decades-old web handling models. In short, the most common type of guide used for correcting web path is called the displacement guide and is shown in Figure 5.4. Alternative guides for other application needs include the unwinder, the winder, and the steering guide. Creativity in deviating from these well-proven designs is discouraged unless one has no other reasonable choice.

Finally, any guide could also oscillate. Oscillation is used to smear the caliper/thickness/weight variations across the width, that is, produce "gauge bands" and valleys that are to some extent unavoidable. As we have already mentioned, the winder will be the fussiest customer to level web thickness profile. If one can oscillate fast enough (<1 cm of diameter buildup on the wound roll) and far enough (about the width of the band or valley), it will be possible to reduce the severity of some winding defects including bagginess, corrugations, and many of the tight and loose defects. However, this increased tolerance to gauge variation comes at a very steep cost; this will be different from the cost of the guide that is only incurred once. The cost will be incurred from increased trim losses that will be incurred "forever." By oscillating 1 cm to help with a 1 cm wide gauge band, one will lose *another* 1 cm of saleable trim width. In this regard, leveling the web is often the best or only solution to many winding defects, especially with narrow web processes.

## 5.8 Slitting and Trim Removal

Slitting is cutting the web lengthwise into narrower lanes as might be dictated by a final customer. Crosscutting is used to cut to length and will not be covered here even though some of the principles are analogous to shear slitting. Strictly speaking, slitting is not a web handling subject because it permanently changes a web. However, it is so common that it must be at least mentioned here. It is not unusual for a web to be slit several times prior to end use. The reader can find additional information on this topic elsewhere [5, 16].

There are many methods for slitting as given in Table 5.3. The most common of the slitting methods are razor, score, and shear. Razor slitting, as its name implies, is not much more than putting a razor into the running web. This is the mode for slitting most films as well as other materials such as vinyl. Score cutting is much like a pizza cutter; it has a rotating circular blade that is not razor sharp as the other cutters desirably are. This blade bears against an anvil roller. Score cutting is used for a variety of products ranging from thin tissue to labels to roofing shingles. Other less common slitting methods are niche to specific applications and are merely listed as follows.

We will discuss the most common of all methods, **shear slitting**, in more detail. This shear slitting method is a circular blade bearing against another circular blade at a small shear angle as demonstrated in Figure 5.5. It is called "shear" slitting because the blades are set at a small camber against each other in a fashion like shear scissors. Indeed, the geometry and physics are a circular analog to the linear blades of scissors or the common paper cutting board. The geometries of all methods are quite fussy. However, when applied well, the shear slitter is capable of cutting materials as varied as tissue and steel and most materials in between.

This capability, however, comes at the cost of simplicity. A simple razor has enough complexity that there is currently no high-level modeling to guide us. The shear slitter is considerably more complicated still, especially regarding

**Table 5.3** Slitting methods.

| |
|---|
| Hot wire |
| Laser |
| Log saw (rolls) |
| Razor (film and plastics) |
| Score |
| Shear |
| Ultrasonic (thermoplastics) |
| Waterjet |

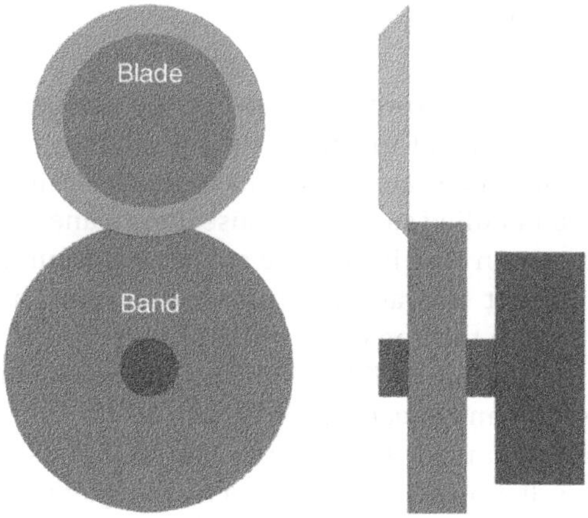


**Figure 5.5** A shear slitter.

**Table 5.4** Shear slitter adjustments and factors.

| |
|---|
| Operator |
| Blade side load |
| Blade overlap |
| Electrical maintenance |
| Band overspeed |
| Mechanical maintenance |
| Blade metallurgy |
| Grinding angles |
| Machine design |
| Shaft/individual mount |
| Wrap or shear |
| Penetration (tangent) |
| Cant (toe-in) angle |
| Rake (shear) angle |
| Auto/manual positioning |

adjustments. The adjustments, as given in Table 5.4, list factors that greatly affect cut quality and blade life. Often, these two concerns are diametrically opposed. For example, increasing the shear angle, as seen in Figure 5.6, will increase cuttability but decrease blade life. Best practices for these adjustments often reflect similar compromises, as well as compromises on web properties.

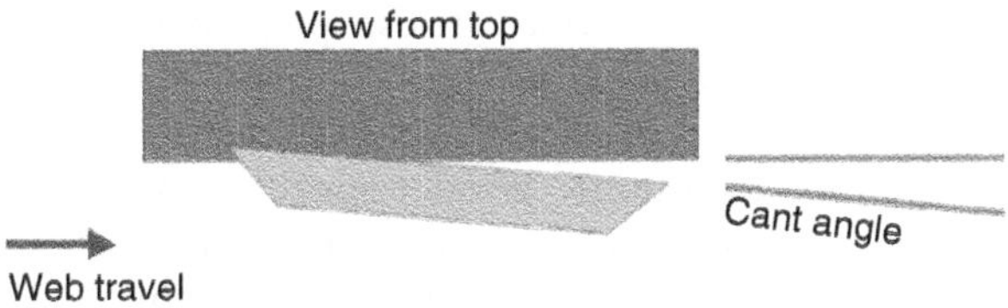


**Figure 5.6** Cant or toe-in angle for the shear slitter blade.

Slitting component suppliers can offer expert guidance on these matters. Indeed, original equipment manufacturers (OEMs) seldom manufacture their own slitting components turning instead to major slitter suppliers for assemblies that are later incorporated into larger machines.

## 5.9 Winding

To store the product developed, winding method is commonly used. Other less convenient options include boxing the web or cutting it into sheets of fixed length. As such, winders are present at the end of almost every web manufacturing line and all R2R lines whether they be paper, film, foil, nonwovens, textiles, or other web products. Winders may be also used to cut wider diameter parent webs into daughter or shipping roll sizes needed by the customer. The winding rolls are also known as slitter rewinders or "slitters," for short, even though "slitting" also refers to a specific component and process, which might create confusion. Finally, small winders called "salvage" winders may be used to rescue some rolls and to do trials or inspection. Thus, just like slitting, a base web may see many winding operations before it reaches the end use customer.

Despite the large number of available winding systems, this operation still carries a level of complexity far greater than any other area of web handling. In some cases, an easy application is possible by buying a winding machine and setting the adjustments in the middle of their range. However, this is seldom the case as winding can be as troublesome as any other manufacturing or converting process, perhaps even more so.

The book by Good and Roisum [14] and *The Ultimate Web and Roll Defect Troubleshooting* [17] are useful guides for winding operations and related defects. The latter covers, in depth, various winding defects and wrinkling problems (see next section). It also reviews paper, film, coating, slitting defects, and much more. This book is the first to resolve the terminology problem by carefully defining proper defect names. For example, the common winding defect known as "telescoping" comprises half-dozen distinctly different defects that share almost nothing in common except the outcome (roll appears to move sideways). To make matters worse, the word "telescoping" has aliases such as scoping, shooting, and more. The book provides common aliases for users to find the proper name, in addition to careful word descriptions, pictures, listing of causes, and possible solutions. Winder-specific training is essential to build winders, rebuild winders, operate winders, run a winder department, and do customer service with a large fraction of complaints related to wound rolls [3]. Furthermore, as winders are rollers and have tension and nip controls, web handling prowess is also required.

One key property of winding machines is the control of wound roll tightness, which, when misused, increases the propensity for some defects. Generally, we can say that wound roll *tightness* is the "*tension* of layers inside the wound roll," or the *pressure* between the layers of the wound roll. The "knobs" or adjustments that control tightness are known as the TNTs of winding. The first T is web tension entering the windup station, which we have already discussed. All winders have this knob. The N is nip, which we have also discussed earlier. However, this time the pressure is between a roll and roll*er* rather than between two roll*ers*. The nip load on a winder tends to be somewhat lighter than other processes. It also tends to be considerably more variable, especially due to caliper variation of the incoming web, which can cause serious problems. Otherwise, we have similar considerations. The second T is an uncommon knob, because it requires two motors, increasing cost and complexity. This knob is formally called "center-surface wind torque differential," often shortened to "torque." However, the abbreviation "torque" does not refer to motor torque but rather the *difference* in torque between two motors. The last knob, small *s*, stands for speed. It is small because only some applications are speed dependent, such as those dealing with air entrainment. Applications that are speed dependent are only observed when the web is simultaneous: fast moving, wide, smooth, and not porous.

A major challenge then becomes the correct and simultaneous adjustment of all knobs. This challenge is even greater because the knobs are not fixed at static set points throughout the build of the roll. Instead, some knobs are intentionally varied as a function of roll diameter or footage, with the most common variation being to "taper" tension. It might also be necessary to vary the pressure on a cylinder to compensate for geometry or roll weight on some winders. We define a set point function (of current roll diameter), given the general term "roll structure," instead of a single set point. The "roll structure" embodies three principles: start tighter at the core, finish looser at the end of the wind, and taper smoothly from tight to loose as the roll builds. Tapering partially contributes to the reduction of certain defects, namely, one type of starring and two types of telescoping.

So, what set point (or set point function) is "best?" The first simplification is that most winders do not have all of the TNT knobs. For example, torque differential is rare, as narrow, slow, or rough products are not speed dependent.

That leaves primarily tension and nip. Also, it is possible to exchange one knob for another to a certain extent, because they all have a similar effect on the wound roll: control tightness (and possibly defects). For example, if tension leads to problems (such as web breaks, necking, etc.), the tension can be decreased while increasing nip load. If nip load is a source of problems, as is the case when the web thickness varies across the width, even slightly, the nip (pressure) can be reduced or perhaps even turned off while increasing tension. If the roll needs tighten, raise both or all knobs.

The next key concept is that almost all the 100+ winding defect types fall into one of four categories, and knowing which one a defect belongs to will guide you to the appropriate correction. *Tight* defects tend to damage the web. Stretching the web into bagginess over gauge (relatively thick) bands is one major tight defect. *Loose* defects tend to damage the wound roll by not keeping it cylindrical enough for the next process that might be storage or unwinding by the customer. Flat spots, egg shapes, and other out-of-round, non-cylindrical rolls are examples of loose winding that may be rejected by customers. *Taper* or *roll structure* defects are few, mainly one type of starring and two types of telescoping, but respond strongly to starting at maximum tightness and tapering (smoothly) to minimum tension. However, the most extensive category is "other," where the defect is not tightness sensitive. These include design, maintenance, and operational issues. An illustrative example would be cutting the wrong width roll because the winder operator misset the slitter blade positions. Accordingly, how classify a defect to respond appropriately? The winding literature [14] can provide some guidelines and solutions. Some of the more common winding defects are listed in Table 5.5. Fortunately, only a few are observed in any application, so the practitioner needs expertise on only those most commonly encountered.

**Table 5.5** Some common winding defects.

| |
|---|
| Baggy webs (winding webs with caliper profile problems) |
| Blocking (layers sticking together) |
| Bulk/thickness loss (tender webs that yield under low pressure) |
| Core crush (three types) |
| Core offset (operational issue) |
| Corrugations (winding webs with caliper profile problems) |
| Out-of-round (including flat spots, egg shaped, etc.) |
| Rough or feathered roll edges (many subcases here) |
| Telescope (five types) |
| Tin canning (film) |

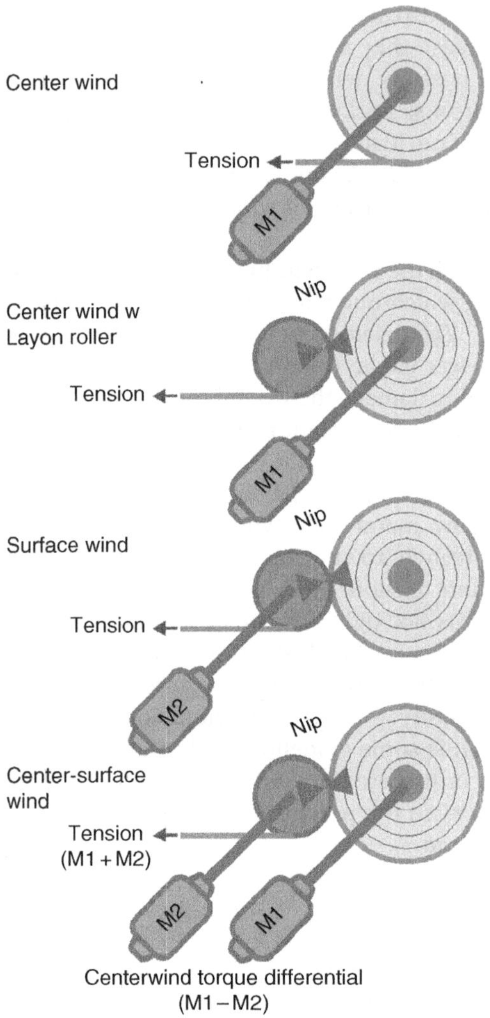


**Figure 5.7** Winder types.

The next key concept is winder "type." The type determines the knobs (aside from speed) used to control tension as seen in Figure 5.7. The most common types for R2R would be a center wind or a center wind with a lay-on roller. The center wind is a one-knob process in which tension is determined by a motor that controls tension for a given set of web properties. The lay-on roller adds another knob for the tension control. The other winder types, surface and center surface, are commonly used in nonwovens, rubber, metal, and paper. It should be noted that web properties also affect tightness, probably even more so than winder type and setting of the TNTs. For example, low modulus

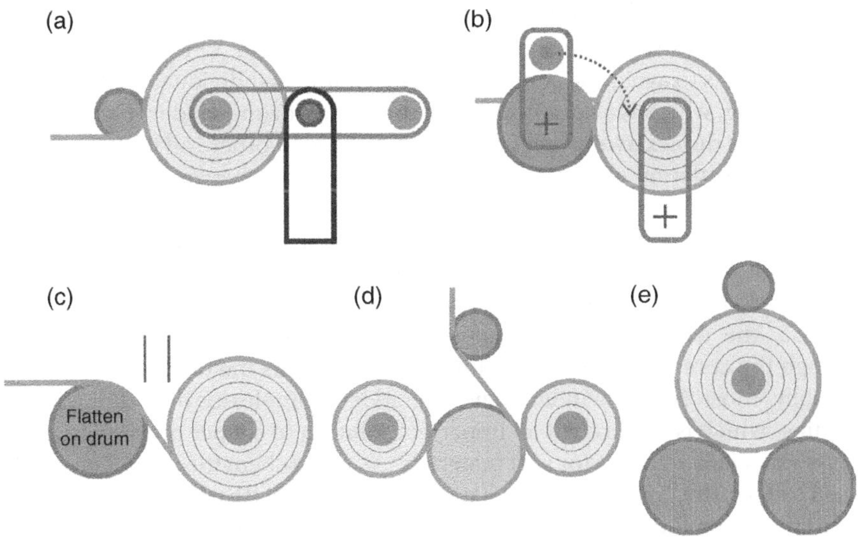


**Figure 5.8** Winder arrangements. (a) Turret winder, (b) reel, (c) two-drum, (d) duplex, and (e) gap winder. Not shown is the slipped core winder that is a variation of the duplex.

products such as rubber and stretchable films generate a significant level of wound roll tightness with little effort. However, stiff materials, such as paper, some films, and certainly any product containing metal, are difficult to wind tightly even at high tension.

Finally, we can discuss winding machinery. Here, whether roll change is done at finite line speed (for a continuous process) or at rest (such as on a rewinder) may influence how a machine is constructed. Other important considerations are roll dimensions (diameter and width), web properties, and whether other functions such as slitting are incorporated. Most winders fall into one of six different variations, called *arrangements*. There are two continuous winders, two rewinders, and two specialty winders, as seen in Figure 5.8.

Note that the *continuous* R2R operation would likely use the turret winder shown in Figure 5.8a. It will build one roll on one spindle and *index* and then transfer the web to next spindle and so on. This index, cut, and splice operation can be disrupted at both the top of the finishing roll and the bottom of the newly started roll. Upsets, though undesirable, are unavoidable to some degree with all winder types, arrangements, design, and operation. The turret winder is in many ways the most capable arrangement, because it can operate as a pure center wind, center wind with lay-on roller, center-surface winder, or gap-winder type. Still, it has limitations beyond cost and complexity. Torque and weight must be borne by a shaft, which limits the diameter of the roll that can be built trouble-free.

For rewinding, cutting rolls to smaller diameter and width, a duplex winder is the norm. This is shown schematically in Figure 5.8d. One reason for this arrangement is to avoid the complexity of spreading of individual lanes to avoid interweaving, which is a common risk with the two-drum winder (Figure 5.8e). The reason a duplex avoids interweaving is that every other roll is alternatively wound on a different shaft, so that there are large gaps between rolls that will keep lanes from tying together. Additionally, with a two-drum winder, the wound roll must be able to support its full weight as a nip, and many materials are too delicate for that (though surprisingly paper can be used if caliper profile is extremely good), thus leaving the duplex winder as the only other recourse. Even so, the duplex winder must support roll weight and tension torque through the core, which presents similar limitations as the turret winder.

Finally, most duplex winders have the added complexity of a slipped core winding, which is required as an imperfect solution to caliper profile variation; this is especially relevant for stiff materials. As seen schematically in Figure 5.9, small caliper/thickness/weight variations across the width of the incoming web translate to small differences in roll diameter. If individual rolls are forced to turn at the same rotational speed, the diameter difference translates into a (surface) speed difference that in turn translates into a tension difference. It is common for the relatively low diameter roll (the relatively lower web thickness position) to have zero tension without slipped core winding. What may be surprising to many in web manufacturing is that for stiff webs, the caliper variation that could cause this problem could be 1% (or even less on metal), which can be considerably beyond the ability of some manufacturing processes to measure, much less control. Therefore, the difficult and limited solution is to turn the shaft at a speed slightly greater than the web speed and transmit torque to the individual wound rolls through slip clutches. A slightly better but considerably more costly option is to have every wound roll use a separate winding station.

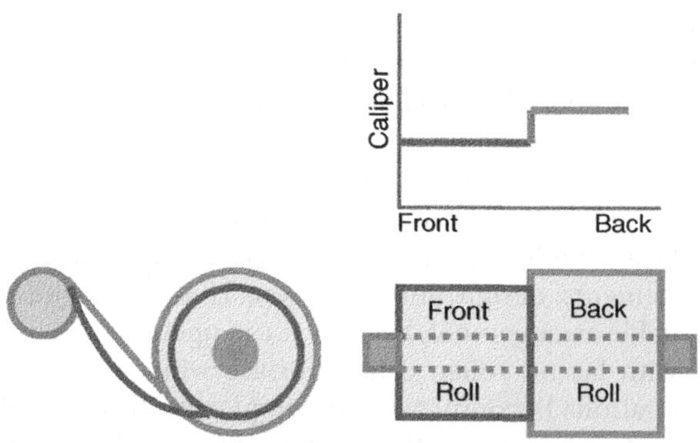


**Figure 5.9** The motivation for slipped core winding.

In conclusion, the main problem for winding operations is inadequate gauge control of the web caused by upstream manufacturing or converting [1]. This will increase the incidence and severity of any tight, loose, and taper-type defects as well as be the source of defects such as baggy webs, corrugations, and many others. This assertion can be validated for any situation by answering a single question. Can the roll be rewound without defect? If a (well-designed/maintained) salvage (or lab) winder cannot process the roll, it is almost certainly not windable. To reduce problems beyond this point would require for the web to be redesigned in nominal properties or made less variable in thickness across the width in upstream manufacturing or converting.

## 5.10 Wrinkling

Another core key area for web handling is wrinkling (insufficient flatness) and its associated topic of spreading. While manufacturing issues tend to be the number one problem in each plant, wrinkling is usually in the top three. Since wrinkling affects nearly all webs and web machines, it generally is the top cause of waste, delay, and customer complaint in the collective web industries.

To get started in wrinkle troubleshooting, we must first be able to observe it, that is, detect a lack of flatness regardless of severity, or judgment with respect to suitability for end use. There are four locations where this lack of flatness observation might be made, as seen in Figure 5.10. The earliest indication of trouble would be a lack of flatness in the open web span that we will call "troughs." If the wrinkling defect is sufficiently large and troughs are sufficiently large, then the web will not lie in complete contact across a roller. At this point, nearly all webs are damaged to the point of rejection. If the bulge gets taller, then it may fold over into a crease. Unfortunately, we are often handed samples of wrinkles, and this would be the last and probably the most difficult place to carry out troubleshooting—in part because we may introduce flatness problems during handling and because we seldom have the provenance or full history of the observed wrinkle. In other words, where was the sample taken from

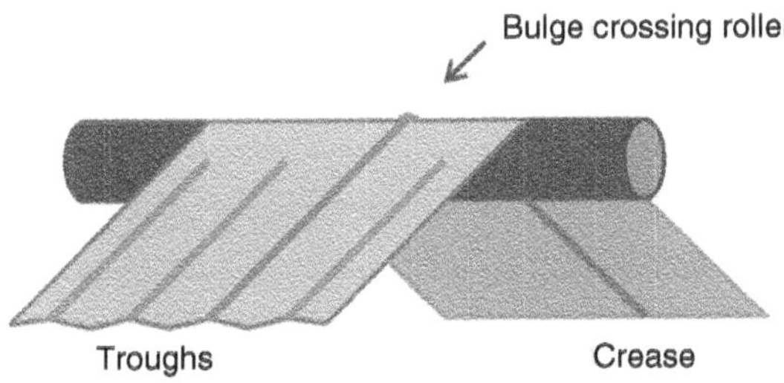


**Figure 5.10** Wrinkle locations.

and, even more importantly, where did the wrinkle form? Finally, it is possible that we may have several different wrinkle types (i.e., “root causes”) in one defective sample.

Though troughs are seldom rejectable, they are the most important for two reasons: First, if the trough is in a forming zone, then the material will likely “remember” that lack of flatness as it cools, dries, cures, and so on. Second, and even more important, is that the open web span is where we do our best troubleshooting. In simple terms, the bulge or crease is the “QA” wrinkle, while the trough is the “troubleshooter” wrinkle. The troubleshooter could and should see the forming wrinkle before becoming a rejectable defect.

The next step in troubleshooting is to note that there are many types and sources of wrinkles in R2R webs. In other words, the lack of flatness is an *outcome* that can come from distinctly different root causes that have very little in common, but must be properly diagnosed. If there were a single crippling problem in wrinkle troubleshooting, it would be not taking the next step and determining the type of wrinkle. All aspects of wrinkle troubleshooting are discussed in several Refs. [3, 5, 17] where major types of wrinkling defects are analyzed by a troubleshooting tree. The top level of the troubleshooting tree taxonomy is given in Table 5.6.

The first three categories in Table 5.6 are related to the orientation of the wrinkle. It is either mostly in the machine direction at a slight/small angle or in the cross direction. Baggy webs are due to uneven formation/converting or, in rare cases, brutish handling. The most common root cause of bagginess, as discussed earlier, is winding material whose caliper/thickness/weight profile across the width is nonuniform. Corrugations, also called rope marks, are also caused by winding a web with a profile problem.

To get to the root cause and thus determine appropriate corrections, we must go at least one layer deeper into the troubleshooting taxonomy. For example, a diagonal wrinkle is always the result of some uneven element, with either a crooked machine or crooked web. To take curl as a final example, we note that subcases, or root causes, are always the result of residual stresses of manufacturing/converting being even across the width (i.e., no “profile problem”), but uneven through the thickness. Root causes of curl include

**Table 5.6** Major wrinkling cases.

| |
|---|
| Machine direction |
| Diagonal shear wrinkle |
| Transverse direction |
| Baggy web |
| Corrugations |
| Curl |

formation, coating, and so on, on one side of the web creating asymmetric structure, laminating, and winding (also called core or roll-set curl).

Once the diagnostics for a situation is complete, it is possible to list options for a solution. Obviously one option would be to eliminate the root cause for that case. Another option would be to spread (active) or flatten (passive) the wrinkle, subjects that are going beyond the scope of this chapter but are well covered in the literature [3, 5]. However, spreading is limited to non-stiff materials. In addition, spreading is limited to relatively wide webs, more than one meter. Both place severe limits on spreading application on many of the new R2R machines. Finally, spreading is also limited in effectiveness. Severe cases are not treatable by spreading. A rare last-resort option is to "anneal" the material to remove stresses before the web is damaged and rejected by customers.

## References

1 Roisum, D.R. Secrets of a Level Process and Product. *Solutions!* August **2002**, pp 45–48 and other earlier venues.
2 Roisum, D.R. *Mechanics of Rollers*; TAPPI Press: Atlanta, **1996**.
3 Roisum, D.R. Web101$^{SM}$ Web Handling, Converting and Winding; self-published course notes: **2016** edition.
4 Roisum, D.R. Roller Alignment: Standards. AIMCAL Web Handling Conference, Prague, Hungary June 2012 and Myrtle Beach Florida October 2012.
5 Abbott, S.A. *AbbottApps*. http://www.stevenabbott.co.uk/AbbottApps/ (Accessed on September 25, 2017).
6 Roisum, D.R. *Web Machine Buying Guide*; DEStech Publications: Lancaster, PA, **2011**.
7 Klassen, C. Drives for Web-handling. AIMCAL Converting School online Shortcourse, **2016**.
8 Roisum, D.R. *Optimization by Integrating Engineering and Business Models*. TAPPI PLACE, Keynote address, Las Vegas, October 26, **2005**.
9 Roisum, D.R. *Web201.16b Dancer Calibration*. http://www.youtube.com/watch?v=4isc_BYxitQ (Accessed on September 25, 2017).
10 Roisum, D.R. The Gold Plated Knob. AIMCAL *Converting Quarterly*, Web Wise column, 4th quarter **2013**.
11 Rucker, L. *Technical Handbook for Elastomeric Roll Covering*; Rubber Roller Group Publication: Springdale, AR, **2000**.
12 Genisot, T. Nip Impressions: a tool for troubleshooting and maintaining quality. *Tappi J.*, vol 77, no 9, pp 246–250, September **1994**.
13 Kennedy, W.B. Nip impressions. *Tappi J.*, vol 74, no 5, pp 277–280, May **1991**.
14 Good, J.K. and Roisum, D.R. *Winding: Machines, Mechanics and Measurements*, Destech: Lancaster, PA, **2006**.

15 Roisum, D.R. Screening Wound Rolls for Bagginess. AIMCAL *Converting Quarterly*, Web Wise column, 4th quarter **2013**.

16 Rumson, D. Slitting. AIMCAL Converting School, **2016**.

17 Smith, R.D. *Ultimate Roll and Web Defect Troubleshooting*, TAPPI Press: Atlanta, **2014**.

# 6

# Polymer Film Substrates for Roll-to-Roll Manufacturing: Process–Structure–Property Relationships

*Baris Yalcin[1] and Miko Cakmak[2]*

[1] *Department of Polymer Engineering, University of Akron, Akron, OH, USA (Present address: The Coca Cola Company, Atlanta, GA, USA)*
[2] *Department of Materials & Mechanical Engineering, Purdue University, West Lafayette, IN, USA*

## 6.1 Introduction

The roll-to-roll (R2R) manufacturing process involves building structures on flexible substrates by additive and/or subtractive process steps to produce flexible and large-area electronics devices such as displays, sensors, solar panels, thin-film solid-state batteries, fibers, textiles, and medical membranes, to name just a few.

The flexible substrate, which is at the heart of the R2R manufacturing process, has the potential to render final products not only bendable or rollable but also low cost due to rapid manufacturing over large areas by unwinding from and to a roll in a continuous manner. Depending on the application, an effective R2R substrate must have many important attributes such as low cost, flexibility/bendability, resistance to chemical attack, thermal stability (especially at elevated temperatures), dimensional stability under thermal cycling, transparency (especially for display and PV applications), stiffness, impermeability to oxygen and moisture (especially for microelectronic and display applications), and smooth surface (microelectronic applications), among other properties.

There are three main classes of flexible substrates in use today, namely, flexible glass, metal foils, and plastic films. Each substrate group has certain advantages and disadvantages.

Thin glass (<100 μm) is flexible and acts as perfect impermeable barrier while at the same time offering superior optical properties, ultrasmooth surfaces, and low thermal expansion coefficients [1]. However, its intrinsic brittle property brings several challenges in the production environment

*Roll-to-Roll Manufacturing: Process Elements and Recent Advances*, First Edition.
Edited by Jehuda Greener, Glen Pearson, and Miko Cakmak.

mostly arising from mishandling and cutting processes. Tiny cracks or stresses arising from cutting glass can cause breakage in subsequent processes. Ultrathin glass [2] can now be chemically strengthened by an ion exchange reaction where smaller sodium ions of glass are replaced by larger potassium ions by simply running the thin sheets of glass through rollers into a bath of molten potassium. In an effort to make thin glass less prone to breaking, a hybrid approach has also been developed where glass foil is coated with a thin polymer layer around the edges. Nevertheless, even with these approaches, the fragile nature of glass raises questions about the final product reliability.

Metal foils can sustain very high temperatures that are required for inorganic material deposition, for example, in the fabrication of inorganic thin-film a-Si, CIS, CIGS solar cells, or silicon-based thin-film transistors (TFTs) [3]. They also provide low coefficient of thermal expansion (CTE), good chemical resistance, and a better barrier to moisture or oxygen ingress compared with plastic sheets, which often require additional special encapsulation for long-lasting operation. Stainless steel and titanium foils are common choices for metal foils. Metal foils have successfully been used as substrates for both amorphous and crystalline silicon-based TFTs to make top emitting active matrix OLED devices. The main drawbacks of metal foils are their high cost, high surface roughness, and non-transparency. Optical clarity with a total light transmission (TLT) of more than 85% over a wavelength range of 400–800 nm is required for bottom emission and electrophoretic displays coupled with a haze of less than 0.7% [4]. For top emissive displays, clarity in the base substrate is not essential, and this opens up the possibility of building display devices on stainless steel substrates. Metal foils with their rough surfaces and high conductivity typically require an insulation spin-on-glass (SOG) planarization layer, such as a chemical vapor-deposited coat of insulating amorphous Si, to ensure flat and nonconductive surfaces that will translate into accurate device registration on subsequent layers.

Plastic films generally fulfill the requirements of low cost, high degree of bendability, transparency, and emissive properties, but there is a reasonable trade-off between physical, chemical, mechanical, optical performance, and cost, which necessitates careful selection of a suitable plastic substrate for a given application [5, 6]. A key challenge of plastic substrates for some advanced applications is the maximum processing temperature (<300°C) and CTE mismatch when combined with other functional layers such as inorganic barrier and planarization layers. For flexible glass and SS steel foil, for example, processing temperatures as high as 550–600 and 900–1000°C, respectively, can be used.

In the following, we will focus on plastic substrate materials due to their high potential for enabling low-cost R2R processing and final flexible electronics products.

Different from other reviews on the subject of substrates, we explore in detail how these polymer films are processed during their manufacture as substrate films/sheets and what kind of structural, thermal, and mechanical properties they are able to achieve during stretching, heat setting, and relaxation modes with real-time mechano-optical studies. Understanding how these materials behave in different, often temporally varying thermomechanical environments will enable readers to choose the right plastic substrates and understand their limitations in a particular R2R processing environment. The substrate manufacturers will also benefit from this survey as the study of uniaxial, simultaneous, and sequential biaxial stretching strategies combined with real-time true stress, true strain, and birefringence provides direct information on molecular mechanisms including orientation, crystallization, and relaxation processes and strain hardening behavior. These mechanisms are most critical to establishing the processing window (temperature, rate, mode of stretching, extent of stretching) needed to obtain the desired morphology (orientation and crystallinity levels and texture) coupled with thickness uniformity and surface roughness controlled by the details of the thermal deformation history and the resulting properties and their anisotropy.

Film casting, whether in solution or melt form, is the first stage of polymer substrate production. In solution casting, the polymer is dissolved or dispersed in solution, coated onto a carrier substrate followed by evaporation of the solvent to create a solid layer on the carrier. The resulting cast layer is then stripped from the carrier to produce a stand-alone film. Before or after stripping, the cast film can be laminated with other webs or coated with other materials to create multilayer (ML) products. High temperature polymer films such as modified polycarbonate (PC) [7] ($T_g \sim 205°C$), polyethersulfone (PES) [8–10] ($T_g \sim 220°C$), aromatic fluorine-containing polyarylate (PAR) ($T_g \sim 340°C$) [11], polynorbornene (PNB) [12, 13] ($T_g \sim 340°C$), and polyimide (PI) [14] ($T_g > 360°C$) cannot be readily melt cast, and therefore the solution processing route is employed. Solution casting process has advantages of casting very thin films and providing, in general, in-plane isotropic substrates with little to no stress accumulated in the structure that hence may not be stretched during manufacture and is not prone to significant relaxation at high temperature processing steps.

Extrusion film casting from the melt, however, has the main advantage of being environmentally friendly due to lack of solvent use in the process. However, most extrusion cast polymer films require subsequent steps such as stretching and heat setting to lock in the morphology that imparts the necessary dimensional and thermal stability. Depending on the initial state of the morphology of the melt-cast films, the processing strategies employed to optimally stretch them are different. For melt-cast polymer substrates, we classify polymers into three categories based on their thermal/crystallization behavior:

I. Amorphous materials that do not crystallize
II. Slow crystallizing polymers that can be quenched into fully amorphous state
III. Fast crystallizing polymers that cannot be quenched into fully amorphous state

*For category I*, following melt casting, the uni- or biaxial stretching is carried out at temperatures between the glass transition temperature ($T_g$) and liquid–liquid transition temperature ($T_{ll}$) [15]. Between $T_g$ and $T_{ll}$ the polymer exhibits heterogeneous structure with soft regions separated by supercooled "rigid" regions, and upon initial deformation, it exhibits elastic behavior due to the presence of these "rigid" regions. The network that sustains the forces exerted by the processing machines on the edges of the films is composed primarily of entanglements in combination with these additional "rigid" segments that help raise the local stresses, leading to strain hardening, resulting in film thickness uniformity. The determination of the $T_{ll}$ has been proposed [16–18] by a number of studies using techniques such as DMA. Mechano-optical techniques in our studies [19] had been shown to reliably measure this temperature for many materials. This fundamental transition determination is critical to establishing the processing temperature window for many noncrystalline materials. Poly(methyl methacrylate) ($T_g = 105°C$), atactic polystyrene ($T_g = 80–100°C$), PC ($T_g = 150°C$), and polyetherimide ($T_g = 218°C$) are included in this category.

*Category II* include aromatic polyesters such as polyethylene terephthalate (PET) ($T_g = 70°C$), polyethylene naphthalate (PEN) ($T_g = 120°C$), polyether ether ketone (PEEK) ($T_g = 145°C$), and their derivatives. Because of their primary aromatic structure, these polymers have relatively stiff chain backbones, and when cast from the melt, they rapidly freeze into amorphous state. A typical film-making process involves melt extrusion and film casting, tenter frame sequential or simultaneous biaxial stretching, and heat setting stages as shown in Figure 6.1. Due to their hygroscopic nature, these polymers need proper drying and moisture-free environments utilizing hopper-drying systems before processing them. Failure to do so leads to significant molecular weight reduction during melt extrusion as the presence of small amount of water molecules in the form of moisture in the melt reverses the polymerization reaction. This is known as hydrolytic degradation. Even following best precautions would not prevent this molecular weight reduction. In order to minimize the degradation, polyesters are typically stabilized against thermal degradation, thermo-oxidative degradation, and hydrolysis through mixing with additives. The most effective method, however, is the proper level of pre-drying prior to melting in the extruders. To ensure the films exhibit high enough molecular weight in the final product, this anticipated reduction is taken into account, and the starting material is selected to have higher molecular weight than is needed in the final film such that it comes down to the desired levels in the final product.

Melt casting is carried out through a sheet casting die onto a chilled roll, and the film contact is typically promoted by electrostatic charge bar to ensure the molten sheet makes a firm contact with the chill roll surface whereby the polymer rapidly quenches into amorphous state. As the differential scanning calorimetry (DSC) diagram indicates in Figure 6.1 inset, the film can be crystallized if the temperature on the chilled roll is raised to $T_{cc}$ (cold

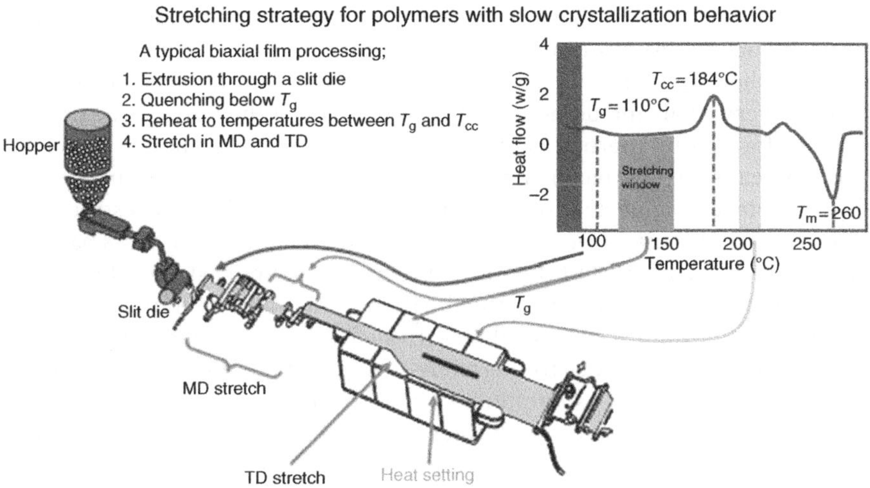


**Figure 6.1** Typical tenter frame film stretching process. Inset shows an example category II polymer and temperature window left on the chill roll to quench the polymer to amorphous state, middle stretching window between $T_g$ and $T_{cc}$, and right heat setting window between $T_{cc}$ and $T_m$.

crystallization) temperature region that is situated roughly between the $T_g$ and $T_m$. Stretching temperatures are typically chosen within the $T_g$–$T_{ll}$ window that is sufficiently away from the $T_{cc}$ (see shaded rectangle for stretching window in the inset of Figure 6.1) in order not to activate thermally induced crystallization (nucleation and growth through chain diffusion to the growing front) prior to stretching. Once this mechanism is activated, these materials lose the ability to be uniformly stretched easily unless they are heated to temperatures where some of these crystals are partially molten. This is usually avoided for a variety of reasons. These films lose rapidly their strain hardening ability when partially molten as these polymers—for example, polycondensates—do not generally have high molecular weights like polyolefins and hence possess low(er) entanglement densities and lesser ability to promote strain hardening during stretching. The activation and promotion of strain hardening mechanism is critical to obtaining uniform thickness films at the end of the stretching processes. In sequential biaxial mode, stretching is accomplished in sequence by first stretching in the machine direction (MD) and then in transverse direction (TD) as shown in Figure 6.1. In the machine direction orientation (MDO) roller systems, the film is increasingly accelerated in the rubbery state ($T_g$–$T_{ll}$ window) between the hot rolls, inducing very high preferential chain orientation to the film. Beyond a critical strain level, the material undergoes orientation-induced crystallization when the polymer chains become parallel to each other and register in crystalline state.

It is important to note that orientation-induced crystallization (also called strain-induced crystallization) is different than thermal crystallization mentioned earlier, which occurs around the $T_{cc}$. Orientation-induced crystals are typically very small and do not affect transparency of the films, while crystals that are formed by thermal crystallization could be large enough to scatter light and cause white (hazy) appearance. After MDO rollers, the material in this next stage is subjected to transverse direction stretching (TD stretching) using the tenter frame stretching process, and this step in the process balances properties in the plane. In both of the previous stretching stages, the level of crystallinity does not typically exceed 25%, and the films contain a great deal of oriented amorphous chains. Films with such structures tend to be unstable above their $T_g$ and in order to thermally stabilize them, they are subjected to constrained annealing—sometimes called heat setting—at temperatures where the thermal crystallization rates are maximum (e.g., for PET ~180°C) that would boost the crystallinity while relaxing some of the oriented amorphous chains, which are responsible for shrinkage that may occur at elevated temperatures. Annealed films remain dimensionally stable nearly up to their crystalline melting range.

*Category III* includes polypropylene (PP), polyethylene (PE), PVDF, and many other polyolefins. The processing strategy is very different for materials fitting into this category. They generally have glass transitions lower than room temperature due to flexibility built in their chain architecture, but they also possess high crystallinities (~60–75% for PP) that melt at high temperatures, meaning the crystalline phase is the dominant phase holding the film together during stretching at high temperatures. Melt casting almost invariably lead to spherulitic crystal growth except maybe at the very surface that touches the chilled roll where quenching may lead to directional crystal growth, but the rest of the film in the interior is generally spherulitic. One of the important morphological dials that these materials need to be optimized is the size of these crystallites (spherulites) to make them suitable for tenter frame biaxial stretching. Generally, if the stretching temperature is too low, then these materials exhibit neck formation that lead to initial nonuniformity of thickness development and void formation. In order to suppress this behavior, the size of the crystallites in the cast film is kept as small as possible either by rapid quenching or by introduction of nucleating agents. This initial "fine" morphology would promote more affine-like behavior at the processing temperatures. These temperatures are generally chosen 1/3 of the way between the onset and the peak of the melting range ($T_{stretching} = T_{m_{onset}} + (T_{m_{peak}} - T_{m_{onset}})/3$). Thus, these materials are deformed in partially molten state where melting and recrystallization take place continuously during the course of stretching. Since melting starts first in small lamellar domains formed at the secondary crystallization stage or at the edges of the spherulites, initial reduction of crystallite sizes promotes more "affine-like" deformation.

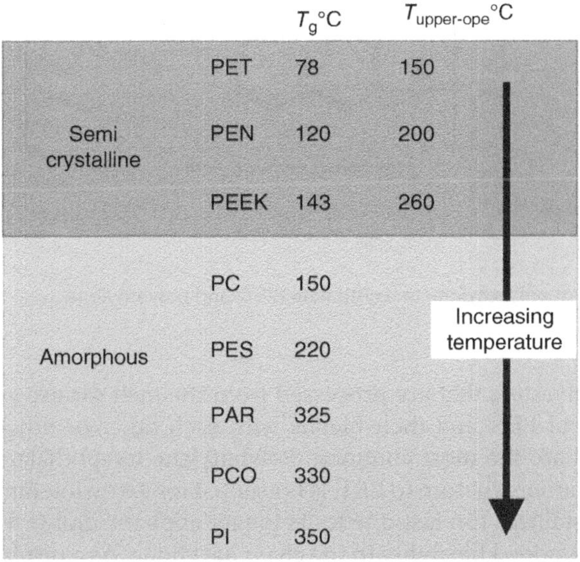


**Figure 6.2** Polymers used in R2R substrates and their transition temperatures.

Most polymer substrates that are investigated for flexible electronics and other advanced applications produced by R2R manufacturing belong to category I (amorphous and solution cast amorphous) and category II (slow crystallizing). These include PC, PES, PAR, polycyclic olefin (PCO) or PNB, and PI for category I and PET and PEN and their blends with PEI and PEEK or other copolymers for category II.

In our review, we will focus on PET and PEN substrates and their blends with PEI and PEEK to give them unique properties as well as high temperature solution cast polymers such as PI.

Figure 6.2 lists some of these polymers in terms of their increasing $T_g$.

## 6.2 Category II: Polyester Films

This class of thermoplastic polymers is formed by condensation polymerization of difunctional monomers of dicarboxylic acid or diesters with aliphatic diols. They can be copolyesters or homopolyesters depending on the type of dicarboxylic acids used. There are aliphatic polyesters, for example, polyglycolic acid, polycaprolactone, and polylactic acid, but their thermal properties are generally much lower than those formed with aromatic dicarboxylic acids, and hence they are not very useful as substrates for R2R manufacturing, which is the subject of this chapter.

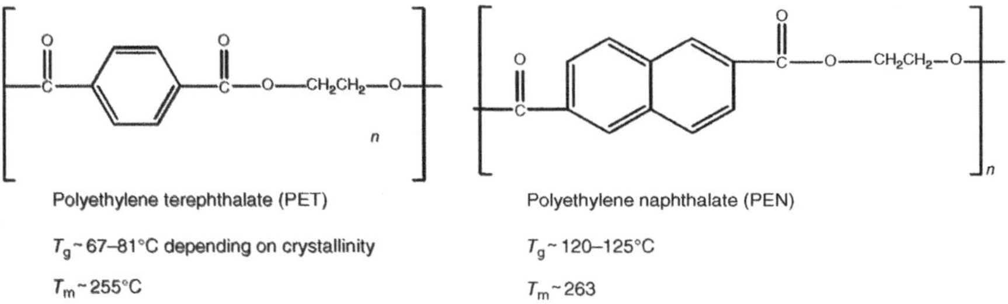


**Figure 6.3** Chemical structure of polyethylene terephthalate (PET) and polyethylene naphthalate (PEN).

Among the aromatic polyesters that are processed from the melt for use as flexible substrates, PET and PEN and their blends with each other or other polymers and copolymers are the most common. Polybutylene terephthalate (PBT), although very similar in structure to PET, is not suited for extrusion film processing because it crystallizes too fast due to its longer aliphatic diol (1,4-butanediol), which imparts added flexibility to the chain backbone. As a result, it is typically used in processes where faster crystallization is desired such as in injection molding.

The chemical structures of PET and PEN are shown in Figure 6.3. The substitution of the phenyl ring of PET by the naphthalene double ring of PEN has little effect on the melting point ($T_m$) but has significant effect on the glass transition temperature ($T_g$), which is about 50°C higher for PEN.

The biaxial stretching (or tenter frame) process is used widely throughout the polymer industry. Its origin dates back to the early 1950s for use in photographic films, laminated films, and flexible disks for storage media, magnetic recording tapes, laminating metallic sheets, thermal transfer ribbons, printing sheets, and packaging as oxygen barrier, insulation films, and heat shrinkage flexible films. Some of these early film applications are predecessors of flexible electronics produced today by substrate-based R2R processes. When producing these biaxially oriented films, processing conditions such as temperature, stretching rate, and sequence of stretching are altered to obtain desired effects on the properties of the films to suit the given application. These properties include optical transparency, optical anisotropy, thermal stability, particularly shrinkage, and balance of in-plane mechanical properties. Typical industrial processes that use multistage multiaxial film processing is graphically summarized in Figure 6.4.

Cakmak et al. have developed laboratory-scale uniaxial [21–24] and biaxial stretching [20, 25–27] and heat setting [28, 29] machines that are equipped with measurement systems that enable real-time measurement of true stress, true strain, birefringence, and IR dichroism [30] of noncrystallizable and crystallizable polymers as they are subjected to deformation fields at industrially

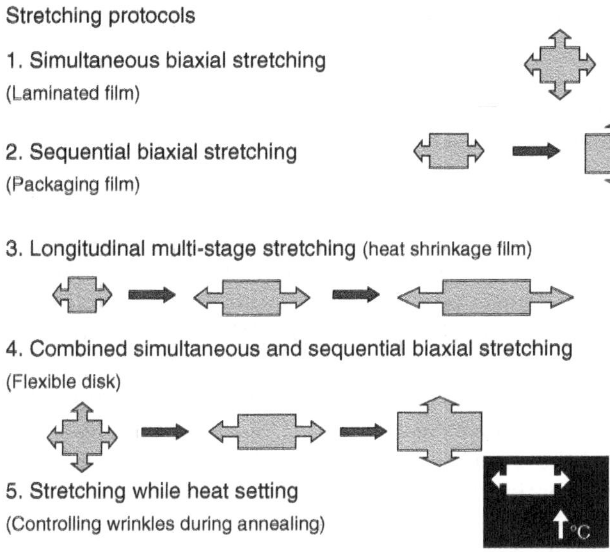


**Figure 6.4** Most common stretching protocols practiced in the film industry [20]. © AIP Publishing LLC.

realistic speeds at elevated temperatures where they are partially molten or in a rubbery state. These machines simulate deformation modes that are typically encountered in the industrial processes including uniaxial free width, uniaxial constant width, unequal simultaneous biaxial, equal simultaneous biaxial, and sequential biaxial. They also apply programmed motion profiles, such as linear motion on both axes as well as user programmable motions that include typically temporal position profiles such as logarithmic, exponential, sinusoidal, sawtooth, and so on. The machine is also capable of receiving deformation temporal profiles from industrial machines and replicates them under well-controlled temperatures.

In processed semicrystalline polymers, birefringence carries combined information about the orientation and crystallinity and to a lesser extent residual stresses. In the case of single-phase polymers such as amorphous polymers, it directly carries information on the orientation and indirectly on the level of residual stresses frozen in the material. Rapid determination of these parameters, together with the "true" mechanical parameters, "true stress" and "true strain," allows an integrated monitoring of strain and stress together with structural (birefringence, IR dichroism) parameters. These "quantitative" relationships between the thermal deformation history and the developed structure imposed by the stretching processes as manifested in optical anisotropy (birefringence, IR dichroism) are essential in uncovering how the polymeric materials behave during various stages of their structural organization in the course of the multistage stretching/heat setting processes.

In the following we will discuss process–structure–property relationships studied with these machines as influenced by the process history on PET and PEN and their miscible blends with each other and other polymers.

### 6.2.1 Polyethylene Terephthalate (PET)

In the rubbery state, when initially amorphous PET is stretched, it exhibits multi-regime behavior—initial linear birefringence true stress regime followed by rapid increase (II) that leads to plateauing III and IV stages as shown in Figure 6.5.

To ascertain what happens in each stage, Hassan et al. stopped the process at strategic strain levels and did offline measurements (DSC, wide and X-ray and Raman spectroscopy), and these studies revealed that the initially amorphous PET that gives linear stress optical behavior at stage I starts crystallizing and the stress optical behavior becomes nonlinear as demonstrated in Figure 6.6. As illustrated in this figure, the increase in crystallinity coincides with transition from stage I to stage II. In this phase, these materials are made up of highly oriented crystals with significant distortions as evidenced by the diffuseness of the WAXS crystalline peaks. Only after annealing (inset) the sharpness of crystalline peaks develops significantly. Offline, Raman data also give additional clue on gauche–trans conformational changes in the films as they

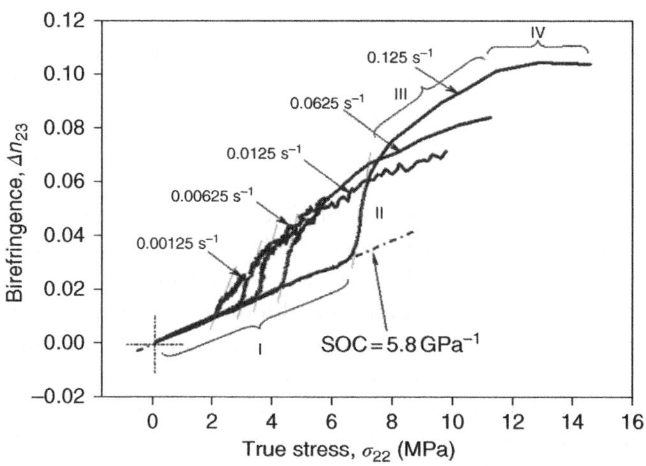


**Figure 6.5** Stress optical behavior of PET at rubbery temperatures: effect of stretching rate [27]. © Elsevier.

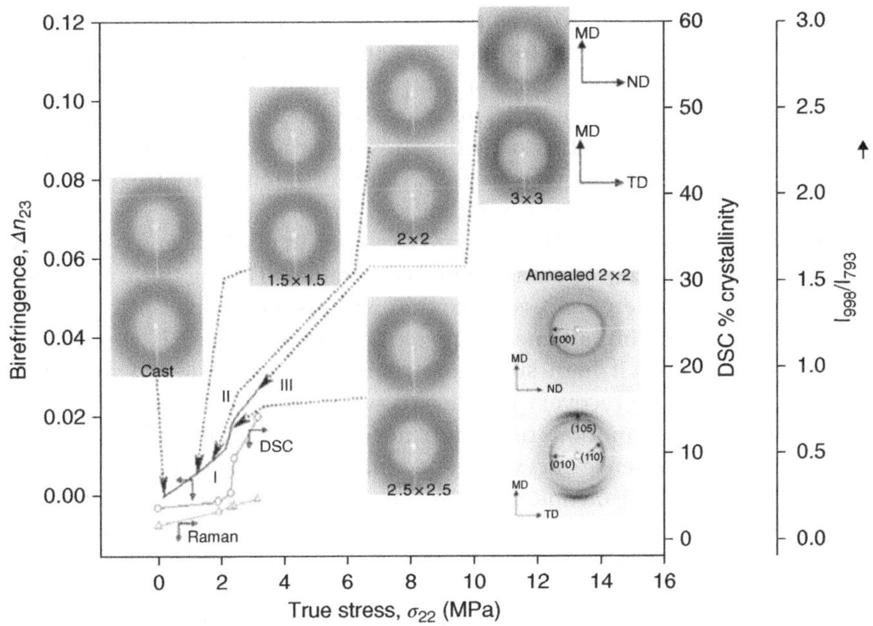


**Figure 6.6** Stress optical behavior of PET combined with offline DSC, X-ray, and Raman spectroscopy [27]. © Elsevier.

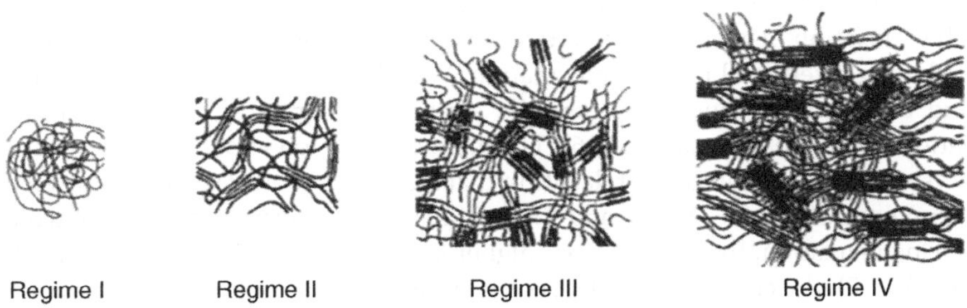


**Figure 6.7** Structural model for simultaneous biaxially stretched films [27]. © Elsevier.

become crystalline. As is well known, only trans conformers exist in the crystalline domains of PET.

The structural evolution during the simultaneous biaxial stretching step is illustrated in Figure 6.7.

In regime I the material remains amorphous. As the polymer chains become oriented in the plane, they start registering side by side with one another (regime II) in parallel orientation that is needed for crystallization. If a sufficient number of them form this side-by-side alignment, they start influencing the birefringence and stress development as the network composed of entanglements is enhanced by the formation of these crystalline "nodes," causing rapid rise in efficiency of orienting polymer chains while suppressing the relaxation processes (regime III). This manifests itself as the rapid increase in birefringence, and beyond this stage the stress optical behavior becomes nonlinear and tends to reach a plateau as the limit of extensibility (regime IV) is reached with rapid increase of crystallinity.

In tenter frame operation, the films are stretched in machine and TD and are transported from station to station while constrained. During these transition stages, materials undergo stress relaxation and associated physical changes. The level of orientation obtained at the end of the stretching is modified by relaxation, especially if high stretching temperatures and low deformation rates are utilized. The recovery that occurs when the acting force is removed has been attributed to motions of the polymer chains in the amorphous regions, particularly in the absence of crystals and entanglements. The oriented chains try to recoil into the isotropic state, as this is thermodynamically the most favorable state. The relaxation process may enhance the crystallization of the polymeric materials particularly when the deformation process creates taut chains with unfavorable orientations adjacent to each other, with substantial amount of local segment chain axes pointing in different directions. During relaxation, these chains tend to orient parallel to each other that is the favorable ordered state, enabling crystallization.

The relaxation behavior of simultaneously and sequentially biaxially stretched PET films was studied at the rubbery state stretching temperatures [24, 31, 32]. The relaxation behavior was found to be dependent on the process prehistory including extent and rate of deformation. WAXD patterns for the as-stretched and the relaxed sequentially biaxially stretched samples at different stretch ratios are shown in Figure 6.8 [26]. The main observation was the appearance of crystalline diffractions after relaxation that were absent after sequential stretch ratio series of 2 ($2 \times 1$, $2 \times 1.5$, and $2 \times 2$), indicating that crystallization occurred during relaxation. In relaxing from higher stretch ratios ($2.5 \times 1$, $2.5 \times 1.5$, $3 \times 1$, and $3 \times 1.5$), the crystalline peaks became intensified as a sign of better orientation and higher crystalline order attained during relaxation. At lower stretching rates $0.0125\,s^{-1}$, the same stretch ratio series ($2 \times 1$, $2 \times 1.5$, and $2 \times 2$) did not show relaxation-induced crystallization, which indicates that relaxation-induced crystallization highly depends on the stretching rate, which defines the state of order of the chains necessary for crystallization.

The relaxation behavior of oriented films is divided into three regimes: In the first regime where the linear stress optical rule applies, the material remains amorphous during stretching and relaxation. At intermediate deformation levels (regime II), the polymer chains may remain amorphous initially. But during the relaxation stage, they begin to exhibit strain-induced crystallization as the adjacent oriented amorphous chains become locally parallel to each other and

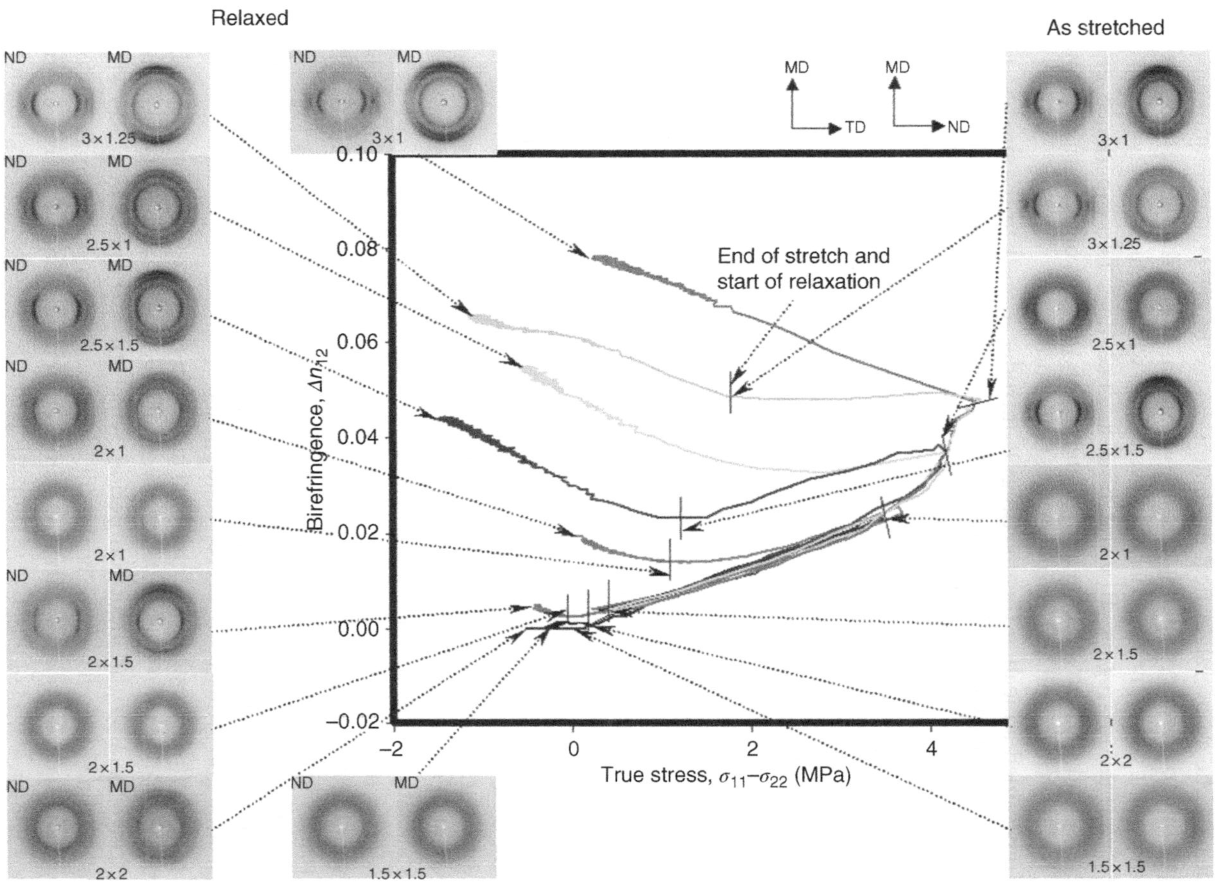


**Figure 6.8** WAXD patterns for the as-stretched and the relaxed sequentially biaxially stretched samples at the rate of 0.125 $s^{-1}$ and different stretch ratios [26]. © Elsevier.

the matrix crystallizes. Once this crystallization occurs, previously decreasing birefringence begins to increase while the stress continues to decrease. There appears to be an overall critical minimum degree of orientation required for the chains to start crystallizing during relaxation. When the relaxation starts from a point where the material has developed a well-connected network during deformation (regime III), the birefringence does not experience any decrease and starts to increase immediately upon the start of relaxation, forming higher orientation and/or higher crystalline order.

Constrained annealing (heat setting) is introduced next to the process to enhance the thermal stability of the stretched films. At this stage constrained films are exposed to elevated temperatures where the polymer exhibits highest rate of crystallization (usually about halfway between $T_{cc}$ and $T_m$(peak). Hassan et al. [29] studied the structural changes that take place during the heat setting of simultaneously and sequentially biaxially stretched films of PET by real-time birefringence, offline Raman spectroscopy, DSC thermal analysis, and X-ray techniques. They reported three different behaviors of the stretched films during the annealing process depending on the orientation and percent crystallinity of the pre-stretched films. These behaviors (categories) are discussed next.

*Category I.* This category is represented by low orientation and crystallinity levels of the pre-stretched films and shown for sample 1.5 × 1.5 in Figure 6.9. In this category the samples experience a little increase in birefringence for a very short time after the induction period then a relaxation during the rest of the annealing process. The high thermal energy transferred to the films by orienting the chains after the induction period, but as there is no physical network to hold that state of orientation, the chains relax very quickly. This behavior is apparent in the low deformation levels prior to the onset of the strain-induced crystallization. The high annealing temperature causes the oriented chains to relax upon heating, as there are no physical junctions or crystalline network to prevent the relaxation of the chains. The low orientation and crystallinity levels results in a loose structure, which, upon heating, the high energy put into the material causes the oriented chains to recoil, resulting in a relaxation process, and if the temperature is high enough, thermal crystallization through nucleation and growth process typically ensues.

*Category II.* This category is observed in the intermediate deformation levels (e.g., 2 × 2 sample in Figure 6.9). This regime starts with a small decrease in birefringence after the induction period, which then starts to increase. The small decline of birefringence is a result of reduction of orientation as part of the oriented chains relaxes at the annealing temperature. The subtle relaxation of the chains helps cooperative organization that leads to parallel registration of the chains and formation of crystalline domains. The high increase in birefringence, crystallinity, and crystallization kinetics in this category is due to the ability of the samples to attain higher crystallinities after substantial amount of

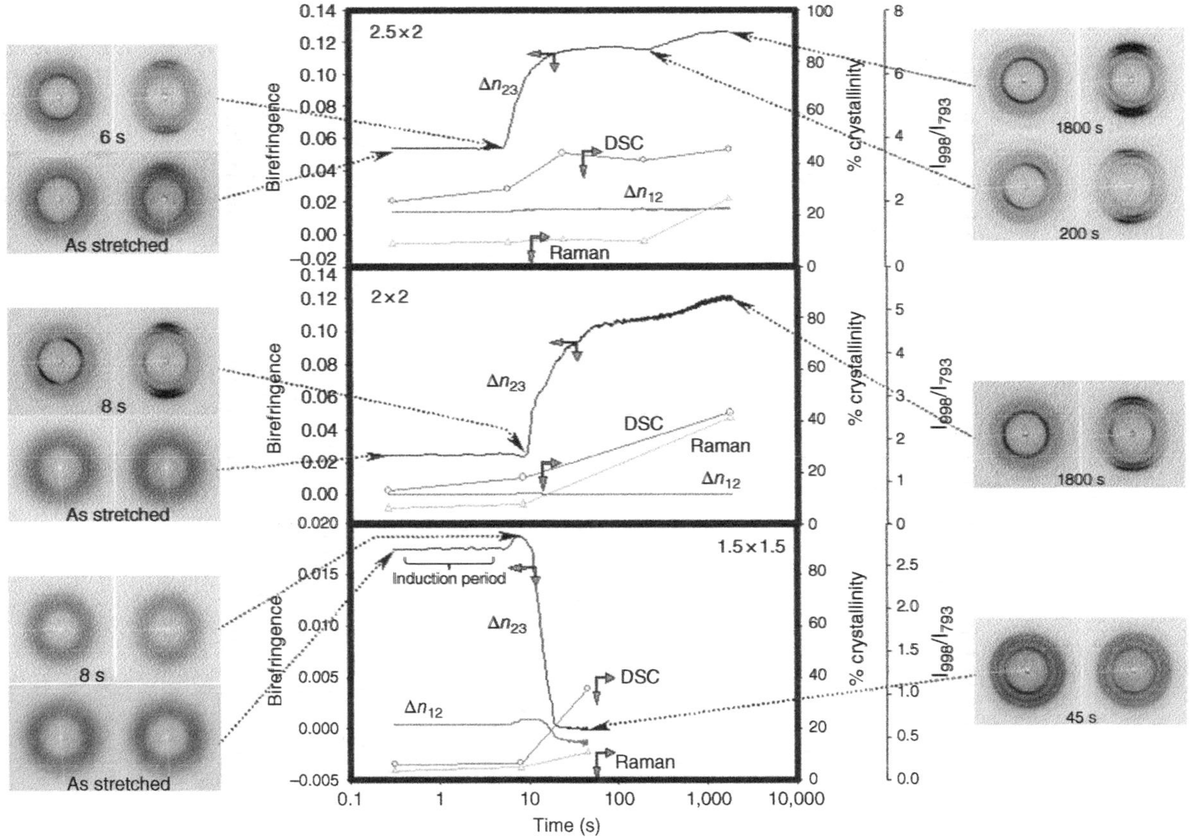


**Figure 6.9** Temporal evolution studies for different sequentially biaxially stretched and annealed samples [29]. © John Wiley & Sons.

oriented amorphous chains were developed during the stretching stage that transform upon annealing.

*Category III*. This category includes the high deformation samples (e.g., 2.5 × 2 in Figure 6.9). These films experience little to no relaxation in the amorphous domains following the induction period followed by the rapid increase in birefringence. Also these samples exhibit a small decline in birefringence before the upturn. The small decline is associated with a decrease in crystallinity as a result of partial melting and reformation of the crystallites. These samples were stretched above the onset of the strain-induced crystallization and exhibit well-formed long-range connected physical networks. In this regime the birefringence and crystallinity increase are lower than in regime II because the previously formed crystalline domains during the stretching process may lock in a large amount of oriented amorphous chains in unfavorable orientations with respect to neighboring chains that prevents the amorphous chains from transforming into crystalline domains, thereby reducing the crystallizability. Another reason for the decrease in crystallizability, which is also the reason for the leveling off behavior in the crystallinity and the trans isomer content at the high annealing times, is that, at these conditions, most of the crystal structure has already been formed either by the stretching or annealing processes, reducing the amount of amorphous chains that could convert during further annealing.

The authors also reported that the sequential mode of stretching provides better environment for the crystallinity development and associated trans conformational transformation than the simultaneous mode of biaxial stretching.

Stretching of the PET films for substrate manufacturing can be performed using nonlinear stretching profiles, which results in the development of crystalline and orientation structures that are different from those that are induced by linear stretching profiles. Martins et al. [33] employed four stretching rate profiles to PET films with different molecular weights (intrinsic viscosities), that is, linear, sigmoidal, logarithmic, and exponential, in uniaxial constrained width mode (Figure 6.10), and found substantial differences in the corresponding mechano-optical behavior (i.e., in-plane birefringence vs. applied stress).

All stages I, II, III, and IV described in Figure 6.5 were still observable in logarithmic and exponential stretching modes. Sigmoidal deformation showed only the first two regimes even though the film was stretched to the same total engineering strain as applied to all profiles.

Figure 6.11 shows mechano-optical behavior of LPET for each type of the deformation profiles studied. The use of logarithmic profile had the largest influence on crystallinity development started earlier. This, in turn, translated into early development of strain hardening that is directly responsible for the achievement of film uniformity at lower strains. In this mode, the use of slower speeds at high strains also helped improve the organization of the crystalline domains by facilitating the partial relaxation of polymer chains that in turn

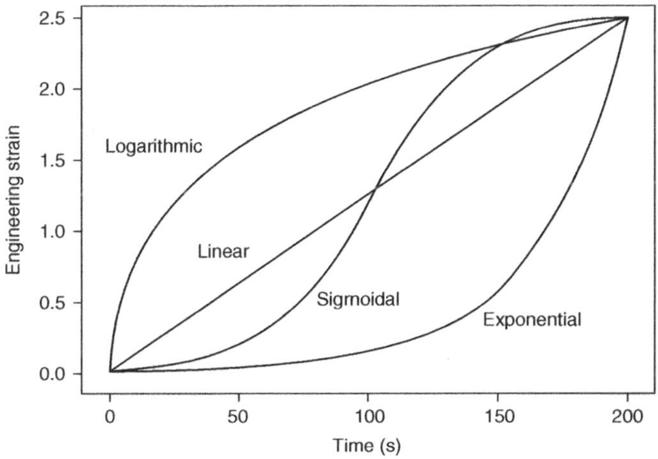


**Figure 6.10** Types of deformation applied to PET film stretching: engineering strain versus time (global deformation history) [33]. © Elsevier.

allowed them to register with oriented crystalline domains, facilitating their growth as well as their perfection.

Exponential profile that imparted the highest strain rate at high strains retarded the strain-induced crystallization as the early stage of deformation was dominated by the relaxation processes. The delay in strain-induced crystallization delayed the strain hardening allowing for an increased orientation of the polymer chains as they were not constrained by the long-range network until high strains were achieved. This, in turn, allowed further development of birefringence. The exponential deformation profile yielded the highest crystallinity for the same total true strain. In other words, the exponential application of strain keeps the structure "looser," while the polymer chains become oriented, preventing the formation of long-range network at intermediate states, thereby facilitating higher strain-induced orientation and consequent crystallinity development at the end of the deformation period. The authors recommended that if a uniform thickness profile is desired at lower strains, the logarithmic deformation profile should be used. However, if higher crystallinity with higher orientation at the end of deformation is desired, it is recommended to use the exponential deformation profile for the same total strain application to the films.

Biaxially oriented PET films can creep over long term even at temperatures below the $T_g$. It is important to understand this behavior when designing flexible devices utilizing PET substrates, especially those that are under constant load. Wang et al. [34] studied the effect of biaxial orientation and crystallinity on the long-term creep behavior of PET below $T_g$. They investigated tensile creep behavior of unoriented films of varying crystallinities and unequal and

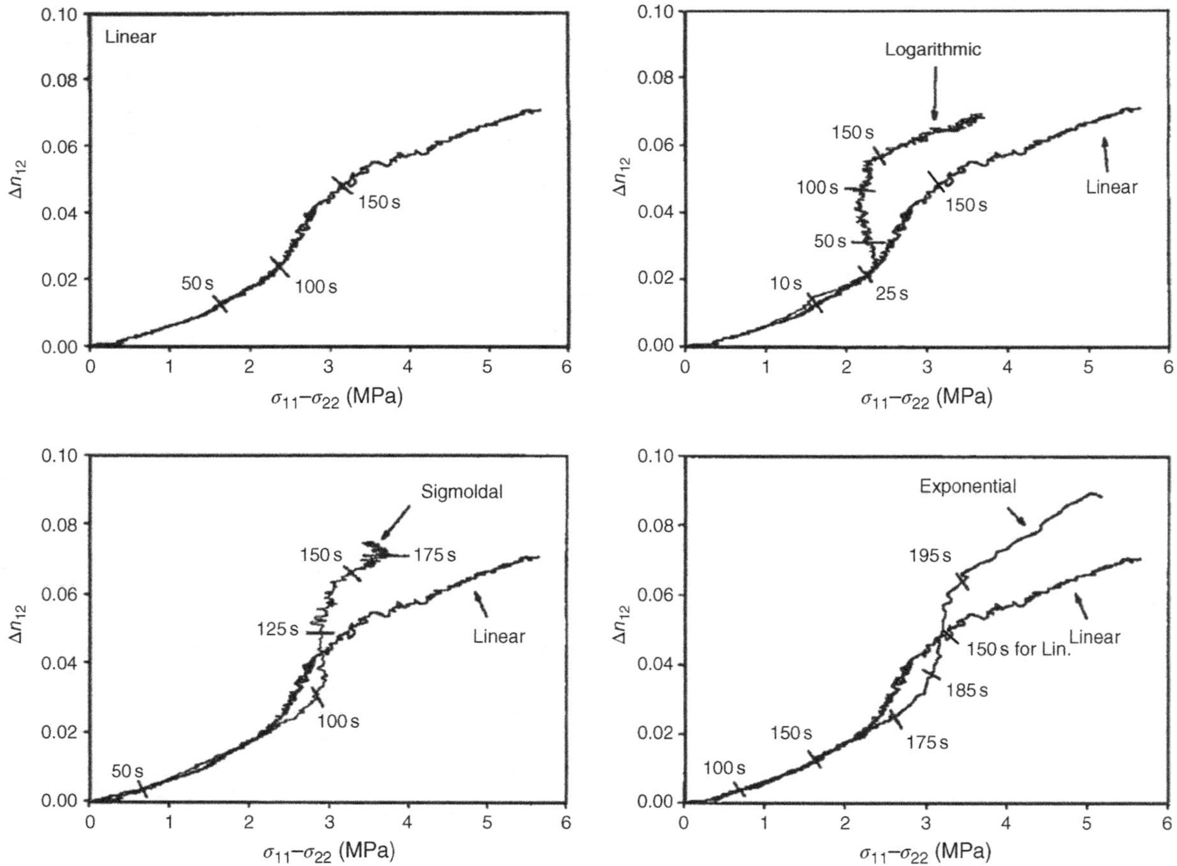


**Figure 6.11** Temporal evolution of mechano-optical behavior of LPET (lower MW PET) for each type of deformation profile [33]. © Elsevier.

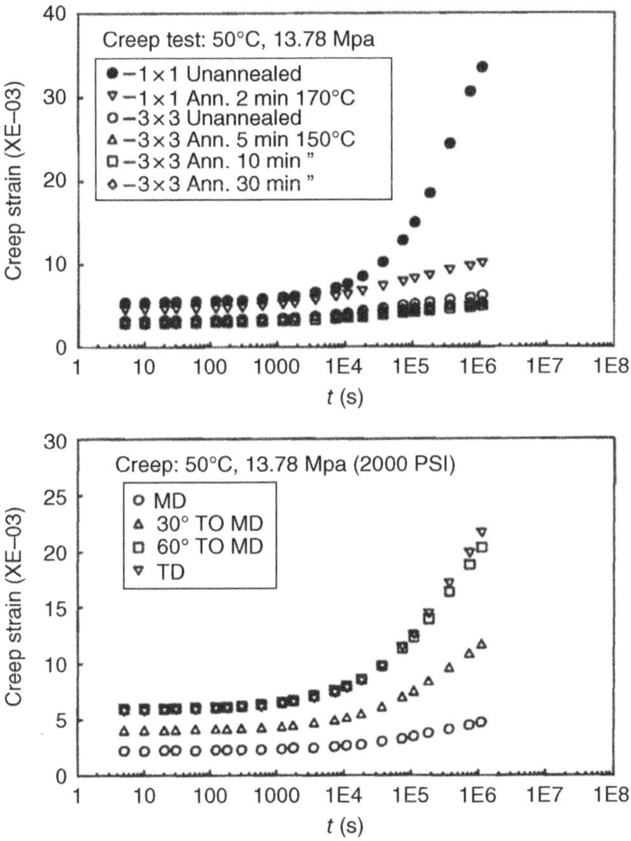


**Figure 6.12** Effect of heat setting on long-term creep behavior of unstretched and 3×3 stretched PET films (top) and effect of in-plane direction on long-term creep behavior of 3×1 stretched PET films (bottom).

equal biaxially oriented PET films and their annealed counterparts. The increase of orientation and crystallinity by stretching and perfecting the crystals by subsequent annealing caused reduction in long-term creep strains (Figure 6.12-top graph). The lowest creep strains were observed in the direction along which the film was stretched to the highest stretch ratio. Unequal biaxially stretched films exhibited in-plane anisotropy in their tensile creep behavior (Figure 6.12-bottom graph). Creep curves exhibited two characteristic regimes, one at short creep times ($<1.10^4$ s) and the other at long creep times. Rapid reduction in creep strain and strain rate with increased crystallinity and orientation (Figure 6.13-top graph) in a given testing direction was particularly apparent at long creep times and at high creep temperatures

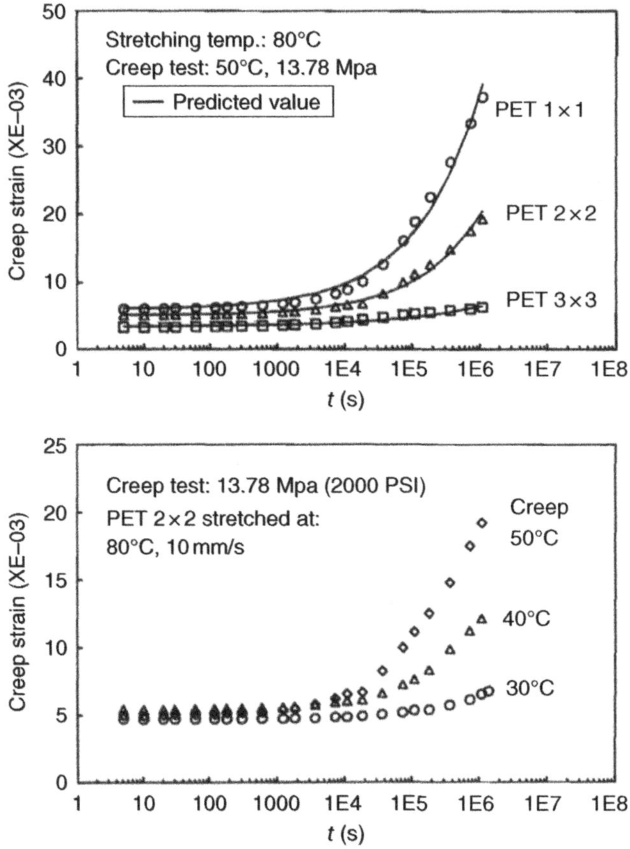


**Figure 6.13** Effect of stretched ratios on the creep behavior of unstretched and 2×2 and 3×3 stretched PET films (top) and effect of temperature on long-term creep behavior of 2×2 stretched PET films (bottom).

(Figure 6.13-bottom graph). They reported that the network structure formed by the crystalline regions acts as an anchor to various portions of the chains by entrapping them in higher density/less mobile crystalline regions.

### 6.2.2 Poly(ethylene Terephthalate) (PET) and Poly(etherimide) (PEI) Blend

The functionality of PET films for use as flexible substrates can be improved by blending with various polymers. One useful blend is formed by compounding PEI into PET. PEI is a linear amorphous high-performance polymer with a high glass transition temperature ($T_g$) of approximately 215°C.

The bulky aromatic imide group provides PEI with the high rigidity, creep resistance, and high thermal stability, while the flexible ether linkage allows for good melt processability. Choi et al. reported the several benefits of PEI in PET rich blends not only for the final utility of the films but also during film stretching and heat setting processes during their manufacture [28, 35]. The increase in $T_g$ of the films due to the presence of PEI (Figure 6.14a) is desirable since it widens the application window of the films by increasing their use temperature. Besides the increase in $T_g$, cold crystallization temperature, $T_{cc}$, also increases as shown in Figure 6.14c. Although the increased $T_{cc}$ is accompanied by the increased $T_g$ with increased PEI concentration, the magnitude of $T_{cc}$ is even larger at a given PEI concentration. Thus, the net result is a wider processing window, defined by the temperature difference between $T_g$ and $T_{cc}$ (Figure 6.14d), in which the process (i.e., deformation, followed by heat setting) operates in a more favorable environment. The crystallization rate of PET in the blend films was dominated by two independently counteractive factors: PEI concentration and preferred orientation. PEI lowers the crystallization rate due mainly to the dilution effect that spatially prevents the crystallizable polymer chains from coming together, while preferred orientation boosts it by reducing the activation energy for transport for crystallization. Although the order of crystalline structure decreases on stretching as the PEI concentration increases, the stress level becomes higher, giving rise to strain hardening at smaller stretch ratios. This deformation behavior has a benefit of being able to strain harden the films with the higher degree of amorphous orientation without locking in the structure by crystallization, which may impart more flexibility to the films during deformation stage, that is, tenter frame processing. The authors found that the increase in PEI concentration led to increase in highly defective crystallites where bulky PEI chains intrudes into the PET crystalline regions (see WAXD patterns in Figure 6.14-bottom). Above 30% PEI, the blend crystallizability disappears. Three categories of distinct structure evolution were observed during stretching of PET/PEI blends with 100/0, 90/10, and 70/30. In 100/0, amorphous orientation takes place rapidly, followed by crystallization. In 90/10, amorphous PET and PEI intermingled molecules steadily orient with suppressed crystallization of PET due to the dilution effect and insufficient stress level to overcome it. At 70/30 PET/PEI composition, amorphous molecules orient significantly even at small stretch ratios due to the high stress level; however, the crystallization is greatly restricted because of the high dilution effect, forming a defective crystalline structure. The slowing down of the crystallization kinetics by the addition of PEI, also shown by the increase in $T_{cc}$ (Figure 6.14c), allows a prolonged deformation process that is conducive to increased molecular order without locking in the structure and hence improved physical and gas barrier properties on which an enhanced utility of the films may be achieved.

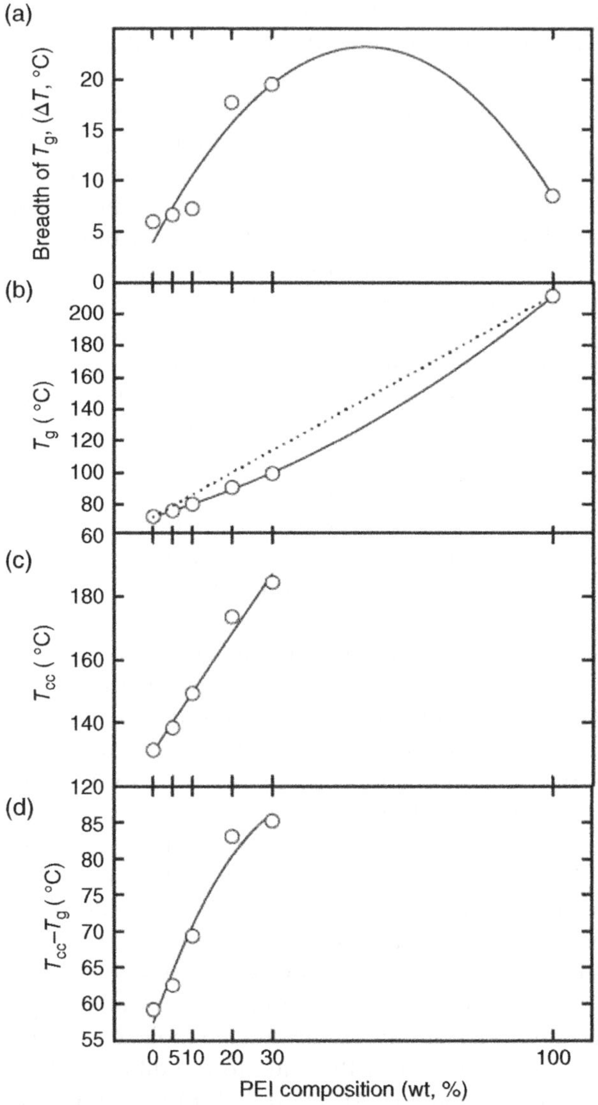


**Figure 6.14** Breadth of $T_g$ range, $T_g$, $T_{cc}$, and $T_{cc}$–$T_g$ for PET/PEI films as a function of PEI composition (top). True stress, degree of crystallinity, and WAXS of PET/PEI films as a function of stretch ratio: (a) 100/0, (b) 90/10, and (c) 70/30 (bottom) [28]. © Elsevier.

Annealing of PET/PEI blends that are stretched to low deformation levels before strain hardening leads to complete elimination of birefringence (Figure 6.15-top) due to relaxation in blends containing as little as 10% PEI [28]. At deformation levels that are beyond the onset of strain hardening (e.g., ~5×), annealing leads to initial relaxation nearly to optical isotropy for 70/30

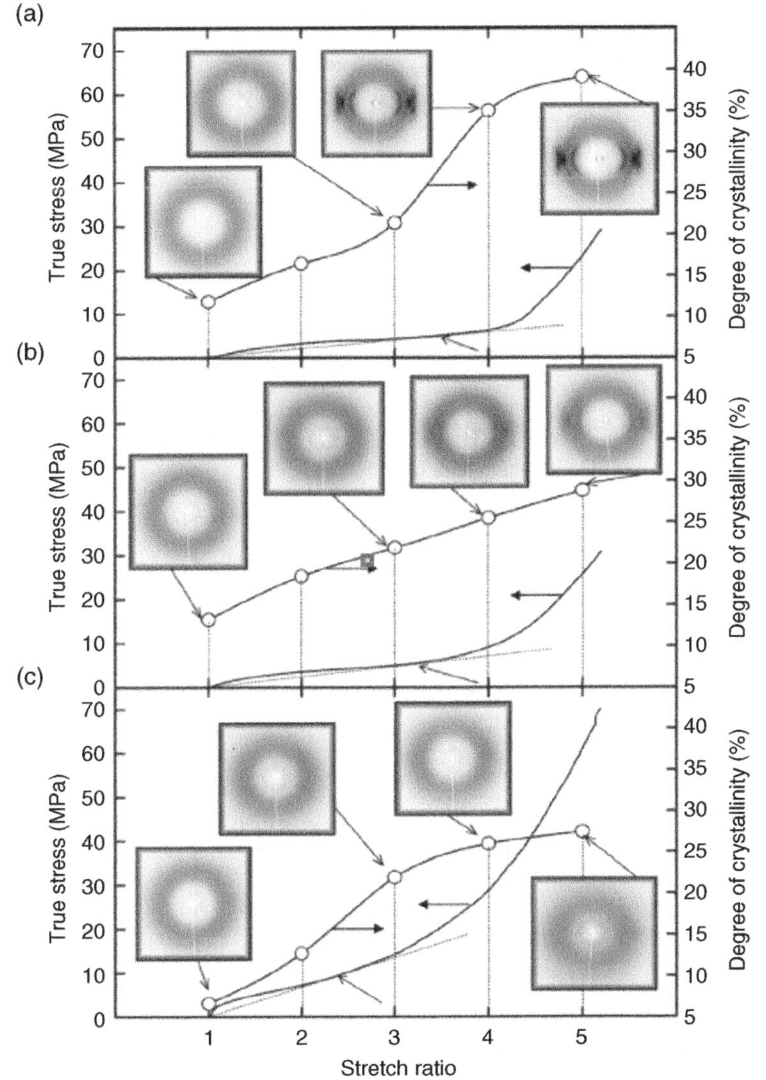


**Figure 6.14 (Cont'd)**

blend and to some extent for the 90/10 blend (Figure 6.15-bottom), but this is followed by renewed increase at later stages of annealing, indicating that oriented nuclei that are formed during deformation survive the initial relaxation phase and dominate the crystallization at longer-term annealing.

Mechanical properties of PET/PEI films stretched to 5× and heat-set for (a) 1 s (b) 5 s (c) 4000 s are shown in Figure 6.16. The final mechanical properties are highly dependent on the structure of the blend (crystallinity and

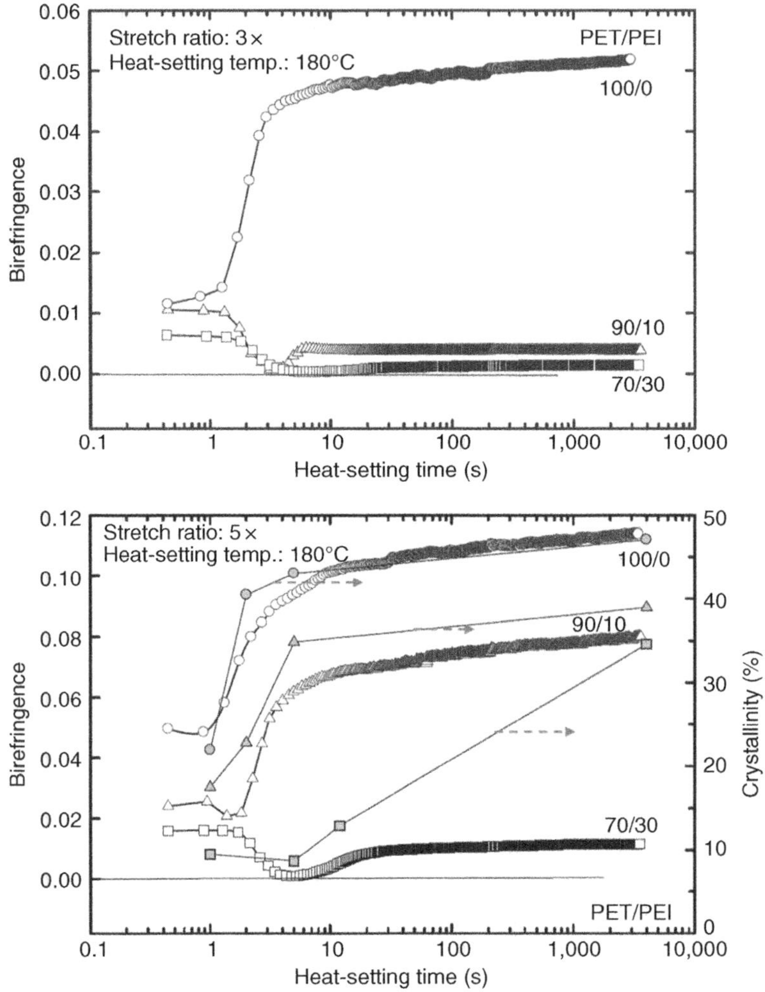


**Figure 6.15** Birefringence of PET/PEI 100/0, 90/10, and 70/30 blends, stretched by 3× (top) and 5× (bottom) at the rubbery state temperatures. Heat setting temperature is 180°C [28]. © Elsevier.

orientation), which is influenced by deformation level and heat setting time. The dramatic increase in toughness, that is, ability of the material to absorb energy and plastically deform without failure, of 70/30 blend at 5 s is attributed to the relaxation process as a result of the suppressed crystallization. However, in 100/0 and 90/10 blends, due to the dominant crystallization process over relaxation, the strength increases with increasing heat setting time, and hence toughness, calculated by the area under the stress–strain curve, tends to

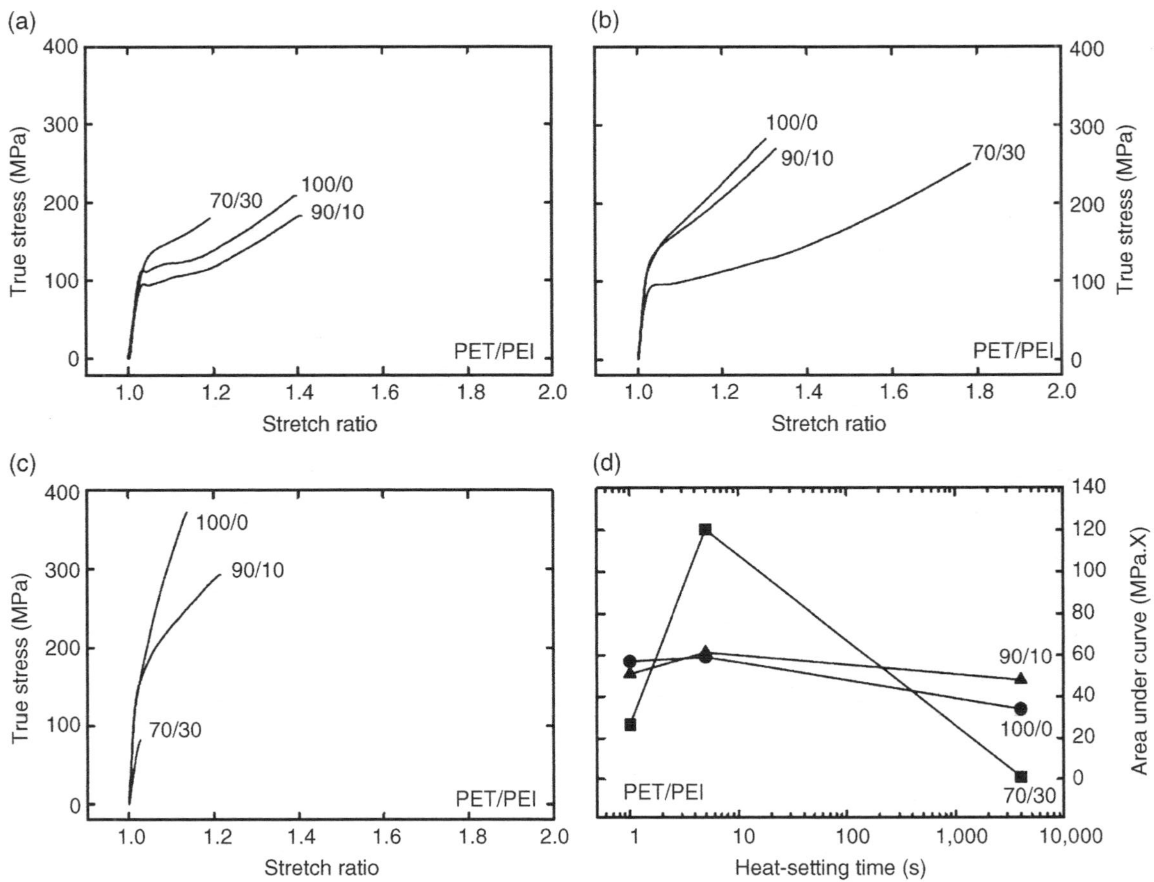


**Figure 6.16** True stress versus stretch ratio of PET/PEI films, stretched by 5× and heat-set for (a) 1 s, (b) 5 s, and (c) 4000 s. (d) Area under the curves. Stretch rate: 300%/min. Stretch temp.: room temp [35]. © Elsevier.

decrease (Figure 6.16d). The brittleness of 70/30 blend (Figure 6.16c) at the end of the heat setting may be attributed to the considerable inhomogeneity induced from the phase separation between PET and PEI in amorphous regions as well as crystallization of PET.

### 6.2.3 Polyethylene Naphthalate (PEN)

PEN, like its close chemical relative, PET (Figure 6.3), is a slow crystallizing polymer that can be quenched into an amorphous state and crystallized either by slow cooling from the melt or by stretching between the glass transition temperature and the cold crystallization temperature.

The structural processes that take place during stretching and relaxation of PEN show similar behavior to PET. Martins et al. [36] reported structural and mechano-optical behavior of amorphous PEN films stretched in the rubbery state as a function of processing temperature, draw rate, and molecular weight ($M_w$). Similar to PET, a three-regime stress optical behavior was identified as shown in Figure 6.17. Regime I corresponds to classical stress optical behavior of amorphous materials where PEN remains completely amorphous during stretching. Oriented crystallization and long-range connected network formation accompany the positive slope change in the stress optical behavior into regime II. In the last regime, regime III, the formation of "muscle-bound" structure begins to level off birefringence as it approaches the finite extensibility of the polymer chains.

WAXD patterns in Figure 6.17 indicate that the material remains amorphous in regime I. Right after the transition to regime II, the amorphous chains are highly oriented but possess low translational order, which prevents them from crystallizing. Further stretching the structure evolves very rapidly to a crystalline structure, with approximately 40% crystallinity that is maintained throughout regime III at the end of which finite chain extensibility prevents further changes in the structure. A model representing the structural mechanisms followed from the amorphous to the crystalline state was suggested by the authors. Both rate and temperature have a significant effect on the structural development of PEN: at intermediate to high rates and low temperatures, a semicrystalline structure is obtained, whereas at very high temperatures and low rates, the structure remains completely amorphous despite the large deformation that has been applied.

Relaxation of uniaxially stretched PEN was found to depend exclusively on the regime from which the relaxation is initiated. If the relaxation is initiated from within regime I, the stress and birefringence decrease traces back the linear stress optical behavior (not clear), while the material remains amorphous. The relaxation of stress and birefringence from the early stages of regime II deviates from the linear stress optical behavior. At long enough relaxation times, birefringence eventually begins to increase, while stress continues to decline (Figure 6.18). This marks the formation of highly localized

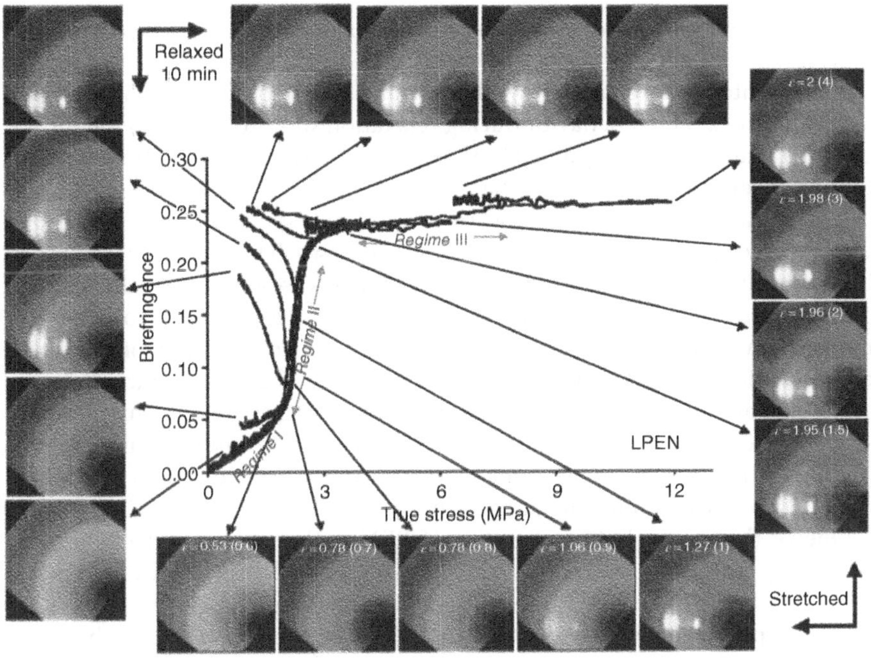


**Figure 6.17** Mechano-optical behavior of Low MW PEN during stretching and relaxation processes with corresponding WAXD pattern. Samples were stretched at 140°C and 0.01 $s^{-1}$ and relaxed for 10 min. The values indicated in the pictures are referring to the true strain and engineering strain (in parentheses) at relaxation time zero [36]. © American Chemical Society.

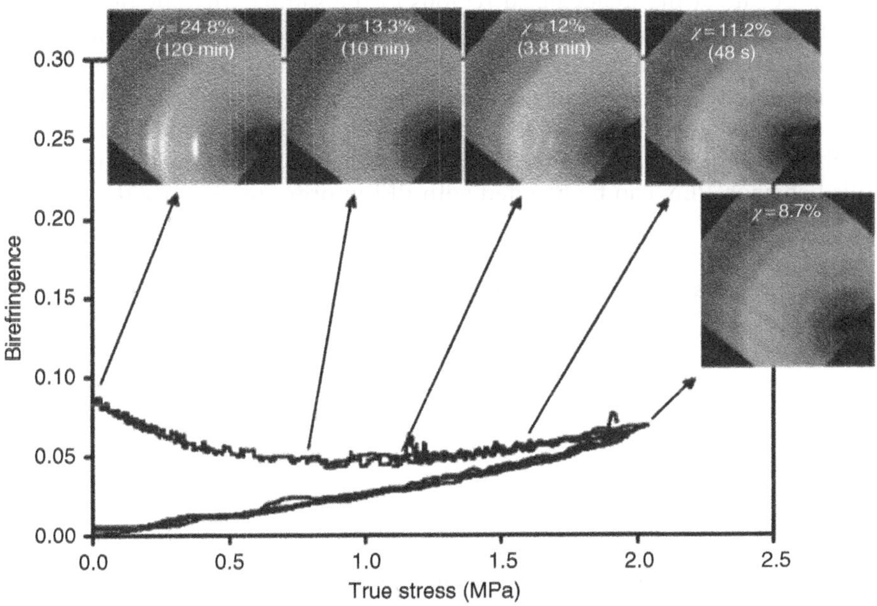


**Figure 6.18** Detailed mechano-optical behavior and WAXD patterns of LPEN strained at 70% (early stages of regime 2) and relaxed for 2 h [36]. © American Chemical Society.

highly oriented ordered regions with low translational ordering along the chain axis. This unexpected behavior during relaxation was attributed to the formation of a long-range physical network that helps arrest the relaxation of orientations generated during stretching.

If the relaxation is initiated toward the end of regime II, a fast increase of birefringence is observed, while the stress decreases. This behavior is found to be the result of a spontaneous deformation process as observed by the increase of local true strain in the absence of macroscopic deformation. However, this mechanism is not a result of relaxation alone, but a combination of relaxation coupled with continuation of the deformation process that started during stretching and did not stop when the relaxation process started. In regime III, stress relaxation accompanies little or no increase in birefringence. In this regime, other than establishment of three-dimensional order in crystalline regions, no other significant change is observed.

One of the unusual characteristics of PEN is that it shows necking upon stretching from the amorphous state above the glass transition temperature [37]. This was shown to be caused by a highly localized cooperative orientation of the naphthalene planes parallel to the surface of the film as shown in Figure 6.19.

This results in uneven thickness distribution that is detrimental to the appearance and the final properties of the product (Figure 6.20).

This behavior resembles an isotropic (random order) to nematic (oriented in parallel but not arranged in well-defined planes) structural transition, which occurs at highly localized regions of the sample, thereby manifesting itself as a neck. A rapid drop in the measured stretching force accompanies the initial neck formation. This suggests that when the naphthalene planes are aligned parallel to the broad surface of the sample, the interchain friction is greatly reduced as a result of this "graphite-like" orientation behavior of the naphthalene planes parallel to the broad surfaces of the film. To counteract this behavior, PEN can be blended with PEI in order to increase the friction

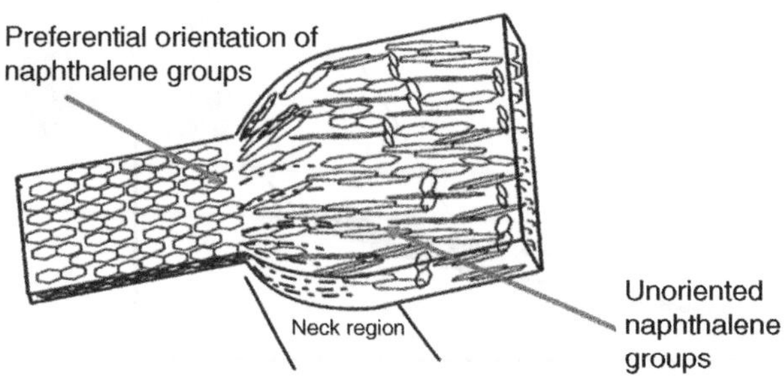


**Figure 6.19** Localized cooperative orientation of the naphthalene planes parallel to the surface of the film [37]. © John Wiley & Sons.

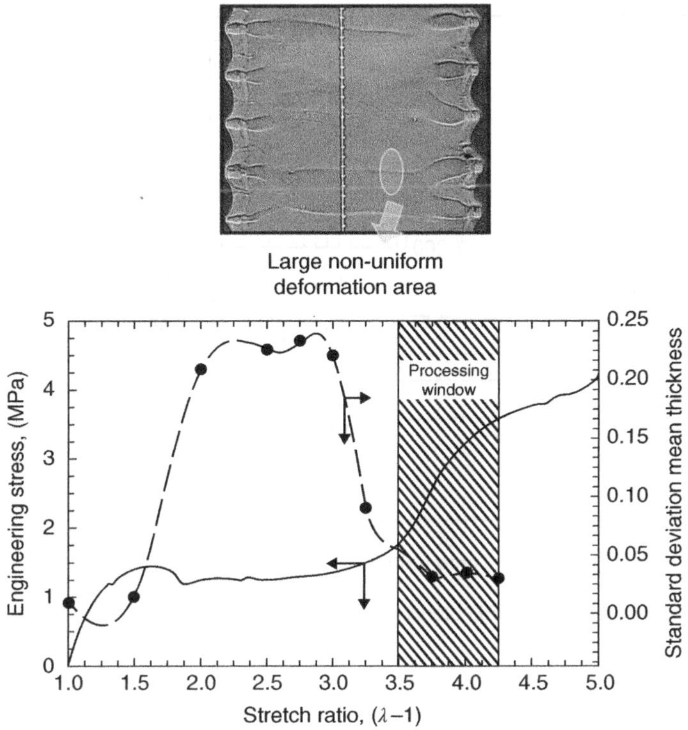


**Figure 6.20** Uneven thickness distribution [37] in PEN film. © John Wiley & Sons.

forces between the PEN chains by introducing stiff and bulky PEI chains into their environment. This molecular-level dispersion of PEI is possible as PEN and PEI exhibit miscibility in the amorphous state at all compositions. In film processing, this neck formation causes the processing window to be smaller by limiting the production to high draw ratios (uniaxial or biaxial) where films become more uniform in thickness once the necks in the films are eliminated. Kim et al. [37] studied the influence of PEI concentration on the neck formation of PEN/PEI films stretched above the $T_g$ from amorphous precursors. Thermal properties of PEN/PEI blends before drawing are shown in Figure 6.21. The glass transition temperature increases slightly with the addition of PEI. The melting temperature was observed to increase first and then decrease as the PEI fraction in the blend increased. The authors attributed this behavior to the relatively low melt temperature (310°C) used for the film casting process as compared with its equilibrium melting point of $T_m = 337$°C. PEN/PEI blends do not follow the normal melting point depression phenomenon until they are melted over 340°C. The presence of PEI increases the viscosity of the melt, which in turn helps the PEN crystals to survive in the form

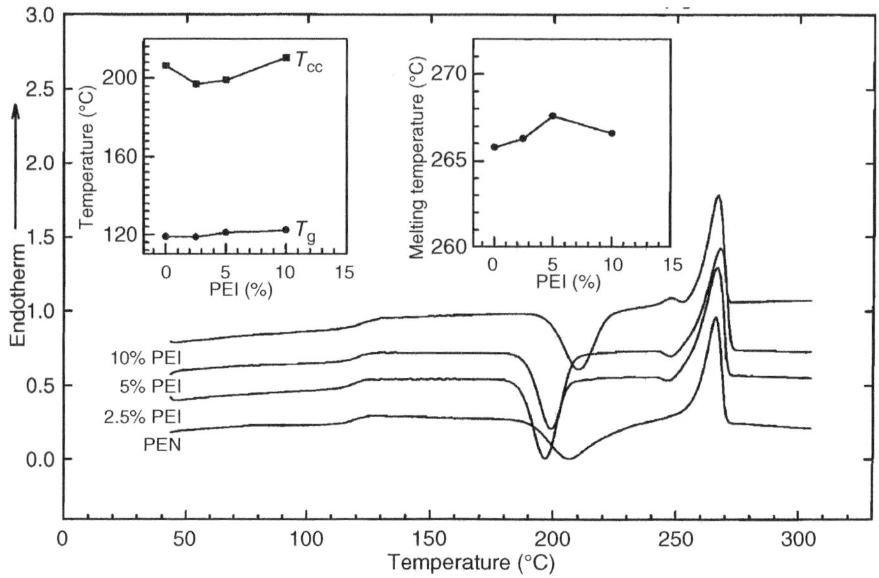


**Figure 6.21** Thermal properties of PEN/PEI blend films before drawing [37]. © John Wiley & Sons.

of small crystallites or frozen ordered state above the normal melting temperature. These remaining crystallites or the frozen ordered states in the quenched as cast films act as nuclei during the subsequent crystallization stage at lower temperatures. Because of this earlier start of crystallization, crystals grow bigger and more perfect, which in turn increases the melting point of the blend. Above 5% PEI concentration, another mechanism, dilution effect, starts to decrease the crystallizability of PEN similar to what is seen with PET/PEI blends as discussed earlier in this chapter. Above 20% PEI, PEN significantly loses its ability to crystallize.

When the films were drawn, the melting temperature of the blends decreased with the increase of the PEI fraction. The yield stress and the draw stress increased as shown in Figure 6.22, which is indicative of bulky and stiff PEI molecules, hindering the motion of the PEN molecules. With a PEI fraction greater than 10%, the necking behavior was eliminated as indicated by the slope of the stress–strain curve between the yield point and the draw point (see Figure 6.22).

A model for structure development in the neck region of PEN was proposed by the authors after detailed WAXD and birefringence studies as shown in Figure 6.23. Crystallization, which starts in the neck region, is eliminated with PEI addition.

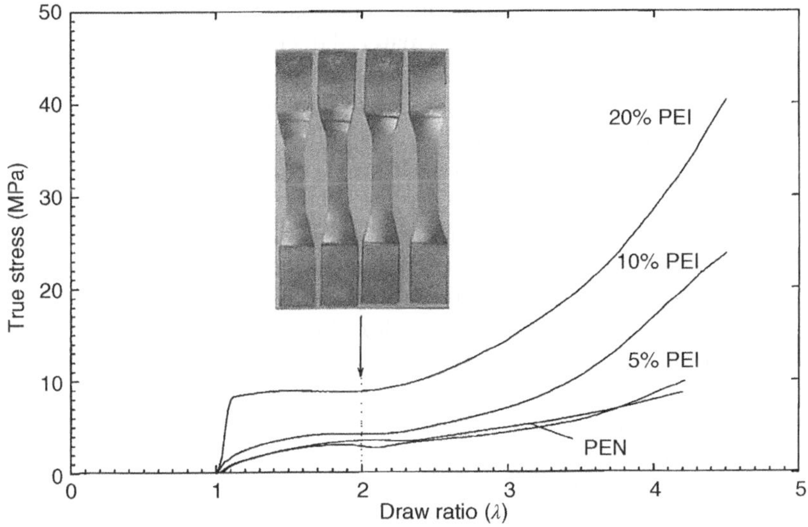


**Figure 6.22** Stress–strain curves of PEN/PEI films and film photographs stretched to draw ratio of 2. Draw temperature: $T_g + 20°C$. Draw rate: 1200%/min [37]. © John Wiley & Sons.

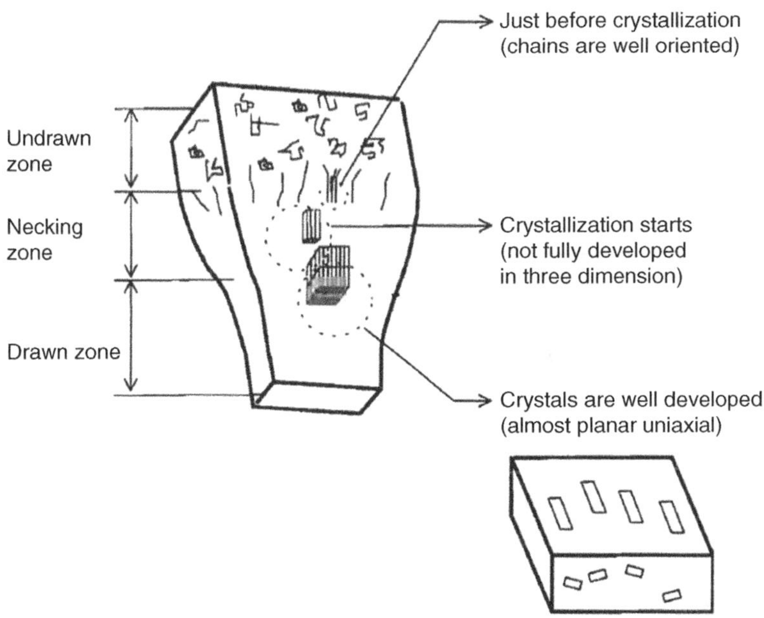


**Figure 6.23** Model structure development in the neck region of PEN/PEI blend film.

Although the addition of PEI into PEN eliminates necking, it has some disadvantages. PEI being an amorphous material decreases the crystallizability of PEN, rendering the material completely amorphous beyond 20% PEI content in the blend due to the dilution effect of PEI and increased viscosity. This absence of crystallinity in the material is detrimental to the processability and final product properties, specifically tensile strength and dimensional stability of the material.

In order to eliminate the neck formation that occurs during deformation in the rubbery state while maintaining the strain-induced crystallizability with increased glass transition temperature, Bicakci et al. [38] blended PEEK with PEN/PEI mixture and investigated the uniaxial deformation [39] of tertiary blends with compositions shown in Figure 6.24.

The influence of stretching temperature on the stress–strain behavior of selected PEN-rich ternary blends is shown in Figure 6.25. At $T_g + 10°C$, the stress–strain curve exhibits distinct yielding followed by stress drop, which accompanies the neck formation. At the later stages of deformation, the rapid rise in stress signifies rapid stress hardening, which occurs as a result of stress-induced crystallization. When the temperature is raised to $T_g + 20°C$, yielding

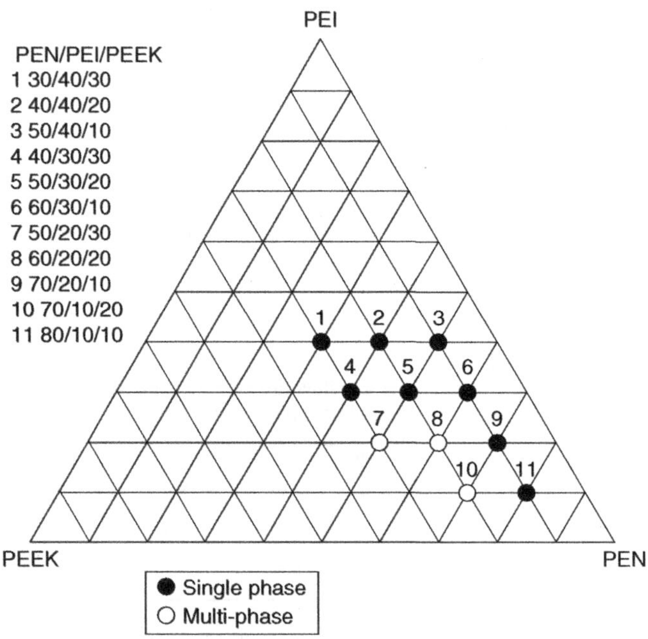


**Figure 6.24** Ternary blend (PEN/PEI/PEEK) compositions investigated [39]. © Elsevier.

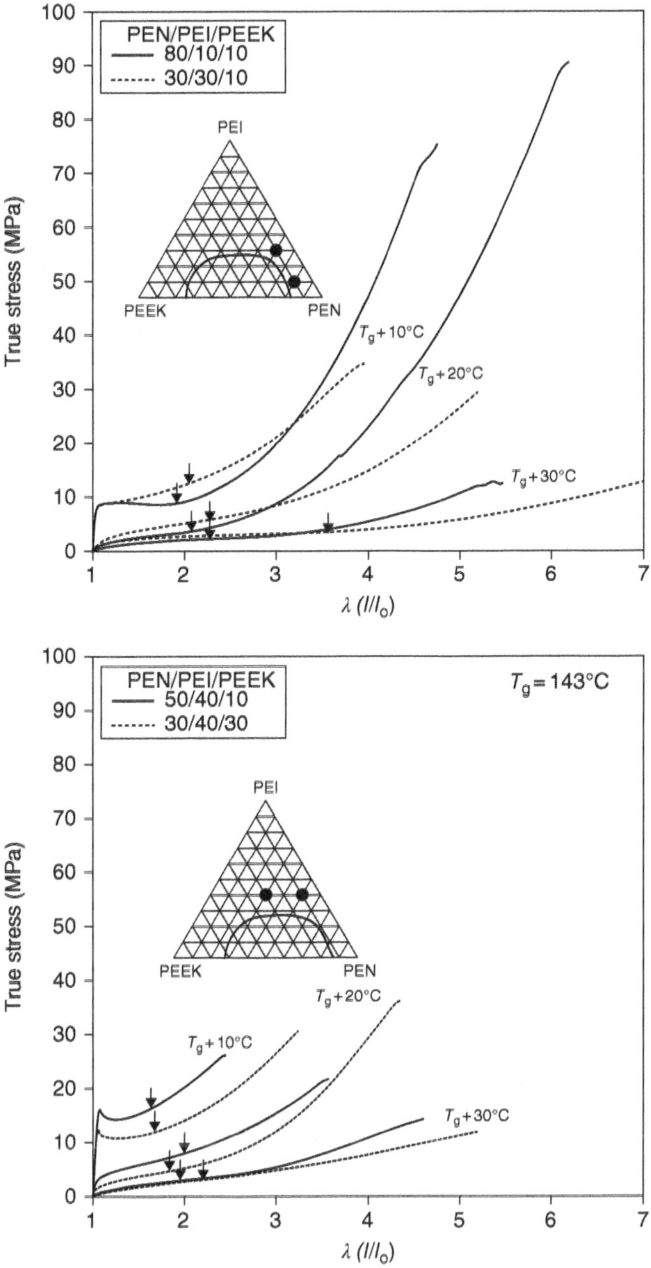


**Figure 6.25** Effect of temperature on the uniaxial deformation behavior of PEN/PEI/PEEK (80/0/10), (60/30/10) films (top) and (50/40/10), (30/40/30) films (bottom) stretched at 250 mm/min. The arrows indicate the onset of stress hardening [39]. © Elsevier.

followed by stress drop is essentially eliminated, and whole stress–strain curve resembles affinely deforming cross-linked rubber. Normally, PEN and PEN/PEI blends exhibit necking under these conditions. The critical strain at which the substantial stress hardening occurs is indicated by arrows on each figure. At $T_g + 30°C$, this increase occurs around 3.5–4 stretch ratio.

When the proportion of PEI is increased at the expense of PEN while keeping the proportion of the PEEK the same (Figure 6.25-top), there are distinct changes in the deformation behavior. One change is that the stress drops after the yield point that was observed in PEN and 80/10/10 PEN/PEI/PEEK films at $T_g + 10°C$ is eliminated, but the elongation at break is also reduced. However, at $T_g + 20°C$ and $T_g + 30°C$, ideal stress–strain curves are observed. The increased proportion of PEI leads to high stresses at low deformations as a result of increased friction between the polymer chains by the presence of stiffer and bulkier PEI chains. But this comes at the expense of reduced stresses at high strains since the stress-induced crystallization process is detrimentally influenced by the increased proportion of noncrystallizable PEI chains. This is clearly shown in Figure 6.25-left comparing stress–strain curves of 80/10/10 and 60/30/10 blends for $T_g + 20°C$ processing temperature. In order to counteract this reduction of stress at high strains, the authors increased the proportion of PEEK to 30%. This is shown in Figure 6.25-bottom at the optimum processing temperature of $T_g + 20°C$. The overall goal of increasing the $T_g$ while reducing the critical strain beyond which substantial stress hardening occurs was thus accomplished. The authors have shown that the addition of PEEK recovered the strain hardening and associated self-leveling behavior while boosting the glass transition temperature of the final films.

At high PEI concentrations of 40%, PEN crystallization was hindered. In these 40% PEI blends, although PEEK could be crystallized with high orientation levels at concentrations as low as 10%, PEN remained amorphous after stretching. The high crystalline orientation in PEEK regions was attributed to the disentanglement effect caused by the presence of PEI, which results in more efficient orientation and lateral packing of the chains in PEEK crystalline domains. Upon constrained annealing, both PEEK and PEN phase-separated from PEI and crystallized into their own domains. PEN crystallized into the unoriented state, while the PEEK crystallized further but with reduced orientation levels.

Another approach to eliminating necking was demonstrated by Kanuga et al. [24, 40] by incorporating small concentrations of clay nanoplatelets—0.5, 1, and 2%—to PEN. They have shown that nanoplatelets help prealign naphthalene planes during the film casting stage, hence avoiding necking during subsequent stretching as evidenced by the absence of stress drop after yielding in Figure 6.26. The authors showed that nanoplatelets help induce early onset of strain hardening (Figure 6.27-top), which aids in the development of films with more uniform thickness profiles at the end of stretching. Although the

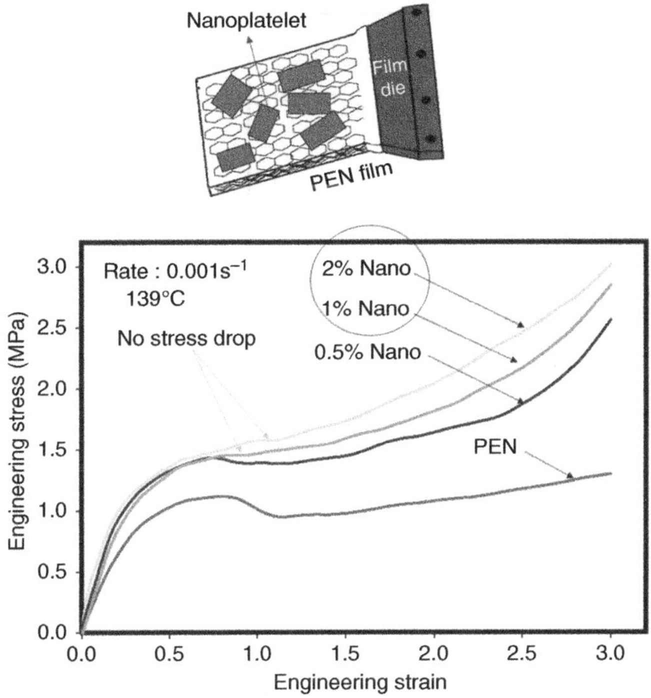


**Figure 6.26** Nanoplatelets and naphthalene moiety prealignment during casting (left) and engineering stress–strain curves of PEN with addition of 0, 0.5, 1, and 2% nanoplatelets (right) [24]. © American Chemical Society.

crystallinity levels were lower with addition of nanoparticles in the PEN films, crystallinity was still shown to develop considerably depending on the stretching rate and temperature (Figure 6.27-bottom).

In this study, the authors have mapped out the final structure and deformation behavior of the nanocomposites as a function of stretching temperature, rate, and nanoparticle content. One example at 2% nanoparticle content is shown in Figure 6.28.

They have identified three distinct types of structural ordering processes that can take place depending on the stretching temperature and rate. At low stretching temperatures and high rates, one can induce “nematic-like” order, which is devoid of significant chain translational order along the stretching direction. At intermediate temperatures and high enough strains, the final state of the films was dominated by orientation-induced crystallization. Finally, by stretching at high enough temperatures, one could induce thermal crystallization.

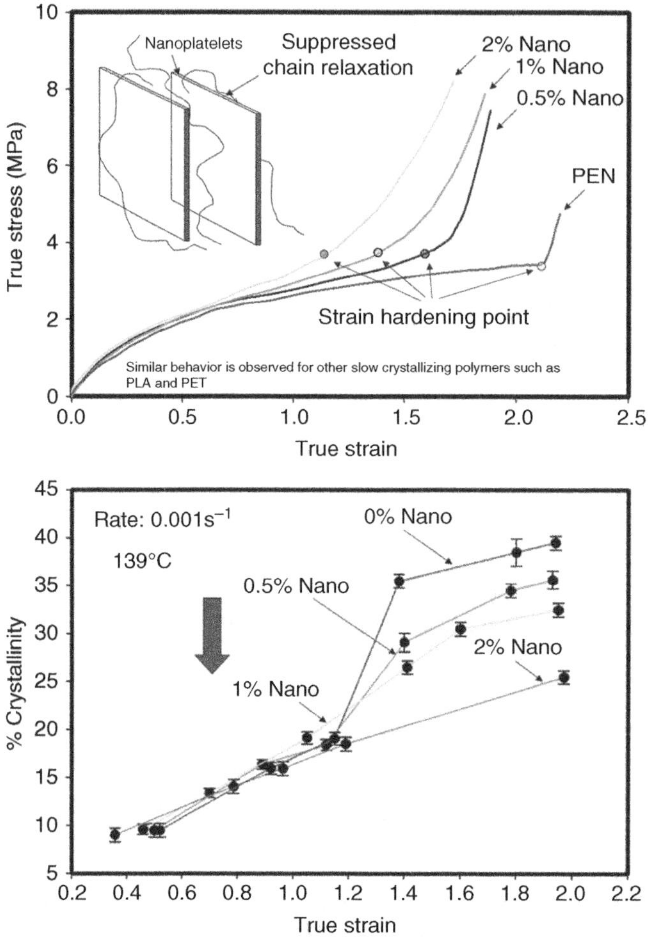


**Figure 6.27** True stress–true strain (top) and % crystallinity as a function of true strain for PEN with 0, 0.5, 1, and 2% nanoplatelets [24]. © American Chemical Society.

## 6.3 Category I: Solvent Cast High $T_g$ Materials

In this section, we will cover high $T_g$ materials that cannot be melt-processed and include aromatic fluorine-containing PAR (e.g., Ferranias Arylite®[1]), PCO (also known as PNB; e.g., Promerus's Appear® 19[2]), and PI (e.g., DuPont's Kapton®[3]).

1 Arylite, Ferrania S.p.A.,Centro Direzionale, Via Rivoltana, 2/d, Edificio A, I 20090 Segrate (MI), Italy.

2 Appear, Promerus LLC, 9921 Brecksville Road, Brecksville, Ohio 44141, USA.

3 Kapton, DuPont High Performance Films, P.O. Box 89, Route 23 South and DuPont Road, Circleville, OH 43113, USA.

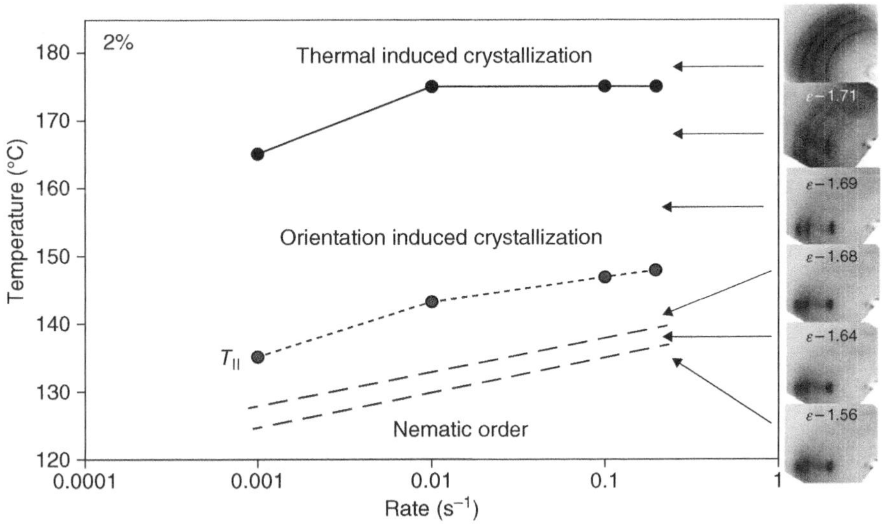


**Figure 6.28** Dynamic phase diagram for PEN with 2 wt % nanoparticle concentration [24]. © American Chemical Society.

### 6.3.1 Polyimides

PI are a class of thermally stable polymers that are often based on stiff aromatic backbones [14, 41]. Most PI films are produced via two-step process [42, 43]. First, a precursor called polyamic acid (PAA) is synthesized in a polar, high boiling point solvent such as *N*-methyl-2-pyrrolidone (NMP), dimethylacetamide (DMAc), and dimethylformamide (DMF). In the second step, PAA is deaerated and cast from the slit die in the form of a continuous film onto the surface of a heated rotating stainless steel drum and gelled by partially evaporating the solvent. This brings the film to a state of green film where it can be handled and lift off from the metal drum. Alternatively, green films can be formed by chemical means by passing the PAA films through a bath containing dehydrating agents such as acetic anhydride and dicyclohexylcarbodiimide. They are then lifted from the metal drum and biaxially stretched, first in the machine direction stretching followed by TD stretching. Finally, the stretched green film is converted to PI via thermal or chemical imidization [44]. During imidization, the film is held in the stretched state in the tenter frame where it is gripped at both transverse edges. Stretching of the PI films improves the stretchability of the final PI films. Without the stretching component, elongations at break of the fully imidized films are typically below 20%, while this value can be improved significantly when stretching is done in the process. For PAA synthesized from pyromellitic dianhydride (PMDA) and 4,4′-oxydianiline (ODA) monomers in NMP solvent, for example, Kapton®, imidization reaction starts at 120°C, and conversion is fastest at 150°C. At 200°C, 90% of the

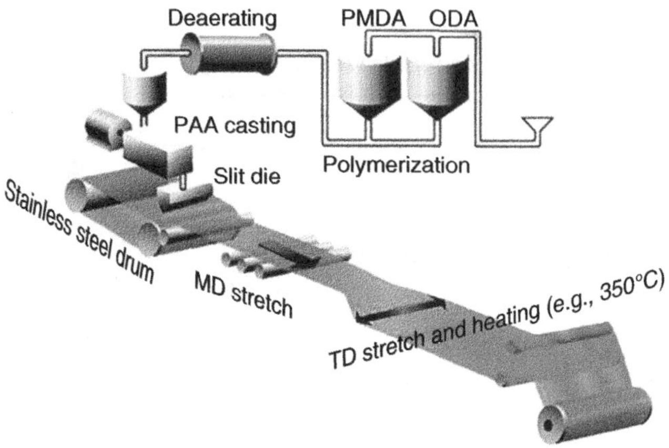


**Figure 6.29** Wholly aromatic polyimide industrial film production.

conversion is achieved, and the temperature needs to be increased from 300 to 450°C for full conversion and complete evaporation of the bound solvent. Industrial film production of wholly aromatic PI is shown in Figure 6.29. Wholly aromatic PI typically have an orange color, which is detrimental for optical transmission in display applications unless very thin films are produced. However, significant advances have been made to produce colorless PI. These colorless PI can be dissolved in dipolar aprotic solvents in their fully imidized form and therefore can be made into film form by a simple solution casting process, which takes place at much lower temperatures than 350°C. In addition, PI solution can still be formed into gel-like state by partial evaporation and stretching to improve physical properties [45].

Imidization reaction is a multistep complex process where solvent evaporation, chain orientation, chemical conversion, and evaporation of water byproduct all take place often concurrently. Because of this complexity, it is challenging to optimize temporal profile of thermomechanical treatment and predict the final properties of PI films as they are affected by these multiple process conditions. Low temperature drying of PAA solution results in the formation of PAA films with bound solvent content of 30–50% by weight. Model compound studies revealed that PAA can form complexes with the polar solvent molecules via hydrogen bonding with the ratios of 4 : 1 (solvent–PAA repeat unit) and 2 : 1 (solvent–PAA repeat unit) depending on the processing conditions. There are four possible hydrogen bonding sites on the repeating unit of the PAA chains—two carboxylic acid groups and two amide groups with two distinct thermal stabilities. When the temperature is increased for imidization reaction, the bound solvent starts to decomplex from the PAA chain, starting from the amide groups, due to lower bonding energy. During decomplexation, the unbounded solvent molecules release the active groups for reaction to occur and cause plasticization as they travel through the thickness to surface and eventually evaporate. The $T_g$ of the medium increases gradually due to evaporation of solvent and imidization reaction that forms more rigid imide groups relative to the amic acid groups. PAA films are generally constrained uniaxially or biaxially to prevent shrinkage during imidization, which causes development of internal stress and spontaneous molecular orientation [46, 47]. This orientation causes PI films to develop mechanical, thermal, and optical anisotropy in the plane or through the thickness direction, depending on the processing conditions and constraints imposed on the films.

In order to determine the details of the physical phenomena taking place during thermal imidization, Unsal et al. studied the solution cast and dried PAA thin films with high solvent content (30–50%) using a custom-built stretcher instrumented with an ultra-rapid-scan FT-IR (URS-FT-IR) equipment and spectral birefringence [30]. This integrated measurement system allowed them to measure the real-time temperature, thickness, mechanical behavior including true strain and true stress, in-plane birefringence and IR dichroism for selected peaks and chemical changes to understand the effects of evaporation of bound solvent and imidization reaction. Time-synchronized rheo-optical true stress and birefringence and URS-FT-IR data (% imidization and % bound solvent) are combined as shown in Figure 6.30, where critical regions are marked with gray-dashed lines and indicated with letters "a"

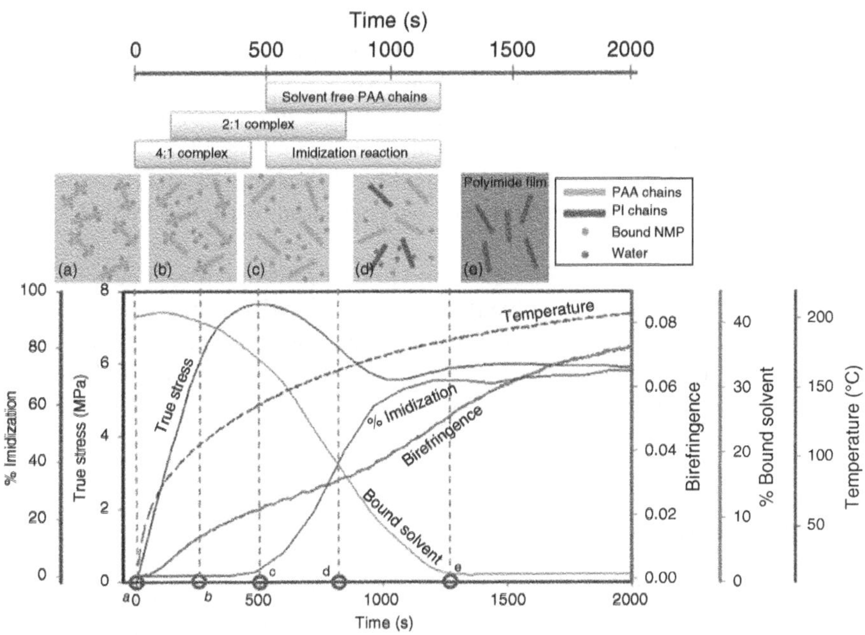


**Figure 6.30** Real-time molecular transformation during thermal imidization through combined mechanical, optical, and spectroscopic data [30]. © AIP Publishing LLC.

through "e." Initially, as-received polyamic acid film contains significant amount of bound solvent (4:1 solvent–PAA repeating unit). From "a" to "b" the temperature increases from room temperature to about 100°C in the first 250 s. At this stage, the bound solvent starts to decomplex from the PAA chains and gradually evaporates from the medium as indicated by the slight decrease in the bound solvent concentration. Since shrinking occurs as a result of solvent evaporation, this translates to rise in stress as the film is constrained uniaxially, and this increases the birefringence. At this stage, imidization does not occur as the temperature is too low to initiate it ($T < 120$°C), and there remains considerable amount of solvent in the film, capping the reactive ends (amide and carboxylic acid groups).

The true stress and birefringence continue to increase up until the onset of imidization (point "c"). As the temperature increases to approximately 130°C, more solvent molecules decomplexed from the PAA chains plasticize the environment and cause a stress relaxation and concurrent temporary reduction on the birefringence curve between points "c" and "d." Birefringence development during the imidization reaction is a combination of two factors: increase in intrinsic birefringence due to the imidization reaction and increase in stress. The reason that the birefringence still increases continuously relative to the stress after point "c" is the continuous increase in the intrinsic birefringence as the imide structure has higher intrinsic birefringence. After point "c," the temperature approaches 150°C where the imidization reaction is fastest. The imidization reaction takes off as the evaporation of solvent accelerates. At point "d," the imidization and bound solvent curves intersect, which indicates the two events take place simultaneously. After point "d," the birefringence increases at a higher rate, due to formation of rigid PI chains and evaporation of solvent, causing increase in $T_g$ and formation of more rigid imide chains, and this also may lead to increase in intrinsic birefringence. PI chains expected to have higher intrinsic birefringence as compared with the PMDA/ODA precursors.

At point "e," true stress, imidization, and bound solvent curves begin to level off. At this point only a 2–3% solvent remains, and imidization also levels off at 80% conversion. The birefringence increases at a slower rate after this point and reaches a plateau region around 2500 s. Higher temperatures (~300–350°C) are required for complete chemical conversion and removal of the remaining bound solvent from the medium.

## 6.4 Summary

We now summarize how well category 1 (amorphous, high-$T_g$ materials) and 2 (polyesters) plastic substrates satisfy the needs of the flexible substrate industry and point out some of their shortcomings.

### 6.4.1 Transparency

All the polyesters discussed earlier such as PET and PEN and their blends satisfy the optical clarity requirements for bottom emissive displays and a TLT of 85% over 400–800 nm coupled with a haze of less than 0.7%. In order for these films to possess measurable levels of haze, the crystals developed during processing should be larger than the wavelength of visible light. Stress-induced crystals of these polymers are typically smaller than the wavelength of visible light and therefore do not scatter light. Precaution needs to be taken during the constrained annealing phase, which could lead to development of larger crystals with time. The polyester blends discussed earlier are completely miscible blends and are therefore transparent. Immiscible blends with phase-separated morphology where the discontinuous phase is larger than the wavelength of visible light could generate haze. In the case of nanoparticle-filled polyester films, similar to domain and crystal size, nanoparticles should be smaller than wavelength of visible light but also have similar refractive index as the plastic phase in order not to scatter light and cause haze development. Another point to keep in mind is that biaxially oriented films of PET and PEN and their blends are birefringent. When used as substrates for liquid crystal display (LCD) applications, they could change the polarization state when used in the front panel and are therefore disadvantaged, but they can be used as substrates in the backlight unit. Birefringence of these films would not be an issue with OLED displays.

Wholly aromatic PI typically have an orange tint originating from the formation of charge transfer complexes (CTC) in their highly conjugated molecular structures [48] and therefore suffer from poor optical properties. However, colorless PI can be produced by inserting fluorine, sulfone, or nonaromatic groups [49]. Commercial colorless PI films such as those from DuPont Teijin, Toray, Mitsubishi, Kolon, and Toyobo exhibit transparencies (87–91%) that surpass 85%.

### 6.4.2 Thermal Properties

The ability to limit if not predict changes in plastic substrate dimensions with temperature is critical in many flexible device manufacturing operations. Thermal and dimensional stability characteristics of plastics need to be taken into account to ensure their robustness in multilayered flexible devices that are often subject to thermal cycling during both their manufacturing and end use. Substrates go through a series of heating and cooling steps during device manufacturing (printing, coating, deposition, etc.). Precise registration of these multilayers highly depends on the substrate on which they are deposited. As discussed earlier, polyester films such as PET and PEN and their various blends have glass transition temperatures ($T_g$) well below the necessary level to impart the thermal and dimensional stability to support flexible display manufacture, for example, flexible OLED display, and other microelectronic devices comprising backplanes and TFTs. Temperatures above the $T_g$ could relax and shrink the oriented amorphous phase of these polymers. However, the crystals that develop during biaxial stretching and heat setting act as anchor points that hold the amorphous phase and hence the entire structure together. In addition, the oriented amorphous phase imparted during stretching relaxes during heat setting and is less prone to shrinkage later on. In this way, $T_g$ effects described earlier are essentially eliminated, and the polyester film remains dimensionally stable even at temperatures higher than $T_g$. For example, heat-set PEN film can remain dimensionally stable up to 200°C, which is within the performance requirements of a flexible substrate for an OLED display.

When designing substrates with well-matched expansion and shrinkage between substrate and other layers, low shrinkage and low coefficient of linear thermal expansion (CLTE) < 20 ppm/°C are desired. When there is a large mismatch in thermal expansion, deposited layers, especially inorganic barrier layers, can crack under thermal cycling. Annealed biaxially stretched PEN film exhibits minimal shrinkage typically less than 0.05% when exposed to temperatures of up to 180°C for 5 min. PET film exhibits minimal shrinkage up to approximately 150°C. Figure 6.31 compares the thermomechanical behavior of 125 μm PEN film and its annealed (heat-set, heat-stabilized) counterpart by thermal cycling at 8–140, 8–160, 8–180, and 8–220°C. As seen in the hysteresis curves, heat-set PEN follows the same expansion and contraction path as the

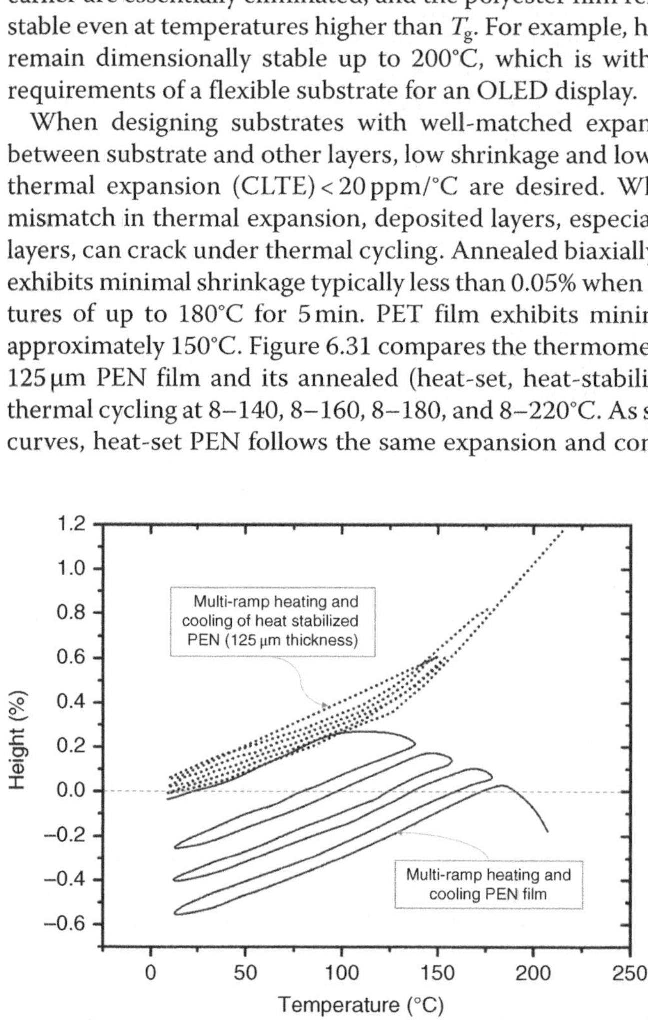


**Figure 6.31** Thermal mechanical analysis of 125 μm thick non-heat-stabilized and heat-stabilized PEN films. *Source:* Data digitized and replotted from Ref. [5].

sample is heated and cooled from 8°C and all the way up to 200°C, whereas PEN that is not exposed to heat setting remains dimensionally stable only up to 120°C above which significant shrinkage is observed.

After annealing, the expansion and shrinkage behavior of PEN and PET are predictable through known CLTE and shrinkage values at the temperature of interest. The degree of CLTE and shrinkage is a function of use temperature and depends on material structural characteristics such as degree of orientation and mode of orientation as well as degree of crystallinity and crystalline structure built in the polymer film after stretching and heat setting. Table 6.1 shows CLTE for biaxially stretched and annealed PEN. One can notice the CLTE is different in the MD and TD direction due to differences in orientation.

Figure 6.32 shows shrinkage of sequential biaxially stretched and heat-stabilized 3 mil PET (Melinex) and 5 mil PEN (Teonex) as a function of temperature. For applications requiring low shrinkage above 140°C, heat-set and biaxially stretched PEN is suitable, whereas PET shrinkage is low only up to about

**Table 6.1** Coefficient of linear thermal expansion (CLTE) for annealed PEN.

| CLTE (PPM) (°C) | −50 to 0 | 0–50 | 50–100 | 100–150 |
|---|---|---|---|---|
| Machine direction | 13 | 16 | 18 | 25 |
| Transverse direction | 8 | 11 | 18 | 29 |

*Source:* Data obtained from Ref. [5].

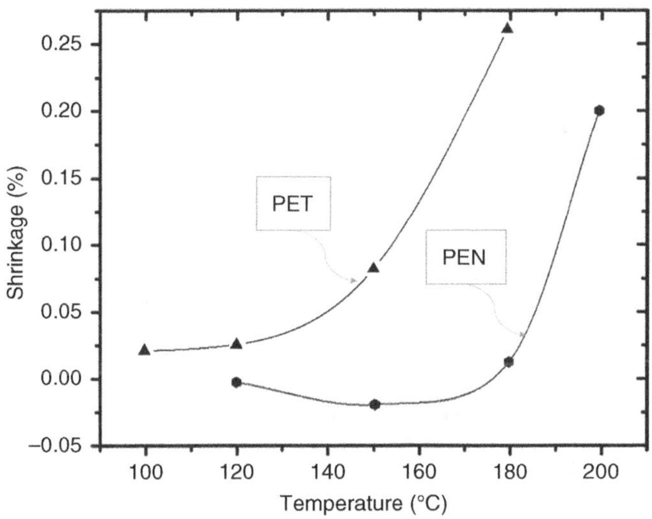


**Figure 6.32** Shrinkage of sequential biaxially stretched and heat-stabilized 3 mil PET (Melinex®) and 5 mil PEN (Teonex®). *Source:* Data obtained from Ref. [5] and redrawn.

120°C. PEN shrinkages of 25 ppm at 150°C during TFT backplane processing after an annealing at 200°C have been reported [50].

Upper processing temperature is another important factor to be considered for flexible substrates such as PET- and PEN-based substrates. Once heat-set, biaxially stretched polyester films remain intact almost up to the onset of the melting range.

PI materials due to their high $T_g$'s, mostly around 330°C and some even above this value, exhibit excellent high temperature resistance when used as a clear substrate. For example, a clear PI substrate from DuPont that has a $T_g$ value of greater than 315°C and CLTE value below 10 ppm/°C was deposited with a-Si TFTs at 250–280°C, showing that such plastic films are getting close to temperatures of 300 and 350°C [51] used in a-Si TFT production on glass substrates.

### 6.4.3 Barrier to Moisture and Gases and Planarization Requirements

It is well known that polyester films are hygroscopic and when used as a substrate for a flexible opto-electronic device, encapsulation is needed for long operating lifetime in order to avoid degradation of the devices by moisture and oxygen. Different types of opto-electronic devices require different level of barrier protection. Oxygen water vapor transmission rate (WVTR) requirements for various opto-electronic applications are shown in Figure 6.33. Devices that use chemically reactive electrodes and active layers such as OLEDs and solar cells are considerably more sensitive than LCDs, RFIDs, and other electrochromic displays. PET films of 50 μm thickness have a WVTR of about 10 g/($m^2$·day) and an oxygen transmission rate (OTR) of 30 $cm^3$/($m^2$·day·bar). By comparison, efficient protection of LCD displays require $10^{-2}$ g/($m^2$·day) for both oxygen and moisture, while OLEDs require a WVTR below $10^{-5}$ g/($m^2$·day) and an OTR below $10^{-5}$ ml/($m^2$·day·bar) [52, 53].

Typical oxygen and water vapor barrier properties of significant polymers are given in Figure 6.34.

These barrier properties can be further enhanced by special coating technologies. These are applied to commonly available films such as PET, PEN, and

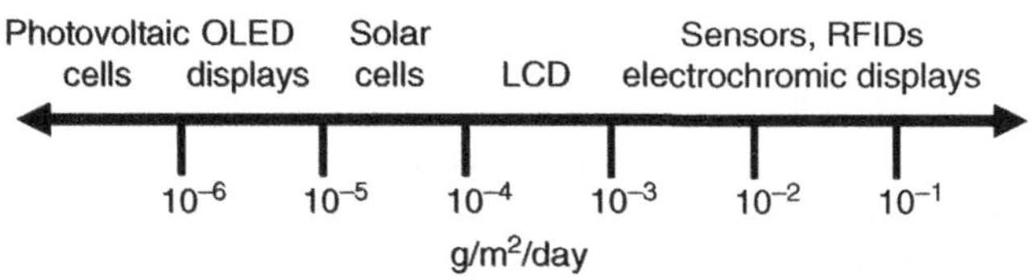


**Figure 6.33** Water permeability requirements for various opto-electronic applications (oxygen permeability requirements are close to water permeability numbers).

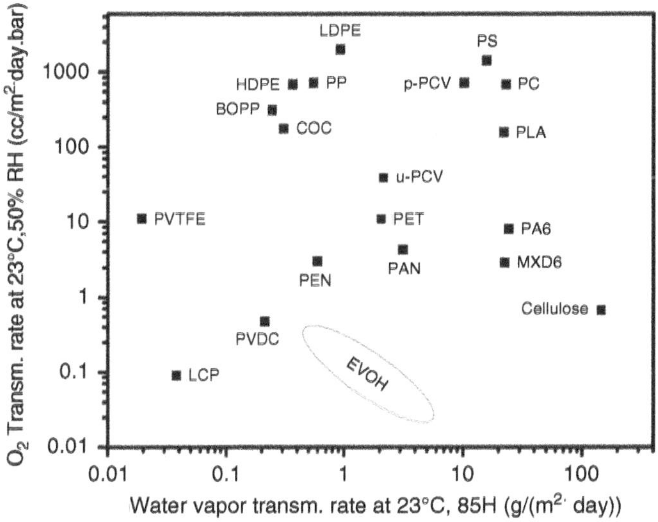


**Figure 6.34** Typical oxygen and water vapor transmission properties of various polymers.

PI substrates. Organic barrier layers include spin-coated cellulose acetate (CA) [54], polyvinylidene chloride (PVDC) [54, 55], or EVOH layers [56] on polyesters. Metallic barriers include thermally evaporated [57, 58] or sputtered metal films, usually aluminum. Ceramic barrier layers are inorganic metal oxides such as $SiO_x$ or $Al_2O_3$, which can be deposited by physical vapor deposition (PVD) [59], chemical vapor deposition (CVD) [60–62], and atomic layer deposition (ALD) techniques. Although barrier technologies are diverse, the choice for opto-electronic devices is limited, especially for OLEDs and solar cells, due to stringent barrier requirements that cannot be met by organic barrier layers such as CA, PVDC, PAN, and EVOH. Another requirement for opto-electronic devices is that the technology used to encapsulate the device should not adversely affect optical transparency at least for one side. This eliminates films coated by metal evaporation. Ceramic barrier layers comprising thin layers of an inorganic film such as $SiO_x$, SiN, or $Al_2O_3$ deposited on the polyester substrates provide the most promising solution in terms of performance and transparency. PVD or CVD techniques were both implemented in an R2R operation to coat polyesters with transparent ceramic layers. Simplified schematics of R2R PVD and CVD processes are shown in Figure 6.35. R2R process typically takes place under vacuum (<1–30 mbar depending on technique). In the PVD process, a deposition source is either evaporated thermally or sputtered by a plasma whereby ions from the source deposit on the web. In CVD process, the web is deposited with a reaction product (e.g., $SiO_x$) of precursor supply gases (e.g., HMDSO, $O_2$), which are fragmented and reacted

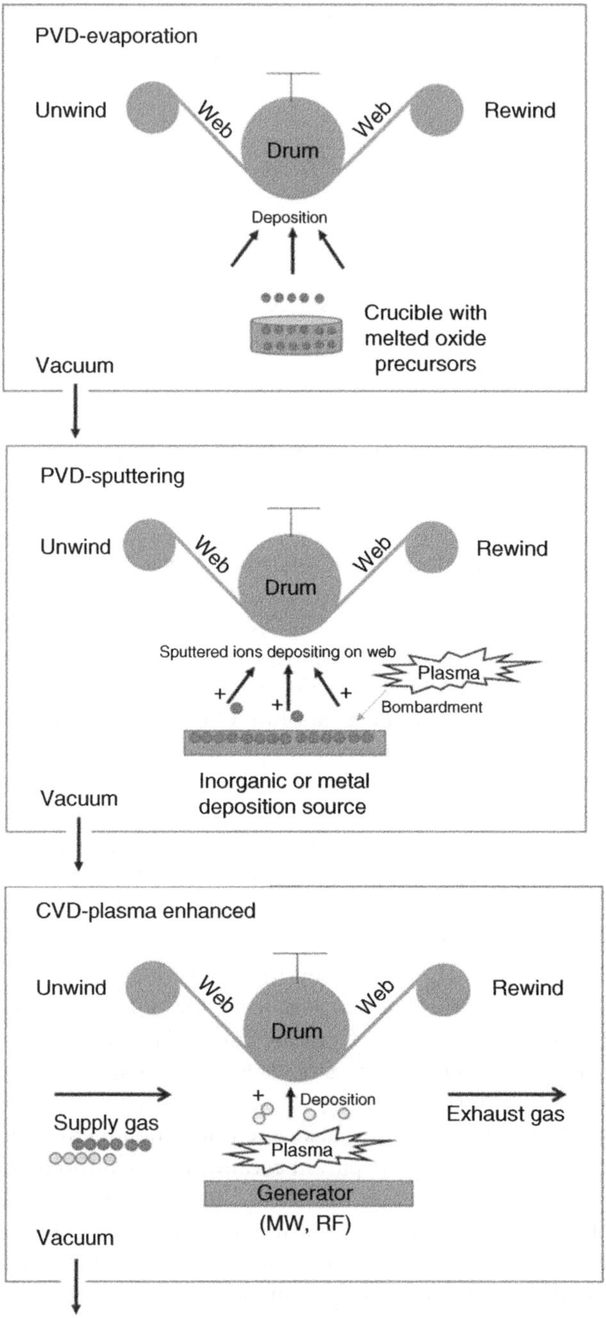


**Figure 6.35** Simplified schematics of R2R PVD or CVD process.

with a high energy plasma environment created by a high voltage RF or microwave generator. More specifically, this process is called plasma-enhanced chemical vapor deposition (PECVD).

Barrier performance of inorganic oxide coatings such as silica on plastic substrates depends not only on the intrinsic coating quality (number of defects, thickness) but also on the adhesion of the coating to the substrate layer. Permeation mechanism in oxide coating layers is governed by defects rather than individual layer permeability and thickness ideal laminate theory (ILT). There are two kinds of coating defects in thin oxide films, micro-defects, and nano-defects. Microdefects can manifest themselves in the form of pinholes that are uncoated regions due to surface roughness or foreign objects or cracks arising from internal stresses due to inherent mismatch of thermomechanical properties with dissimilar materials, for example, PET and $SiO_x$. The latter also influences adhesion and delamination of the coating from the substrate. Nano-defects, on the other hand, result from nonequilibrium thermodynamic nature of the PECVD process and resultant quenched disordered state in the oxide coating. Roberts et al. [63] proposed that nano-defects can result from boundaries between grain-like structures of the $SiO_x$ coatings.

Several studies were published to show barrier improvement via eliminating defects or improving adhesion. Multi-layered (ML) and graded barrier structures were shown to be the most promising solutions to eliminate defects and improve interfacial adhesion. Interfacial adhesion and flexibility of the coating is especially important for flexible and bendable devices. An ML approach can be used to improve adhesion by providing a layer that has a higher chemical affinity to the polymeric substrate before applying the actual barrier layer. Thickness of the coating also influences the coating adhesion and integrity. Although thicker coatings provide more coverage on pinholes thereby improving barrier properties, they are also more prone to delamination and cracks. Therefore, there is a critical coating thickness above which the flexibility of the coating suffers and below which a large number of pinhole defects exist. ML coating approach with alternating inorganic ceramic layers and organic layers, called dyads, were shown to exhibit ultra-barrier close to the requirements of OLEDs. In such an arrangement, first organic layer provide adhesion to the substrate and the subsequent ones provide planarization and smoothing layers that cover pinhole defects and provide a tortuous path for $O_2$ and $H_2O$ migration. GE researchers demonstrated this approach in 1996 with what they called the polymer multilayer (PML) process, which involved vacuum flash evaporation of acrylic monomers with conventional oxide sputtering and evaporation processes to produce ML barrier coatings in a vacuum web coating process [64]. Another example of such a barrier film is Barix™ film provided by Vitex Systems and used by Samsung. The film consists of 5–6 alternating 30–100 nm $Al_2O_3$ ceramic layers between 250 nm thick UV-cured acrylate layers on a PEN substrate [52]. In these structures, $Al_2O_3$ ceramic layers were deposited by DC reactive sputtering, while the organic layer was deposited by first depositing precursor monomer followed by UV curing. The films had WVTR below in the range of $10^{-5}$ g/(m$^2$·day).

The Fraunhofer POLO alliance demonstrated the synergistic combination of sputtered oxide layers with inorganic organic hybrid polymers (ORMOCERS®) for an improved barrier foil that can be used to encapsulate several types of flexible electronic devices [65]. The ultra-barrier performance is achieved by covalent bonding between the oxide and the ORMOCER® layer.

A further method to deposit ML $SiO_x$ coatings with tailored carbon and hydrogen content on PET and PC containers was demonstrated by Walther et al. of Schott Glaswerke company [66]. They tailored the carbon and hydrogen content in the $SiO_x$ coating by manipulating the high energy pulse duration during PECVD process. Pulsed PECVD is required to minimize the thermal load on plastic substrates during deposition. With shorter pulses below 1 ms, they have shown that organometallic precursor molecules are only partially cracked and results in $SiO_xC_yH_z$ or $TiO_xC_yH_z$. With higher energy and longer pulses, precursor molecules fragment to a greater extent, yielding $SiO_x$ layer. Other researchers have shown organometallic precursor to gas, for example, $O_2$, modulation to create graded structures during PECVD deposition [67, 68]. ALD is a subclass of CVD process based on the sequential use of a gas-phase chemical process. DuPont demonstrated permeation barrier performance of WVTR values as low as $10^{-7}$ g/(m$^2$·day) from $Al_2O_3$ grown on PEN at a thickness of 25 nm [69]. $SiN_x$ coatings with Cat-CVD deposition on PET and COC were found to exhibit moisture barrier property (0.0045 g/(m$^2$·day)) [70].

The substrates for R2R manufacturing of electronic devices are required to be free of asperities caused by manufacturing defects such as belt marks, dusts, and scratches to prevent device defects such as shorts. For this purpose, materials used as substrates PET, PEN, and PI are coated with "planarization layer" to cover over these defects to obtain functionally smooth surfaces. For example, HP developed proprietary UV-curable oligomer technology [71] for PI coating that significantly improved the surface quality of the PI. There are a number of factors that must be kept in mind when selecting appropriate chemical architecture as planarization material. It must flow readily (low viscosity), form uniform film, and adhere well to the substrate that is being smoothed and show good adhesion to the functional layers that are deposited on it and withstand the chemical or thermal conditions the subsequent deposition techniques impose on it. HP's technology appear to have solved the problem for PI as shown in the SEM images of PI planarization coating figure of their article [71]. There are a variety of other planarization technologies mostly mentioned in patent literature including polymer or hybrid polymer inorganic nanoparticle coatings [72], UV-curable polymers [73], thermally cured prepolymers and curing [74], silicon oxide derivatives by PECVD or CVD process [75, 76], and diamond coating [77].

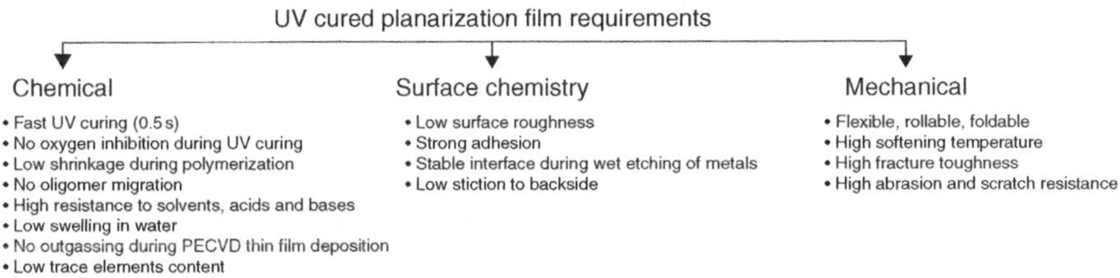


**Figure 6.36** UV-cured planarization film requirements.

Almanza et al. classified additional requirements for UV-cured planarized films. This is shown in Figure 6.36.

Clearly UV-curable resins have a number of advantages in this application. They start as low viscosity fluids, which allow them to go into cracks and crevices in the original carrier film. Having generally fast kinetics, they rapidly cure into solid form without having to worry about solvent entrapment typical of solution coatings. For thermoplastic polyesters, PET, PEN, with the use of high temperature processing during deposition of functional layers, the surface roughens due to migration of smaller molecules (e.g., oligomers) to the surface [78, 79]. Planarization layers used to coat these films tend to prevent such roughening process by acting as barriers to these smaller molecules.

## References


1 Weber, A.; Deutschbein, S.; Plichta, A.; Habeck, A. Thin glass polymer systems as flexible substrates for displays. *SID Dig.* **2002**, 33, 53–55.

2 Horvatitsch, T. On the Way to Super Strong Glass. The Schott Technology magazine. SCHOTT solutions > Research and Technology (Online). **2013**, no 1. http://www.us.schott.com/magazine/english/sol113/sol113_08_strong-glass.html (Accessed on February 28, 2017).

3 Shin, H.S.; Koo, J.B.; Jeong, J.K.; Mo, Y.G.; Chung, H.K.; Cheon, J.H. et al. 4.1 inch top-emission AMOLED on flexible metal foil. *SID Dig.* **2005**, 36, 1642–1645.

4 Crawford, G.P. *Flexible flat panel display technology*. New York: John Wiley & Sons, Inc., **2005**.

5 MacDonald, W.A. Engineered films for display technologies. *J. Mater. Chem.* **2004**, 14, 4–10.

6 Choi, M.C.; Kim, Y.; Ha, C.-S. Polymers for flexible displays: From material selection to device applications. *Prog. Polym. Sci.* **2008**, 33, 581–630.

7 Yan, M.; Kim, T.W.; Erlat, A.G.; Pellow, M.; Foust, D.F.; Liu, J.; Schaepkens, M.; Heller, C.M.; McConnelee, P.A.; Feist, T.P.; Duggal, A.R. A transparent, high barrier, and high heat substrate for organic electronics. *Proc. IEEE* **2005**, 93, 1468–1477.

8 Loan, S. *Functionalized polysulfones: Synthesis, characterization, and applications*. Boca Raton: CRC Press, **2015**.
9 Jang, J.; Han, S.H. High-performance OTFTs on flexible substrate. *SID Dig.* **2005**, 36, 10–13.
10 Choi, Y.-K.; Yarimaga, O.; Kim, T.W.; Jung, Y.K.; Park, H.G. Flexible display device with organic composite film. *IMID/IDMC/Asia Disp.* **2008**, 8, 1233–1236.
11 Angiolini, S.; Avidano, M. Polyarylate films for optical applications with improved UV-resistance. *SID Dig.* **2001**, 32, 651–653.
12 Choi, M.-C.; Hwang, J.-C.; Kim, C.; Ando, S.; Ha, C.-S. New colorless substrates based on polynorbornene-chlorinated polyimide copolymers and their application for flexible displays. *J Polym. Sci. A Polym. Chem.* **2010**, 48, 1806–1814.
13 Choi, M.-C.; Nagappan, S.; Lee, W.-K.; Han, M.; Ha, C.-S. Polynorbornene/fluorosilica hybrids for hydrophobic and oleophobic coatings. *Polym. Bull.* **2013**, 70, 619–630.
14 Ghosh, M.K.; Mittal, K.L. *Polyimides: Fundamentals and applications.* New York: Marcel Dekker; **1996**. p. 891.
15 Keinath, S.E.; Miller, R.L.; Reike, J. *Order in the amorphous state of polymers.* New York: Plenum Press; **1987**.
16 Lobanov, A.; Frenkel, S.Y. The nature of the so-called "liquid-liquid" transition in polymer melts. A review. *Polym. Sci. USSR* **1980**, 22, 1150–1163.
17 Warner, C.M.; Boyer, R.I. $T_{ll}$ from differential scanning calorimeter (DSC) cooling traces for atactic polystyrene. *J. Polym. Sci. B Polym. Phys.* **1992**, 30, 1177–1181.
18 Stadnicki, S.J.; Gillham, J.K.; Boyer, R.F. The $T_{ll}$ (>Tg) transition of atactic polystyrene. *J. Appl. Polym. Sci.* **1976**, 20, 1245–1275.
19 Kanuga, K.D. A Fundamental Study on the Non-Linear Mechano-Optical Behavior of Polyethylene Naphthalate, Its Blends with Polyetherimide and Its Nanocomposites. Ph.D. Dissertation, University of Akron, Akron, OH, **2006**.
20 Cakmak, M.; Hassan, M.; Unsal, E.; Martins, C. A fast real time measurement system to track in and out of plane optical retardation/birefringence, true stress, and true strain during biaxial stretching of polymer films. *Rev. Sci. Instrum.* **2012**, 83(12), 123901/1–123901/10.
21 Jiang, F.; Cakmak, M. Real time optical and mechano-optical studies during drying and uniaxial stretching of polyetherimide films from solution. *Polymer* **2015**, 68, 168–175.
22 Wang, E.; Batra, S.; Cakmak, M. A real time study on drying and the mechano-optical behavior of polyvinyl alcohol films in solid and swollen state. *Polymer* **2015**, 67, 200–207.
23 Offenbach, I.; Gupta, S.; Chung, T.C.M.; Weiss, R.A.; Cakmak, M. Real-time infrared-mechano-optical behavior and structural evolution of polypropylene and hydroxyl-functionalized polypropylene during uniaxial deformation. *Macromolecules* **2015**, 48(17), 6294–6305.

**24** Kanuga, K.; Cakmak, M. Role of nanoparticles and relaxation on strain-induced crystallization behavior in uniaxially stretched polyethylene naphthalate films: A mechano-optical study. *Macromolecules* **2013**, 46(15), 6300–6308.

**25** Hassan, M.K.; Cakmak, M. Strain induced crystallization during relaxation following biaxial stretching of PET films: A real time mechano optical study. *Macromolecules* **2015**, 48(13), 4657–4668.

**26** Hassan, M.K.; Cakmak, M. Mechanisms of structural organizational processes as revealed by real time mechano optical behavior of PET film during sequential biaxial stretching. *Polymer* **2014**, 55(20), 5245–5254.

**27** Hassan, M.K.; Cakmak, M. Mechano optical behavior of polyethylene terephthalate films during simultaneous biaxial stretching: Real time measurements with an instrumented system. *Polymer* **2013**, 54(23), 6463–6470.

**28** Choi, J.; Serhatkulu, T.; Cakmak, M. Temporal evolution of structure in uniaxially stretched PET/PEI blends during constrained annealing: A real time birefringence study. *Polymer* **2016**, 97, 569–579.

**29** Hassan, M.K.; Cakmak, M. Rapid structural changes that take place during constrained annealing of biaxially stretched pet films. *Polym. Eng. Sci.* **2016**, 57, 550–562.

**30** Unsal, E.; Nugay, I.I.; Offenbach, I.; Gross, M.; Manning, C.; Cakmak, M. Rapid integrated rheo-optical and polarized Fourier-transform infrared spectrometry measurement system for polymer films undergoing chemo-mechanical changes. *Rev. Sci. Instrum.* **2013**, 84(7), 073901/1–073901/10.

**31** Daver, F.; Blake, A.; Cakmak, M. Stages of structural ordering leading to stress induced crystallization of PEEK films: A mechano-optical study on deformation, Relaxation and Retraction. *Macromolecules* **2009**, 42(7), 2626–2633.

**32** Martins, C.; Cakmak, M. Role of relaxation on strain crystallization of uniaxially stretched poly(ethylene naphthalate) films: A mechanooptical study. *Macromolecules* **2006**, 39(14), 4824–4833.

**33** Martins, C.I.; Cakmak, M. Control of the strain-induced crystallization of polyethylene terephthalate by temporally varying deformation rates: A mechano-optical study. *Polymer* **2007**, 48(7), 2109–2123.

**34** Wang, Y.D.; Cakmak, M. The effect of biaxial orientation and crystallinity on the long-term creep behavior of PET films below glass transition temperature. *J. Appl. Polym. Sci.* **1990**, 41, 1867–1890.

**35** Choi, J.; Cakmak, M. Morphological evolution during thermal and strain induced crystallization in poly (ethylene terephthalate)/poly (ether imide) blend films. *Polymer* **2016**, 84, 10–20.

**36** Martins, C.I.; Cakmak, M. Large deformation mechano-optical and dynamical phase behavior in uniaxially stretched poly(ethylene naphthalate). *Macromolecules* **2005**, 38(10), 4260–4273.

37 Cakmak, M.; Kim, J.C. Necking mechanism and its elimination in uniaxially drawn films of poly(ethylene naphthalate) (PEN)/polyetherimide (PEI) blends. *J. Appl. Polym. Sci.* **1997**, 65(11), 2059–2074.

38 Bicakci, S.; Cakmak, M. Phase behavior of ternary blends of poly(ethylene naphthalate), poly(ether imide) and poly(ether ether ketone). *Polymer* **1998**, 39(17), 4001–4010.

39 Bicakci, S.; Cakmak, M. Uniaxial deformation behavior of ternary blends of poly(ethylene naphthalate), poly(ether imide) and poly(ether ether ketone) films. *Polymer* **1998**, 39, 5405–5420.

40 Kanuga, K.; Cakmak, M. Dynamic phase diagram derived from large deformation non-linear mechano-optical behavior of polyethylene naphthalate nanocomposites. *Polymer* **2007**, 48, 7176–7192.

41 St Clair, A.K.; St Clair, T.L. Process for preparing highly optically transparent/colorless aromatic polyimide film. U.S. Patent 4,603,061, July 29, **1986**.

42 Walter, M.E. Polyamide-acids, compositions thereof, and process for their preparation. U.S. Patent 3,179,614, April 20, **1965**.

43 Laszlo, E.A. Aromatic polyimide particles from polycyclic diamines. U.S. Patent 3,179,631, April 20, **1965**.

44 Bell V. Process for preparing thermoplastic aromatic polyimides. U.S. Patent 4,094,862, June 13, **1978**.

45 Fay, C.C.; Stoakley, D.M.; St Clair, A.K. Molecularly oriented films for space applications. *High Perform. Polym.* **1999**, 11, 145–156.

46 Hsu, T.-C.J.; Liu, Z.-L. Solvent effect on the curing of polyimide resins. *J. Appl. Polym. Sci.* **1992**, 46, 1821–1833.

47 Echigo, Y.; Iwaya, Y.; Tomioka, I.; Yamada, H. Solvent effects in thermal curing of poly(4,4′-oxybis(phenylenepyromellitamic acid)). *Macromolecules* **1995**, 28(14), 4861–4865.

48 Ke, F.; Song, N.; Liang, D.; Xu, H. A method to break charge transfer complex of polyimide: A study on solution behavior. *J. Appl. Polym. Sci.* **2013**, 127, 797–803.

49 Ni, H.-J.; Liu, J.-G.; Wang, Z.-H.; Yang, S.-Y. A review on colorless and optically transparent polyimide films: Chemistry, process and engineering applications. *J. Ind. Eng. Chem.* **2015**, 28, 16–27.

50 Sarma, K.R.; Chanley, C.; Dodd, S.; Roush, J.; Schmidt, J.; Srdanov, G.; Stevenson, M.R.; Wessel, J.I.; Yu, G.; O'Regan, M.; Macdonald, W.A.; Eveson, R.; Long, K.; Gleskova, H.; Wagner, S.; Sturm, J.C. Active matrix OLED using 150°C a-Si TFT backplane built on flexible plastic substrate. *Proceedings from SPIE Aerosense, Technologies and Systems for Defense and Security*, Orlando, FL, April 22–25, **2003**.

51 Long, K.; Kattamis, A.Z.; Cheng, I.C.; Gleskova, H.; Wagner S.; Sturm J.C. Stability of amorphous-silicon TFTs deposited on clear plastic substrates at 250°C to 280° C. *IEEE Electron Device Lett.* **2006**, 27, 111–113.

52 Burrows, P.E.; Graff, G.L.; Gross, M.E.; Martin, P.M.; Shi, M.K.; Hall, M.; Mast, E.; Bonham, C.; Bennett, W.; Sullivan, M.B. Ultra barrier flexible substrates for flat panel displays. *Displays* **2001**, 22, 65–69.

53 Langowski, H.C. Flexible barrier materials for technical applications. *Proceedings of the 46th Annual Technical Conference of the Society of Vacuum Coaters*, San Francisco, CA, May 3–8, **2003**, 559–565.

54 Rendu, P.L.; Nguyen, T.P.; Carrois, L. Cellulose acetate and PVDC used as protective layers for organic diodes. *Synth. Met.* **2003**, 138, 285–288

55 Mohr, J.M.; Paul, D.R. Comparison of gas permeation in vinyl and vinylidene polymer. *J. Appl. Polym. Sci.* **1991**, 42(6), 1711–1720.

56 Granger, V.M.; Tigani, M.A. Permeation properties of moisture sensitive barrier materials under end use conditions. *Tappi J.* **1993**, 76, 98–102.

57 Prins, W.; Hermans, J.J. Theory of permeation through metal coated polymer films. *J. Phys. Chem.* **1959**, 63, 716–720.

58 Jamieson, E.H.H.; Windle, A.H. Structure and oxygen-barrier properties of metallized polymer film. *J. Mater. Sci.* **1983**, 18, 64–80.

59 Felts, J.T.; Grubb, A.D. Commercial-scale application of plasma processing for polymeric substrates: From laboratory to production. *J. Vac. Sci. Technol. A* **1992**, 10, 1675–1681.

60 Nelson, R.J.; Chatham, H. Double Sided QLF Coatings for gas barriers. In: *Society of Vacuum Coaters, Proceedings of the 34th Annual Technical Conference, Philadelphia, March 17–22, 1991*. Albuquerque: The Society; **1991**, 113–117.

61 Izu, M.; Dotter, B.; Ovshinski, S.R. High Performance Clear Coat Barrier Film. *Society of Vacuum Coaters, Proceedings of the 36th Annual Technical Conference*, Dallas, TX, April 25–30, **1993**, 333–340.

62 Klemberg-Sapieha, J.E.; Martinu, L.; Küttel, O.M.; Wertheimer, M.R. Transparent gas barrier coatings produced by dual-frequency PECVD. *Society of Vacuum Coaters. Proceedings of the 36th Annual Technical Conference*, Dallas, TX, April 25–30, **1993**, 445–449.

63 Roberts, A.P.; Henry, B.M.; Sutton, A.P.; Grovenor, C.R.M.; Briggs, G.A.D.; Miyamoto, T.; Kano, M.; Tsukahara, Y.; Yanaka, M. Gas permeation in silicon-oxide/polymer (SiOx/PET) barrier films: Role of the oxide lattice, nano-defects and macro-defects. *J. Membr. Sci.* **2002**, 208, 75–88.

64 Affinito, J.D.; Gross, M.E.; Coronado, C.A.; Graff, G.L.; Greenwell, I.N.; Martin P.M. A new method for fabricating transparent barrier layers. *Thin Solid Films* **1996**, 290–291, 63–67.

65 Charton, C.; Schiller, N.; Fahland, M.; Holländer, A.; Wedel, A.; Noller, K. Development of high barrier films on flexible polymer substrates. *Thin Solid Films* **2006**, 502, 99–103.

66 (a)Whalter, M.; Heming, M.; Spallek, M. Multilayer barrier coating system produced by plasma-impulse chemical vapor deposition (PICVD). *Surf. Coat.*

*Technol.* **1996**, 80, 200–202; (b)Walther, M.; Heming, M.; Spallek, M.; Zschaschler, G. Method of making a vessel having a wall surface having a barrier coating. U.S. Patent 5,900,285, May 4, **1999**.

67 Lamendola, R.; d'Agostino, R.; Fracassi, F. Thin film deposition from hexamethyldisiloxane fed glow discharges. *Plasmas Polym.* **1997**, 2(3), 147–163.

68 Patelli, A.; Vezzù, S.; Zottarel, L.; Menin, E.; Sada, C.; Martucci, A. Costacurta, S. SiOx-based multilayer barrier coatings produced by a single PECVD process. *Plasma Process. Polym.* **2009**, 6, S665–S670.

69 Carcia, P.F.; McLean, R.S.; Reilly, M.H.; Groner, M.D.; George, S.M. Ca test of Al2O3Al2O3 gas diffusion barriers grown by atomic layer deposition on polymers. *Appl. Phys. Lett.* **2006**, 89, 031915.

70 Saitoh, K.; Kumar, R.S.; Chua, S.; Masuda, A.; Matsumura, H. Estimation of moisture barrier ability of thin SiNx single layer on polymer substrates prepared by Cat-CVD method. *Thin Solid Films* **2008**, 516(5), 607–610.

71 Almanza-Workman, M.A.; Jeans, A.; Braymen, S.; Elder, R.E.; Garcia, R.A.; De La Fuente Vornbrock, A. et al. Planarization coating for polyimide substrates used in roll-to-roll fabrication of active matrix backplanes for flexible displays. *Proc. SPIE 8323 Altern. Lithograph. Technol. IV* **2012**, 8323, 83231K.

72 Woody, J.W.; Kolb, W.B. Substrate with planarizing coating and method of making the same. W.O. Patent WO2010078233 A2, July 8, **2010**.

73 Adam, R.; Eveson, R.W.; Macdonald, W.A. Coated and planarized polymeric films, an electronic device comprising the composite film and processes for fabricating coated films. W.O. Patent 2009127842, October 22, **2009**.

74 Economy, J.; Flandera, M.A. Providing a layer of polymeric material on a substrate. European Patent Application EP 55819 A2, July 14, **1982**.

75 Premkumar, P.A.; Starostin, S.A.; de Vries, H.; Creatore, M.; Koenraad, P.M.; van de Sanden, M.C.M. Morphological description of ultra-smooth organo-silicone layers synthesized using atmospheric pressure dielectric barrier discharge assisted PECVD. *Plasma Process. Polym.* **2013**, 10(4), 313–319.

76 Coclite, A.M.; Gleason, K.K. Global and local planarization of surface roughness by chemical vapor deposition of organosilicon polymer for barrier applications. *J. Appl. Phys.* **2012**, 111(7), 073516/1–073516/7.

77 Wang, D.-Y.; Chen, F.-K. Characterization of hydrogen-free diamond-like carbon film deposited on cyclic olefin copolymer. *Diamond Relat. Mater.* **2008**, 17(4–5), 822–825.

78 MacDonald, W.A.; Looney, M.K.; MacKerron, D.; Eveson, R.; Adam, R.; Hashimoto, K.; Rakos, K. Latest advances in substrates for flexible electronics. *J. SID* **2007**, 15(12), 1075.

79 Macdonald, W.A.; Looney, M.K.; Mackerron, D.; Eveson, R.; Rakos, K. Designing and manufacturing substrates for flexible electronics. *Plast. Rubber Compos.* **2008**, 37(2–4), 41–45.

# 7

# Curl Effects in Roll-to-Roll Operations

*Jehuda Greener*

*J. Greener Consulting, Rochester, NY, USA*

## 7.1 Introduction

The vast majority of products manufactured by roll-to-roll (R2R) operations have planar, two-dimensional film-like configuration, and many must meet stringent dimensional requirements. Dimensional changes in the product that contribute to deviations from geometric specifications are often associated with thermal or moisture-induced effects. These changes can be isotropic and homogeneous in nature, leading to a simple increase or decrease in physical dimensions, or they can be anisotropic, giving rise to a shape distortion and, in particular, deviation from planarity. One very distinct shape distortion, commonly observed in R2R-based products, is manifested as curl or curvature in the nominally flat product and is typically associated with the *core-set curl* effect [1–3]. This mechanically induced distortion is closely related to the fact that the production web is wound up on rigid cylindrical cores or bent around conveyance rollers during the manufacturing process, and the final film product is often stored in roll form for extended times. The corresponding non-planar distortion is manifested as film curl—the nominally flat film has finite curvature whose magnitude is generally dependent on the winding history and storage conditions of the film product. This curvature (= curl) can be severe enough to render the product defective and out of spec. This particular geometric distortion is important in many advanced applications, especially in various microelectronics and display components, that require high levels of planarity and dimensional conformity, but other applications also require the web to be as flat as possible during conveyance in the R2R machine and later during functional use.

*Roll-to-Roll Manufacturing: Process Elements and Recent Advances*, First Edition.
Edited by Jehuda Greener, Glen Pearson, and Miko Cakmak.

We note that the curvature produced by the core-set curl effect is oriented along the winding direction (= machine direction, MD), although when flattening tension is applied along the MD of the curled film, the MD curl reverts to a transverse direction (TD) curl associated with the Poisson effect. Indeed, this TD curl scales with the Poisson's ratio of the film and is often observed in R2R operations as the curled web is conveyed under finite MD tension. This curl can adversely impact conveyance, thickness uniformity of deposited layers and process robustness in general. In some applications involving relatively large-area film products, as the weight of the film causes some flattening, the curl usually manifests as edge curvature ("edge lift").

Deviations from planarity may be caused by other physical mechanisms such as nonuniform drying or UV curing, and nonuniform temperature- or humidity-induced dilation associated with the so-called bimetal effect [4], observed mostly in multilayered films (see Section 7.5 for further discussion). But core-set curl is by far the prevalent mechanism leading to deviations from planarity in film products produced in R2R operations. In this chapter we review the phenomenology and mechanism of core-set curl and discuss ways to mitigate and control its deleterious effect on film planarity. Other curl-inducing mechanisms are also briefly reviewed in Section 7.5.

## 7.2 Core-Set Curl

The phenomenology of core-set curl (sometimes referred to as roll-set curl or bending recovery (BR)) is well known [1, 2] and is observed to a greater or lesser extent in most polymeric films that are wound around cylindrical cores. In a typical core-set test, a flat film is wrapped around a rigid cylinder (core) of radius $R$ for time $t = t_w$ (the winding time). At $t = t_w^+$ the film is abruptly unwound from the cylinder by removing the imposed bending moment, and its curvature along the winding direction (MD) is recorded as a function of time. The curvature $\chi = 1/\rho$ at $t > t_w$ can be scaled as

$$\mathrm{BR} = \frac{R}{\rho} \tag{7.1}$$

This dimensionless curvature or curl is referred to as core-set curl or *BR*. We will refer to the intervals $\{0, t_w\}$ and $\{t_w^+, t > t_w^+\}$ as the winding and recovery intervals, respectively. It is generally observed that the curvature is maximum immediately after release ($BR^+$ at $t = t_w^+$), and it gradually decreases thereafter during the recovery interval, ultimately approaching some asymptotic limit $BR_\infty$, with $BR_\infty < BR^+ < 1$. The actual curvature of the unwound film ($\chi$ at $t > t_w$) depends in some way on the core radius $R$, the temperature(s) during the winding and recovery intervals, the viscoelastic properties of the polymeric film

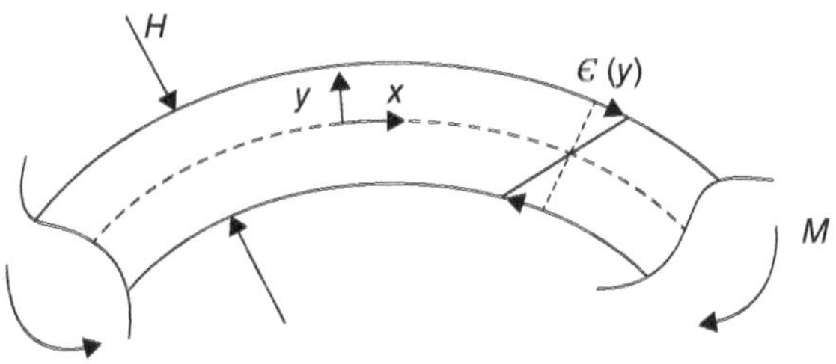


**Figure 7.1** Strain distribution across a bent (curled) film. H is the film thickness and M is the applied bending moment.

and, to a lesser extent, the thickness of the film. When the film undergoes a series of winding steps, as is usually the case in various R2R production scenarios, the dependence of the final film curl on a given winding history can be quite complex. This particular problem is discussed in Section 7.4.

The underlying mechanism of core-set curl is discussed in detail elsewhere [2], but, briefly, this effect is a result of stress relaxation occurring during the winding interval followed by strain recovery when the applied bending moment is removed and the film is allowed to change unconstrained during the recovery interval. When a bending moment $M$ is applied to a thin polymeric plate (film), a linear strain profile is produced across its thickness as shown in Figure 7.1. The strain function is given by

$$\epsilon = \frac{y}{R} \tag{7.2}$$

The $y$ coordinate is the cross-thickness direction with its origin at the mid-plane of the film or, more precisely, at its neutral axis. It is thus noted that the strain varies linearly across the film with the "outer" band being in a state of tension (+), while the "inner" band is in compression (−). Given the viscoelastic nature of most polymeric materials, we expect the bending-induced stress to relax during the winding interval followed by strain recovery during the recovery interval after the bending moment (and stress) is removed, in accordance with a given viscoelastic constitutive law.

Qualitative stress and strain histories during the core-set test are shown in Figure 7.2.

With strains in typical core-set scenarios being relatively small (typically < 1%), it is reasonable to assume that the polymeric material obeys the linear viscoelastic (LVE) constitutive equation [5]:

$$\sigma(t) = \int_{-\infty}^{t} E(t-\xi)\cdot\dot{\epsilon}(\xi)d\xi \tag{7.3}$$

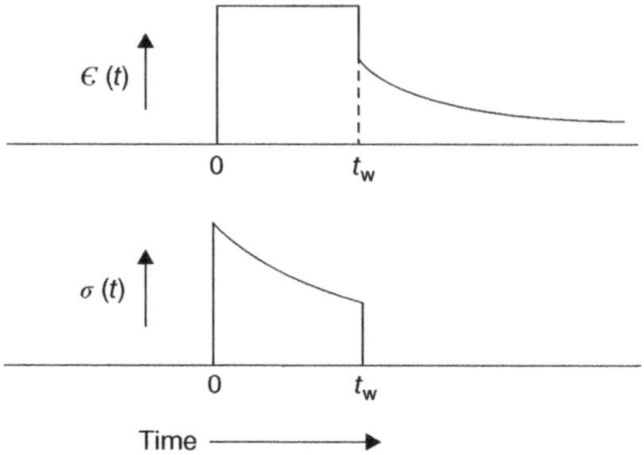


**Figure 7.2** Typical strain and stress histories during the core-set test.

where $E(t)$ is the relaxation modulus of the film, a time-dependent material function, and $\dot{\epsilon}$ is the strain rate.

By applying the LVE model to the plate-bending problem, it is possible to estimate the evolution of stress during the winding interval as

$$\sigma(y,t) = \frac{y}{R}\frac{E(t)}{(1-\nu^2)} \tag{7.4}$$

where $\nu$ is Poisson's ratio. From Eq. (7.4) it follows that the core-set curl immediately after release from the core (@ $t = t_w^+$) is given by

$$\mathrm{BR}^+ = \frac{1}{2}\left[1 - \frac{E(t_w)}{E_0}\right] \tag{7.5}$$

where $E_0$ is the initial (instantaneous) modulus of the film and $E(t_w)$, its relaxation modulus at the end of the winding interval, can be expressed by a variety of time-dependent empirical forms. Thus, the instantaneous core-set curl is strictly related to the relaxed portion of the modulus at $t_w$. One complication in the LVE analysis is that the relaxation behavior in the outer tensile band of the film and the inner compressive band may not be the same, in which case Eq. (7.5) must be rewritten as follows:

$$\mathrm{BR}^+ = \frac{1}{2}\left[1 - \frac{E_t(t_w) + E_c(t_w)}{2E_0}\right] \tag{7.6}$$

where $E_t(t)$ and $E_c(t)$ are, correspondingly, the tensile and compressive relaxation moduli of the film. $E_t(t)$ is easily measured using standard tensile testing methods [5], while $E_c(t)$ can be estimated by shifting the tensile relaxation spectrum to longer times using an effective compressive–tensile shift factor $a_c^2$. A shift factor $a_c$ of 13,000 seems to fit the data well for several film types. It is noted that the use of a tensile–compressive shift factor is an *ad hoc* correction to account for differences in relaxation time between tensile and bending modes for common polymeric films. Another possible explanation for this discrepancy is time dependence of Poisson's ratio [5, 6], but this property is not readily attainable, so the use of an apparent $a_c$ is a practical way to circumvent this problem.

The evolution of curl (strain) during the recovery interval requires solution of the following integral equation for the LVE film:

$$0 = \epsilon_0 E(t) + \int_{t_w}^{t} E(t-\xi)\cdot\dot{\epsilon}(\xi)d\xi \tag{7.7}$$

where $\epsilon_0$ is the initial and constant strain during the winding interval. Equation (7.7) is a Volterra integral equation of the first kind for $\epsilon(t)$, the strain history during the recovery interval, which can then be converted to curl. An analytical solution of Eq. (7.7) is described elsewhere [2]. From inspection of Eq. (7.7), and in the context of the LVE approximation, it is seen that the dimensionless curl (*BR*) depends strictly on the relaxation modulus of the film, but not on the core radius. So prediction of the dimensionless core-set curl requires knowledge of $E(t)$ for the given film material and making some assumptions about $a_c$, the compressive–tensile shift factor.

Tensile relaxation moduli of several common materials are shown in Figure 7.3 [2]. These materials exhibit very different relaxation rates at room temperature, with the nylon film (MXD) being the slowest to relax, while polypropylene (PP) is the fastest. The corresponding relaxation functions can be used to predict the

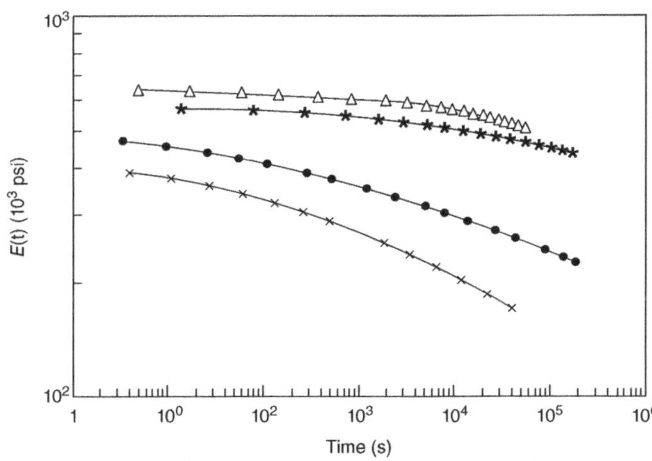


**Figure 7.3** Tensile relaxation moduli for several film types at 21°C: cellulose acetate (CA) (●), polyethylene terephthalate (PET) (*), nylon (MXD) (Δ), and polypropylene (PP) (×) [2]. © John Wiley & Sons.

evolution of curl in a core-set test. Figures 7.4 and 7.5 show the instantaneous core-set curl (curl immediately after release at $t = t_w^+$) as a function of winding time for two commonly used materials: polyethylene terephthalate (PET) and cellulose acetate (CA). In both cases model results are compared with experimental data, showing good agreement. Curl recovery (core-set curl evolution during the recovery interval) results for the same materials are shown in Figures 7.6 and 7.7. Both sets of results show that CA, which is faster to relax,

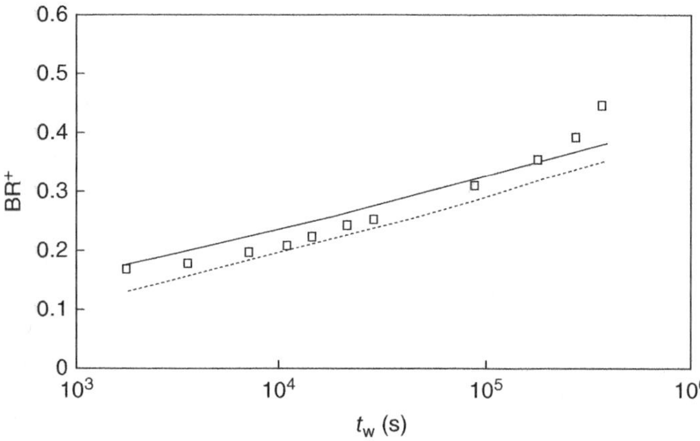


**Figure 7.4** Instantaneous core-set curl versus winding time for CA at 21°C. Data compared with analytical predictions for recovery times of 0 (solid) and 10s (dotted) after unwinding [2]. © John Wiley & Sons.

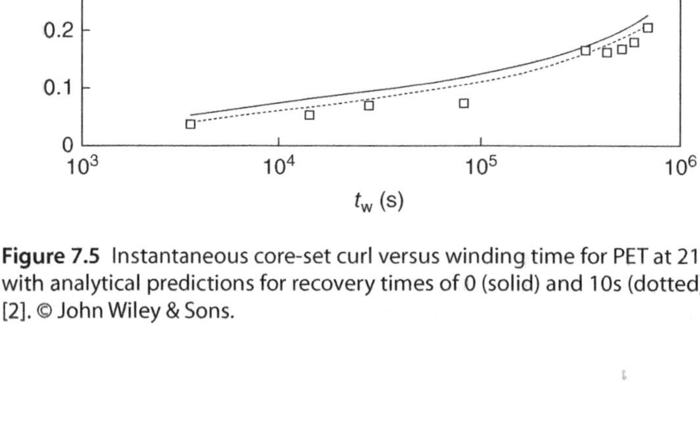


**Figure 7.5** Instantaneous core-set curl versus winding time for PET at 21°C. Data compared with analytical predictions for recovery times of 0 (solid) and 10s (dotted) after unwinding [2]. © John Wiley & Sons.

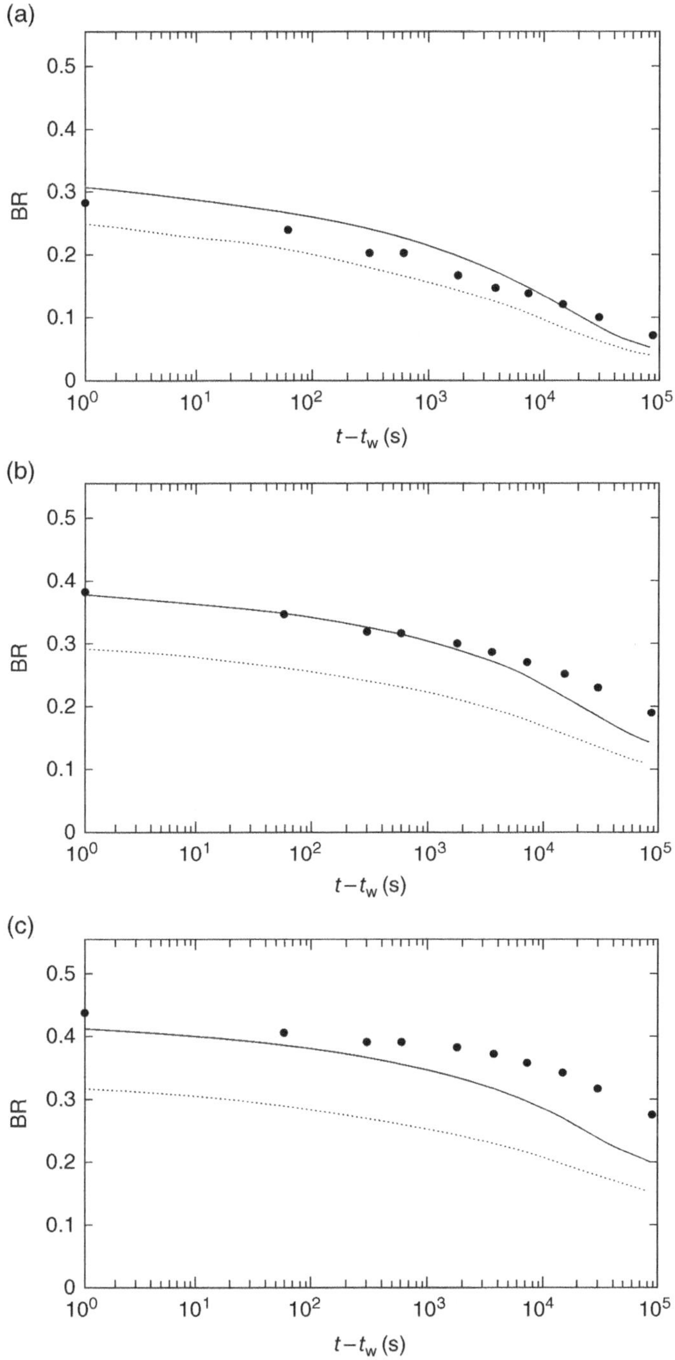


**Figure 7.6** Core-set recovery of CA film at 21°C for three winding times: (a) 24 h, (b) 120 h, and (c) 220 h. Curves are model predictions for $a_c = 13{,}000$ (solid) and $a_c = \infty$ (dotted) [2]. © John Wiley & Sons.

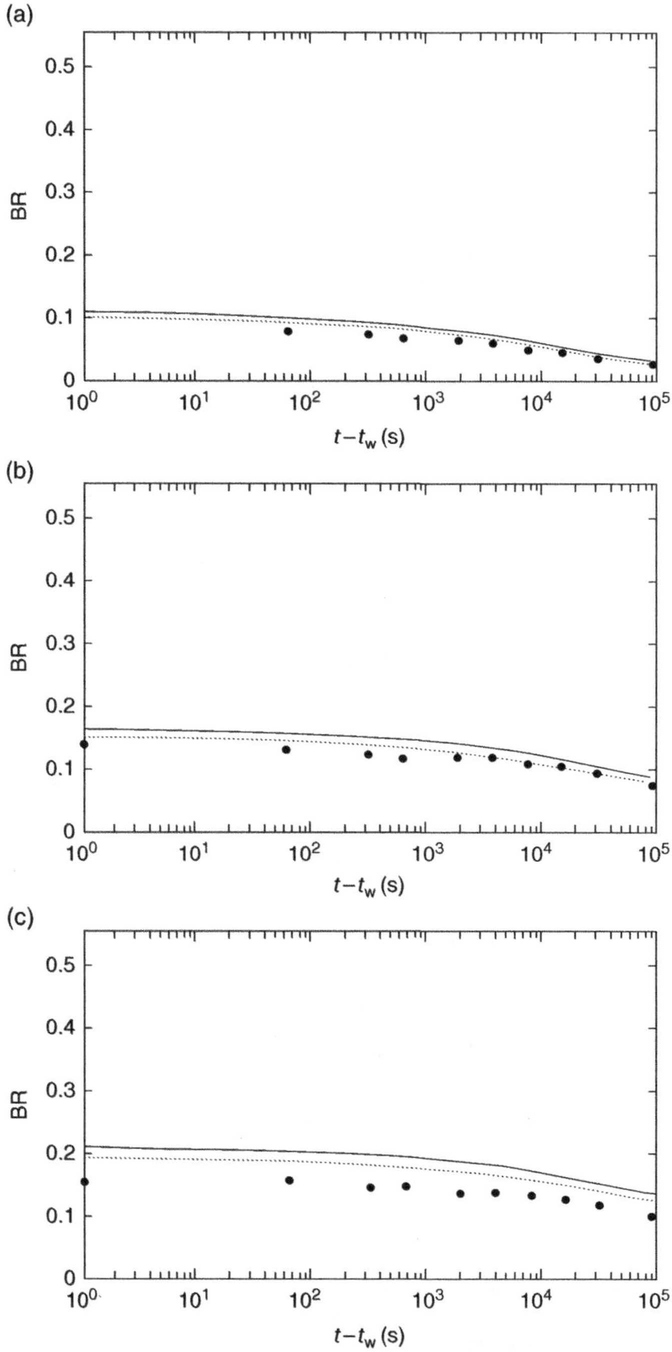


**Figure 7.7** Core-set recovery of PET film at 21°C for three winding times: (a) 24 h, (b) 96 h, and (c) 196 h. Curves are model predictions for $a_c = 13,000$ (solid) and $a_c = \infty$ (dotted) [2]. © John Wiley & Sons.

produces higher instantaneous curl levels under similar conditions, but the curl is also faster to decay during the recovery interval. A general illustration of the effect of winding time on the core-set recovery function for CA is shown in Figure 7.8. Clearly, long winding (storage) times under ambient conditions can have a significant effect on the curl function—it is generally shifted upward for longer times. The effect of core size ($R$) is implied from the scaling of the problem (Eq. 7.1). The dimensionless core-set curl ($BR$) is independent of core radius and film thickness in the context of the LVE assumption, but the actual curl ($1/\rho$) is proportional to $1/R$ based on the scaling of the problem. Thus, winding the film around smaller cores produces proportionately larger curl levels and more distorted films. Film thickness, however, will play a role only when the LVE assumption breaks down, that is, when the film is thick relative to the core radius ($H/R > \sim 0.008$) [5] and the material behaves nonlinearly or when the film is stored in large production rolls wherein the effective winding radius depends on film thickness, as discussed in Section 7.4.

One parameter with a potentially significant effect on the curl function is temperature. The temperature ($T$) enters the problem through its effect on the relaxation spectrum of the material, which can be expressed by a given time–temperature shift factor, for example, the Arrhenius equation:

$$a_{\mathrm{T}} = \exp\left[\frac{\Delta E}{\mathcal{R}}\left(\frac{1}{T} - \frac{1}{T_{\mathrm{r}}}\right)\right] \tag{7.8}$$

where $\Delta E$ is an activation energy, $\mathcal{R}$ is the gas constant, and $T_{\mathrm{r}}$ is a reference temperature. With the given shift factor, it is possible to predict the effect of temperature on the core-set function. It is useful to define two temperatures for the standard core-set test—one for the winding (rolled storage) interval and the other for the recovery interval—after the film is unwound and stored in its unconstrained state. The effect of temperature on the bending recovery function is illustrated in Figures 7.9 and 7.10 for CA film. The storage temperature during the winding phase has a significant effect on the curl immediately after release ($BR^{+}$) and some effect on the rate of recovery of curl after release—the instantaneous curl and the curl throughout the recovery interval appear to increase substantially with temperature. The temperature during the recovery interval influences only the rate of recovery after release with the rate increasing with temperature, that is, curl decays faster when the film is exposed to elevated temperatures after it is unwound. These results suggest that curl minimization requires rolled storage at low temperatures and/or exposure of the unwound film to high temperatures.

We finally note that the rate of core-set buildup and recovery can be influenced not only by temperature but also by other environmental factors that

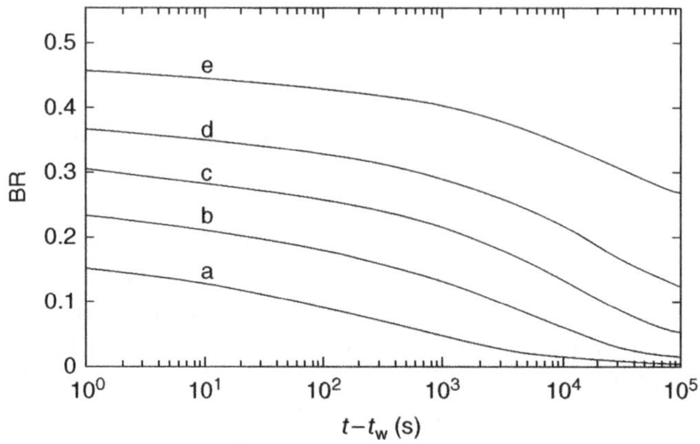


**Figure 7.8** Effect of winding time on core-set recovery for CA film at 21°C: (a) 0.5 h, (b) 4 h, (c) 24 h, (d) 1 week, and (e) 4 weeks [2]. © John Wiley & Sons.

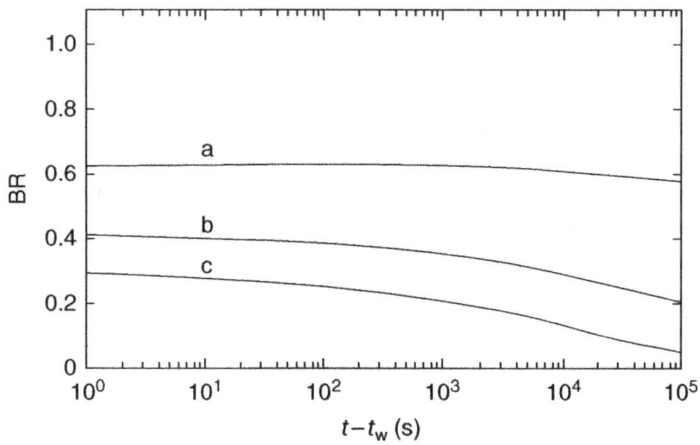


**Figure 7.9** Effect of temperature during the winding interval on core-set recovery of CA film at 21°C for $t_w = 24$ h: (a) 38°C, (b) 29°C, and (c) 21°C [2]. © John Wiley & Sons.

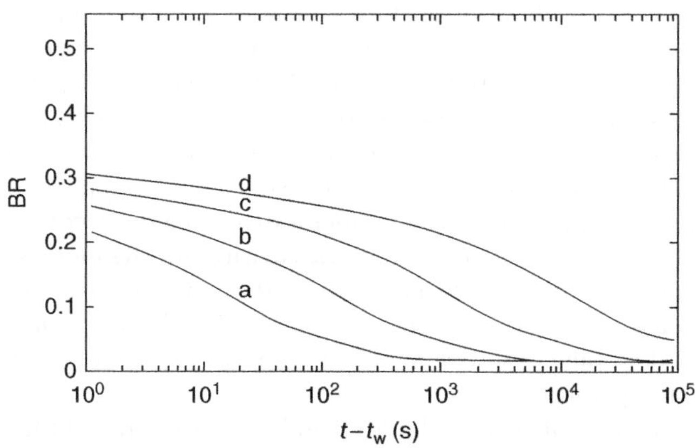


**Figure 7.10** Effect of temperature during the recovery interval on core-set recovery of CA film. Winding temperature = 21°C, $t_w = 24$ h: (a) 47°C, (b) 38°C, (c) 29°C, and (d) 21°C [2]. © John Wiley & Sons.

contribute to the relaxation spectrum of the material. For example, for some hydrophilic and hygroscopic materials, exposure to a high humidity environment can shorten the relaxation time of the material and thus increase the rate of core-set buildup and recovery. Another important parameter is the age of the sample, to be discussed in the following section.

## 7.3 Physical Aging Effects

Film products are often stored in roll form for extended times, either before the R2R operation (substrate only) or after the operation (coated or processed product) and before or after various finishing steps. This extended storage can modify the mechanical response of the film in a significant way because of the physical aging of the polymeric material at temperatures below the glass transition temperature ($T_g$) [7]. The main effect of physical aging is to retard the mechanical relaxation of the polymer by shifting its relaxation spectrum to longer times. A simple formalism to describe this effect was proposed by Struik [7, 8]:

$$\lambda(t, t_e) = \int_0^t a_{te}(t_e, t') \cdot dt' \tag{7.9}$$

where $\lambda$ is the relaxation time of the aged polymer, $t_e$ is the aging time, and $a_{te}$ is a characteristic time–aging time shift factor that can be expressed empirically by [7, 8]:

$$a_{te} = \left( \frac{t_{e,r}}{t_e} \right)^{\mu} \tag{7.10}$$

where $t_{e,r}$ is some arbitrary reference aging time and $\mu$ is the so-called aging rate parameter, a temperature-dependent material characteristic. Based on Eqs. (7.9) and (7.10), the effective relaxation time of the aged sample becomes

$$\lambda(t, t_e) = \frac{t_e}{1-\mu} \left[ \left( 1 + \frac{t}{t_e} \right)^{1-\mu} - 1 \right] \tag{7.11}$$

With this updated expression for the effective relaxation time, the formalism of Section 7.2 can now be extended to account for the effect of physical aging in estimating the curl response of the film to a particular winding history [9].

Figure 7.11 shows the effect of aging time on the evolution of the instantaneous core-set curl with winding time for a PET film. Predictions of the updated LVE model are compared with data for two aging times using an aging rate parameter of $\mu = 0.85$ with reasonably good agreement. In the core-set test represented by Figure 7.11, the PET film is pre-aged before the winding interval at

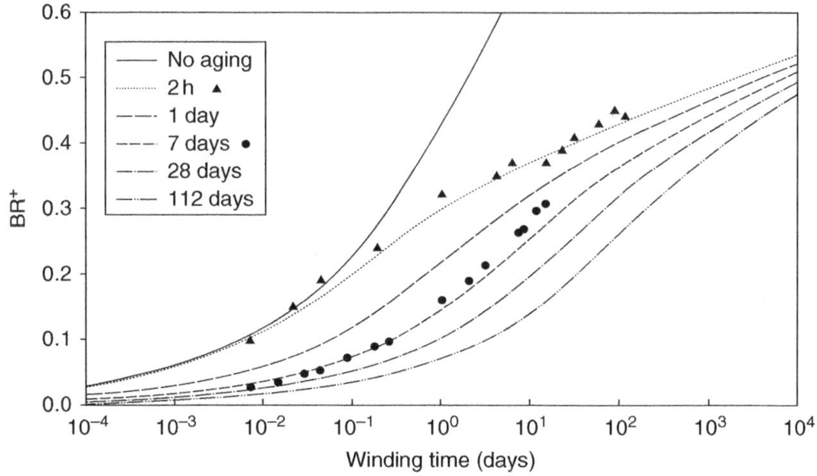


**Figure 7.11** Instantaneous core-set curl for PET film versus winding time for various pre-aging times. Data (points) are compared with predictions of the LVE core-set model modified to account for physical aging effects [9]. © John Wiley & Sons.

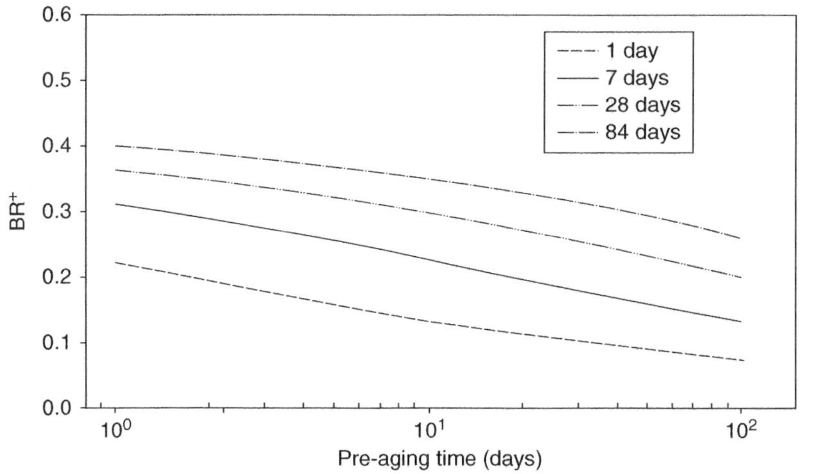


**Figure 7.12** Instantaneous core-set curl for PET film versus pre-aging time for various winding times [9]. © John Wiley & Sons.

ambient temperature, and aging during the winding interval is assumed to be negligible. Model predictions of the instantaneous core-set curl (curl immediately after release) as a function of pre-aging time for a PET film are shown in Figure 7.12 for different winding times. The results in Figures 7.11 and 7.12 underscore the strong retarding effect of physical aging on the ability of the

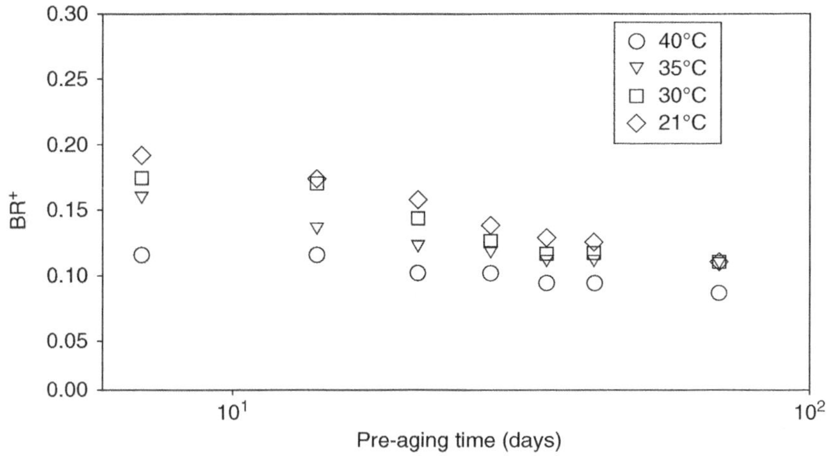


**Figure 7.13** Instantaneous core-set curl versus pre-aging time for various pre-aging temperatures. Data for PET film, with winding temperature of 21°C and winding time of 1 week [9]. © John Wiley & Sons.

film to acquire curl. This suggests that a properly pre-aged film will be less prone to acquire curl in a particular winding scenario. However, long pre-aging times may be impractical and costly.

One way to speed up the physical aging process is to pre-age the sample at elevated temperatures below $T_g$. It is known that aging rate peaks at some elevated temperature below $T_g$ [7], so by pre-aging the film close to this peak temperature, pre-aging times can be considerably shortened. The effect of pre-aging temperature on dimensionless core-set curl for PET film is shown in Figure 7.13, clearly illustrating the enhanced effect of physical aging when the film is pre-aged at elevated temperature. It is generally expected that the rate of aging will peak at 15–20°C below $T_g$ [10(a)] and will start decreasing above this temperature. In any case, pre-aging when properly exercised can provide another approach for minimizing curl in a film for a particular winding scenario [1]. Pre-aging of the film is best applied when the film is flat to minimize any acquisition of curl during the pre-aging step itself. If the aging process is sufficiently rapid (when conducted at the optimum temperature), it should be possible to pre-age the film (the bare substrate or the coated substrate) in a properly designed R2R annealing operation whereby the film passes through an annealing oven at a suitable line speed [10(b)]. But if such "flat" annealing treatment is not possible or cost-effective, the unaged film can be annealed in roll form at the optimal temperature for sufficiently long times to modify its relaxation characteristics. However, in such a "rolled" annealing step, it is important to use large-diameter cores to minimize the curl acquired during the pre-aging step, and selection of appropriate annealing times should take into account the considerable thermal lag associated with heating large wound production rolls.

## 7.4 Core-Set Curl in R2R Operations

The mechanical history of the film in a typical R2R manufacturing operation is considerably more complex than the history of the simple core-set test discussed in Section 7.2 and as represented by the strain history of Figure 7.2. R2R operations often involve multiple winding and recovery steps with varying conditions (temperature, tension, humidity) applied in each step to satisfy certain operational and quality requirements. Moreover, the history within a single production roll is not uniform because of positional variation along the wound film of the effective winding (core) diameter. In principle, the LVE analysis discussed in Section 7.2 can be extended to any arbitrary strain history corresponding to a particular R2R winding scenario, but the analysis is considerably more involved requiring spatial mapping of the evolution in strain of each film element along the production roll. In general, the curl is a function of position along the roll, that is, the final processed roll will have a characteristic "curl profile" that is expected to depend on the particular winding history, storage conditions, core size, film thickness, roll length, and viscoelastic properties of the film material(s).

Figure 7.14 depicts schematically a localized strain history for a film element along a production roll subjected to a multistep winding process. By rewriting

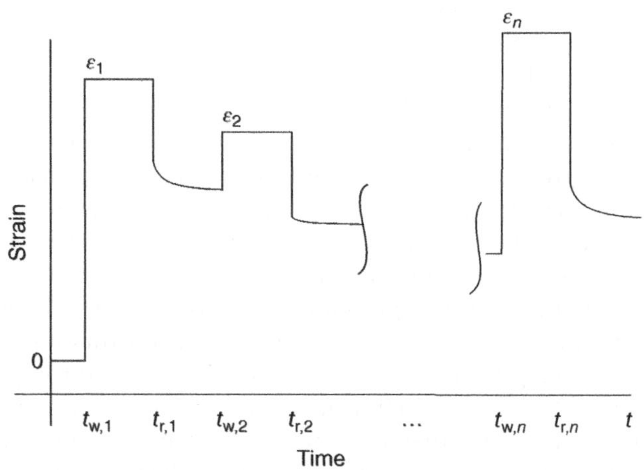


**Figure 7.14** Localized schematic strain history for a film element along a production roll undergoing a multistep winding process.

Eq. (7.2), the imposed strain during a given winding interval for a film element $\Xi$ takes the form

$$\epsilon_i(\Xi) = \frac{y}{R_i(\Xi)} i = 1, 2, \ldots, n \tag{7.12}$$

where $R_i$, the radial position of the element during the $i$th step, depends on the axial position of the element along the production roll, the step number, the winding mode, and the core radius of the roll. From simple geometric considerations, the radial position for the $n$th (last) step (effective core radius at position $x$ along the roll) is given by

$$R_n(x) = \sqrt{R_{C,n}^2 + \frac{H}{\pi} x} \tag{7.13}$$

where $H$ is the thickness of the film (may vary for different steps if coating is applied) and $R_{C,n}$ is the core radius for step $n$. In Eq. (7.13) we use the convention that the origin of the axial coordinate $x$ is arbitrarily set as the core of the roll for the nth (last) winding step. For the $(n-1)$th step two possible rewinding modes may be considered as shown in Figure 7.15: "normal" ($N$) and "reverse" ($R$) rewinding.

For $N$ rewinding (rewinding in the same direction as in the $n$th step), the radial position function for the $(n-1)$th step takes the form

$$R_{n-1}(x) = \sqrt{R_{C,n-1}^2 + \frac{H}{\pi}(L - x)} \tag{7.14a}$$

and for $R$ rewinding (rewinding the film in the opposite direction) the radial position function becomes

$$R_{n-1}(x) = -\sqrt{R_{C,n-1}^2 + \frac{H}{\pi}(L - x)} \tag{7.14b}$$

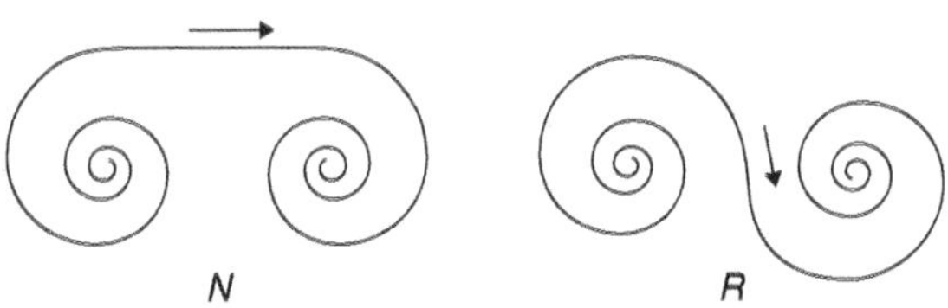


**Figure 7.15** Rewinding modes: normal ($N$) and reverse ($R$).

where $L$ is the total length of the rolled film and $H$ is the film thickness. The radial position function for the $(n-2)$th step is given by

$$R_{n-2}(x)=\begin{cases}\sqrt{R_{C,n-2}^2+\dfrac{H}{\pi}x} & \text{for "Normal" rewinding}\\ -\sqrt{R_{C,n-2}^2+\dfrac{H}{\pi}x} & \text{for "Reverse" rewinding}\end{cases} \tag{7.15}$$

This scheme can be generalized for any $(n-k)$th step in a given winding scenario. Equation (7.14) is used to describe the radial position if $k$ is odd and Eq. (7.15) is used if $k$ is even. Once the radial position functions $R_i(x)$ are mapped for all $n$ steps, the discrete strains during the winding intervals can be computed as a function of axial position along the roll using Eq. (7.12). The strain during the recovery intervals in Figure 7.14 is unspecified and must be computed independently based on the condition that the stress is always zero during the recovery phase (see Section 7.2). For the general strain history shown schematically in Figure 7.14, Eq. (7.7) can be rewritten as

$$\sigma(t\geq t_{w,n})=0=\int_{-\infty}^{t}E(t-\xi)\cdot\dot{\epsilon}(\xi)d\xi \tag{7.16}$$

This integral equation must be solved for $\epsilon(x,t)$, the axial strain of the film element at time $t$ (for $t\geq t_{w,n}$), by substituting the specified strain history. The strain can then be converted to curl during the final ($n$th) recovery phase, and the curl profile ($\chi(x,t)$) for the full roll can be estimated from the relation,

$$\chi(x,t)=2\frac{\epsilon(x,t)}{H} \tag{7.17}$$

This formalism is now illustrated for a typical R2R process flow shown schematically in Figure 7.16. This particular operation involves three major steps and corresponding winding events. The first step (S1) represents the substrate-making process, and it is followed by the first winding event (W1) whereby the substrate is stored wound on a roll with a given core radius for a given duration ($t_{w1}$). In the next step (S2) the substrate is fed into an R2R machine wherein it is processed (e.g., coated) in one or multiple sub-steps. After exiting the R2R machine, the processed film undergoes a second winding event (W2) whereby

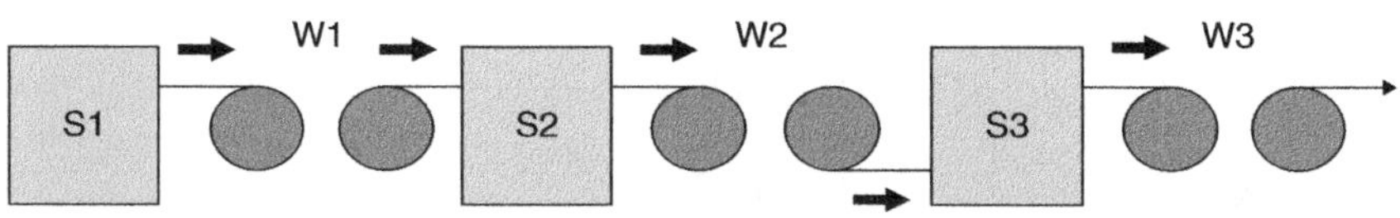


**Figure 7.16** A schematic multistep R2R process flow with three winding events.

it is stored on a second roll with the same or different core radius for a duration $t_{w2}$. In this case the film can be rewound either in $R$ mode or $N$ mode (see Figure 7.15). Finally, in the third step (S3) the processed film is fed into a slitter whereby the film in the production master roll is slit down into narrower rolls. This is followed by the third winding event (W3) where the slit film is stored on a third roll for a duration $t_{w3}$. In this winding step the rewinding mode can again be either $N$ or $R$. The slit roll is finally unwound, cut into the final nominally flat product, and curl is measured along its length to establish the *curl profile* for the entire finished roll. Curl measurement can be done at some arbitrary, fixed recovery time after unwinding ($t_r$). This winding scenario represents a fairly typical process flow for many R2R operations.

Using the process of Figure 7.16, we will examine the effect of various process parameters on the final curl profile along the finished roll. The curl profile is calculated using Eqs. (7.16) and (7.17) and assuming a 175 μm thick PET substrate with a total roll length of 14,000 ft and a given relaxation modulus for the PET film (see Figure 7.17 and Table 7.2). It is further assumed that any layer applied in the R2R step (S2) and the conditions during this step have negligible effect on the relaxation properties and total thickness of the processed film. We also assume that web tension and winding pressure during the three winding events have an insignificant effect on the relaxation properties of the web material. The parameter space covered in this analysis is summarized in Table 7.1. The parameters examined are the rolled storage times for the three winding events, recovery time (time after unwinding the slit rolls, following the last winding event, when curl is measured), core diameter (assuming the same core sizes for the three events), rewinding modes for

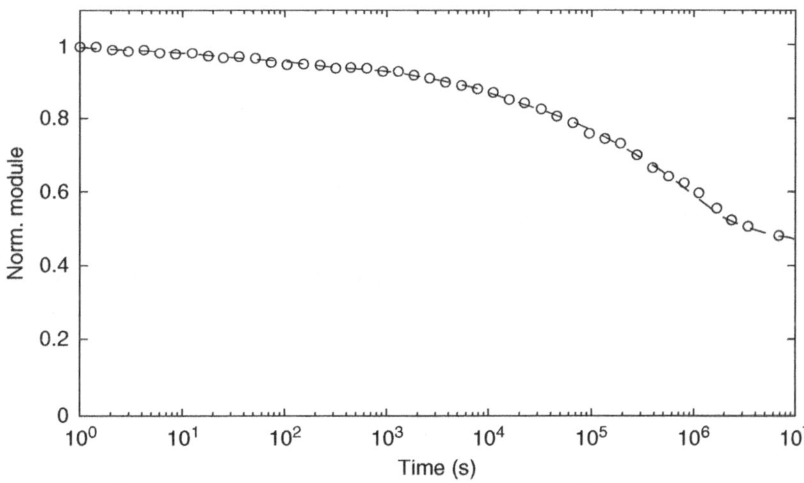


**Figure 7.17** Normalized relaxation modulus of unaged PET film. Curve is best fit of Maxwell function (Eq. 7.18).

**Table 7.1** Winding conditions for sensitivity analysis.

| Variable | | Values[a] |
|---|---|---|
| Winding time for W1 ($t_{w1}$) | Days | 7, 14, 21, 28 |
| Winding time for W2 ($t_{w2}$) | Days | 7, 14, 21, 28 |
| Winding time for W3 ($t_{w3}$) | h | 2, 4, 8, 16, 24 |
| Recovery time for W3[b] ($t_r$) | h | 1, 3, 6, 10, 12, 24 |
| Core diameter | in. | 6, 12, 18, 24 |
| Rewinding mode and sequence | | *NN*, *RN*, *RR* |
| Storage temperature for W2 ($T_{W2}$) | °C | 25, 21, 17 |

[a] Underlined values are baseline (nominal) condition.
[b] Recovery times for W1 and W2 are assumed to be negligibly short.

**Table 7.2** Viscoelastic parameters for PET film (Eq. 7.18) at 25°C.

| *i* | $E_i/E_0$ | $\lambda_i$ (s) |
|---|---|---|
| 1 | 2.360E−2 | 0.100E+2 |
| 2 | 2.490E−2 | 0.100E+3 |
| 3 | 1.960E−2 | 0.100E+4 |
| 4 | 7.190E−2 | 0.100E+5 |
| 5 | 0.107 | 0.100E+6 |
| 6 | 0.252 | 0.100E+7 |
| 7 | 0.213 | 0.100E+9 |
| 8 | 0.285 | 0.100E+10 |

the second and third winding events, and storage temperature for the second winding event. Storage temperature for all events is ambient (25°C) unless otherwise stated. The underlined values in Table 7.1 are the nominal (baseline) values implied when examining the effect of a given parameter. The relaxation modulus of the PET substrate used in the analysis is given in Figure 7.17. This time-dependent modulus is expressed by the following empirical Maxwell function:

$$E(t) = E_0 \sum_{i=1}^{n} E_i \exp(-t/\lambda_i) \tag{7.18}$$

where $E_i$ and $\lambda_i$ are material parameters representing the relaxation spectrum of the film under ambient conditions. The corresponding empirically fitted parameters are listed in Table 7.2, and, as shown in Figure 7.17, the empirical

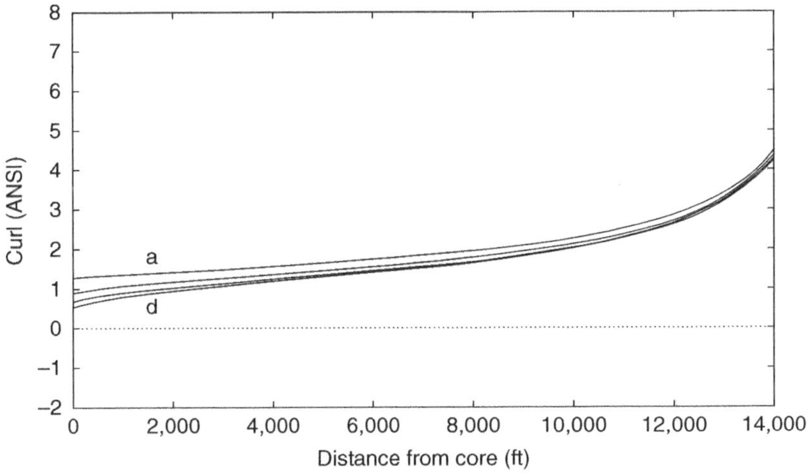


**Figure 7.18** Effect of winding time for the first winding event ($t_{w1}$) on the curl profile for the nominal production flow: (a) 7, (b) 14, (c) 21, and (d) 28 days.

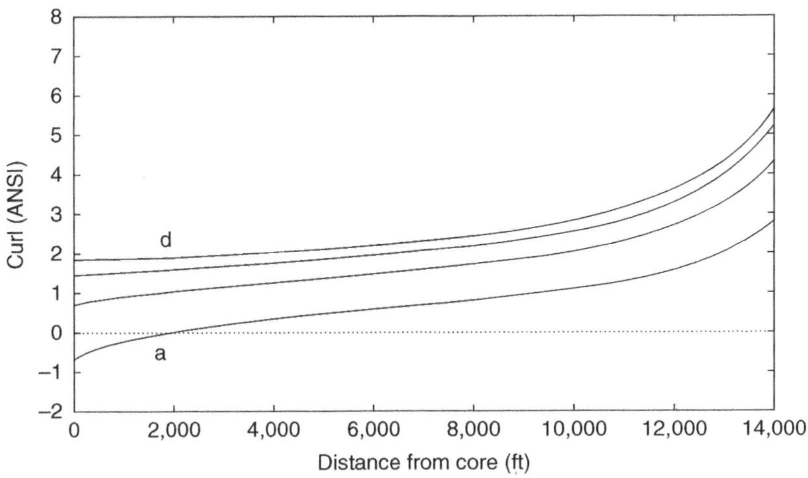


**Figure 7.19** Effect of winding time for the second winding event ($t_{w2}$) on the curl profile for the nominal production flow: (a) 7, (b) 14, (c) 21, and (d) 28 days.

relaxation function agrees well with data for the PET film analyzed. The results of the analysis, expressed as curl profiles (curl vs. axial position along the slit roll following the last winding event), are shown in Figures 7.18, 7.19, 7.20, 7.21, 7.22, 7.23, and 7.24 for the different winding scenarios covered in Table 7.1, with the curl given in ANSI units (100/in.).

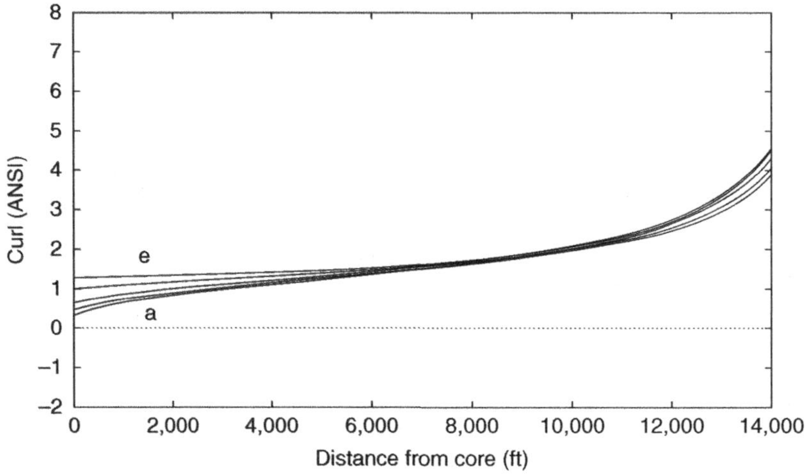


**Figure 7.20** Effect of winding time for the third winding event ($t_{w3}$) on the curl profile for the nominal production flow: (a) 2, (b) 4, (c) 8, (d) 16, and (e) 24 h.

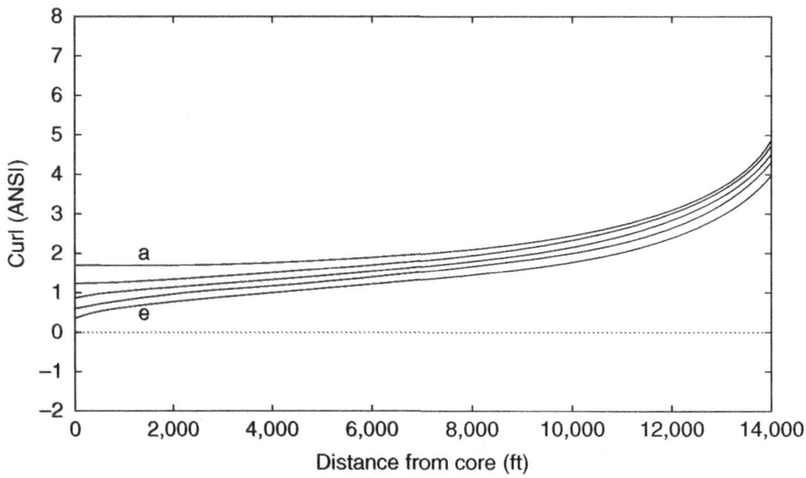


**Figure 7.21** Effect of recovery time after the third winding event ($t_r$) on the curl profile for the nominal production flow: (a) 1, (b) 3, (c) 6, (d) 12, and (e) 24 h.

The effect of the three rolled storage times is shown in Figures 7.18, 7.19, and 7.20. In all cases and for all storage times, the curl is lowest near the core (for the roll in the last winding event, W3), and it grows steadily with axial position away from the core with the curl at the outer position peaking at approximately two to six times the core value. This result seems counterintuitive, but it is consistent with the fading-memory effect of the viscoelastic web

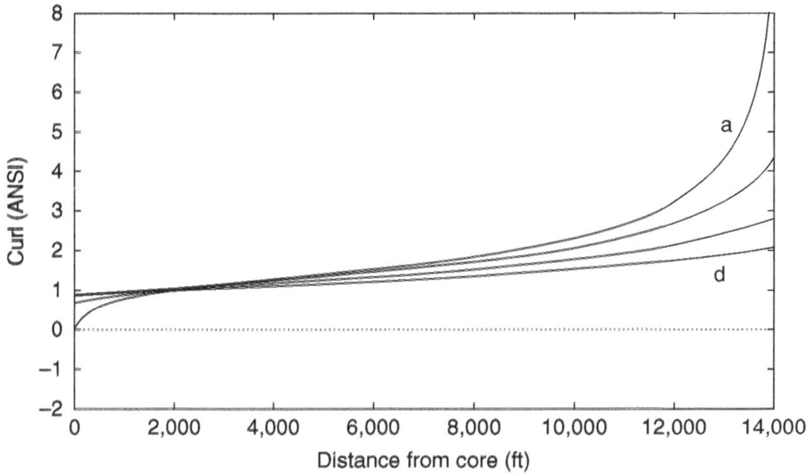


**Figure 7.22** Effect of core diameter on the curl profile for the nominal production flow: (a) 6, (b) 12, (c) 18, and (d) 24 in.

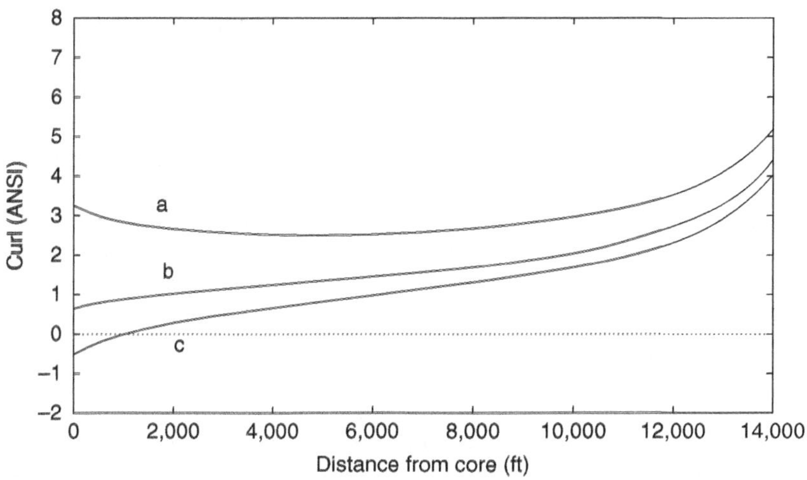


**Figure 7.23** Effect of winding mode and sequence on the curl profile for the nominal production flow: (a) *NN*, (b) *RN*, and (c) *RR*.

material. The storage (winding) time in the last winding event, $t_{w3}$, is relatively short compared with that in the previous events (W1 and W2) so it has a lesser effect on the final curl profile, but, of the two longer winding events, W1 and W2, the more recent event (W2) seems to dominate the response. Figure 7.19 shows a significant and consistent increase in curl throughout the roll as $t_{w2}$ increases from 1 to 4 weeks. Clearly, reducing storage time for this winding

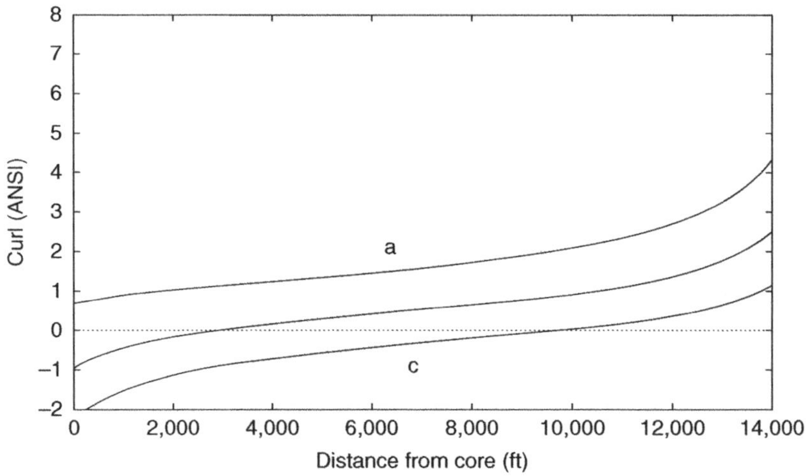


**Figure 7.24** Effect of storage temperature during the second winding event ($T_{W2}$) on the curl profile for the nominal production flow: (a) 25, (b) 21, and (c) 17°C.

event can facilitate a significant overall reduction in curl. By comparison, the contributions of the other rolled storage times, $t_{w1}$ and $t_{w3}$, are substantially less important. As noted previously, the diminished effect of $t_{w1}$ is due to the fading memory of the material, while the relatively small effect of $t_{w3}$ is due to the relatively short storage times for this step. The effect of the recovery time is depicted in Figure 7.21. This time represents the interval between unwinding the last (slit) roll and the instant of curl measurement. The result in Figure 7.21 shows a consistent drop in curl over time across the entire length of the roll due to viscoelastic strain recovery. However, as noted in Section 7.2, the curl is expected to reach asymptotically a finite value and remain unchanged indefinitely. Apparently, this asymptote is not reached within the times covered in this analysis.

Figures 7.22, 7.23, and 7.24 illustrate the effects of core size, winding mode, and storage temperature during the second winding event on the curl profile of the finished roll. The effect of core diameter (assumed identical for all three winding events), Figure 7.22, is very significant near the end of the slit roll, which corresponds to the core position in the second winding event. The curl at that position drops proportionately by a factor of ~4 as the core diameter increases from 6 to 24 in. However, the effect of core diameter on the curl profile at positions closer to the core is fairly small. By contrast, the winding mode and sequence (Figure 7.23) have a pronounced effect on curl throughout the length of the roll. Of the three options considered (*NN, RN, RR*), *RR* appears to produce the lowest curl throughout the roll, that is, by reverse winding the film twice, it is possible to lower the curl substantially. Finally, Figure 7.24 illustrates the effect of storage temperature during the second winding event on the final curl profile. As noted in Section 7.2, the effect of temperature can be represented by a suitable time–temperature shift factor such as the Arrhenius equation (Eq. 7.8). Also, in assessing the effect of rolled storage temperature, the considerable thermal lag associated with the relatively sluggish heat transfer in a large production roll is neglected. This thermal lag would retard the temperature effect to some extent requiring longer storage times to make the effect of storage temperature practically beneficial. However, within these assumptions, the results in Figure 7.24 show a very significant reduction in curl throughout the roll by storing the wound roll at 4–8°C below ambient temperature.

The results of the winding analysis clearly illustrate the multitude of options available for reducing the core-set curl acquired during a typical R2R process flow through judicious control of key process parameters such as rolled storage time and temperature, core radius, and winding mode. Of course, selecting materials, especially substrates, with longer relaxation times can also help lower core-set curl, but this option may not be practically feasible in most cases because of other considerations in material selection, including cost.

## 7.5 Other Curl Mechanisms and Curl Mitigation Strategies

We have shown in Sections 7.2–7.4 that a nominally flat film product produced by R2R operations can become significantly curled (non-planar) as a result of the various winding, bending, and rolling steps to which the processed web is being subjected during the operation itself or during rolled storage before or after the operation. This deviation from planarity is associated with the core-set curl effect and the viscoelastic nature of the polymeric materials used, and it can render the product defective and out of spec. Excessive film curl can also negatively impact the robustness of the R2R process itself, for example, by increasing coating variability or by affecting conveyance, thereby producing a less uniform and less robust product. To be sure, there are other mechanisms leading to non-planarity of processed films such as nonuniform thermal or humidity-induced dilation of multilayered films as well as drying and curing stresses produced during the film manufacturing process [11–18], but core-set curl is typically the leading cause of curling in films produced by R2R operations.

In order to minimize curling effects in film products, it is necessary to first identify the underlying curling mechanism that drives the deviations from planarity. If core-set curl is identified as the primary cause for non-planarity, then the discussion in Sections 7.2–7.4 can provide a useful guideline for mitigating the problem. Some obvious curl-reducing steps include the following: increase core diameter(s), shorten rolled storage times as much as possible before or after the R2R operation, consider the possibility of flat storage, store wound rolls at low temperatures (below ambient), select web materials (especially substrates) with relatively long mechanical relaxation times, and, if possible, preage (anneal) the substrate at elevated temperatures to increase its effective relaxation time and, thus, reduce its propensity to acquire core-set curl. It was noted in Section 7.3 that the most effective annealing temperature (where the physical aging process is the fastest) is typically ~15–20°C below the glass transition temperature of the substrate material [10(a)]. It is also important to note that the annealing step must be conducted when the film is flat or when it is wound in rolls with relatively large core sizes to minimize acquisition of curl during the annealing (pre-aging) step. Finally, as illustrated in the winding analysis and case study of Section 7.4, the curl in large production rolls can be substantially reduced by implementing an effective "campaigning" strategy, that is, by judiciously selecting a winding sequence and mode as well as rolled storage times aimed at minimizing the curl level throughout the roll.

Curl can also be induced in multilayered films with unbalanced (asymmetric) structure by temperature or humidity changes as a result of the so-called "bimetal" effect [4, 11, 12]. In this case curl minimization requires changes in the layered structure of the film and/or selection of materials with more closely matched thermal expansion or humidity expansion coefficients. The extent of the problem will depend on the environment to which the film product is likely to be exposed during its functional lifetime. If the film is expected to function in extreme temperatures and humidities, then this curling effect can be significant and it must be addressed by necessary design changes to the film product. Otherwise, if the film is expected to function under near-ambient conditions with little changes in temperature and humidity, this effect may be inconsequential. A common design solution to this problem is addition of a backing layer on the opposite side of the film from the functional layer, whose main function would be to counter the thermal or humidity expansion effect of the functional layer. To provide an effective counterbalance, this so-called *pelloid* layer must have a thermal (or humidity) expansion coefficient and thickness that match closely those of the functional layer(s) on the topside of the film structure. Of course, the cost of adding this pelloid layer must be carefully weighed against the implications to film performance from thermal or humidity-induced curl effects. Finally, the general effect of mismatched thermal or humidity expansion coefficients in the film structure can be minimized by stiffening the substrate layer through increase in its thickness and/or its Young's modulus.

Another potential source of curling in film products is residual stress induced in a deposited layer(s) during the R2R process. Examples of such stresses include drying stresses generated during a drying step [13–16, 19] or curing stresses induced during a UV (or other radiation) curing step [17, 18]. In both cases residual tensile stresses are locked in a deposited matrix as it transitions from a liquid to a solid during the R2R operation. These stresses apply only in the deposited layer and could lead to curling in the finished film product if the film structure is not properly balanced. The extent of curling in this case depends on the magnitude of the induced stresses and the layered structure of the film. The main approach to mitigate the problem is to change the process conditions, especially for the drying or curing steps, such that the level of residual stress is minimized. For example, lower residual drying stresses can be achieved by applying less "harsh" drying conditions, that is, slowing down the drying process by lowering the drying temperature or reducing the line speed [16], which may necessitate increasing the length of the drying station. This, of course, can have consequences for the economics of the operation [20], but it must be weighed against the problems associated with the corresponding curl induced in the film. Similarly, residual stresses generated in a curing step can be lowered by delivering a similar radiation dosage distributed over a longer curing station. Again, this solution may be limited by cost and design constraints. In the latter case it is also important to ensure that the curing level is uniform across the web and through the thickness of the solidified layer as gradients in the extent of cure could also lead to a non-planar product. Finally, as for the thermal and humidity expansion effects discussed previously, stiffening the substrate layer by increases in thickness and/or Young's modulus could also minimize the effect of residual stresses on the extent of curling [14–17].

All in all, deviations from planarity are a significant quality issue in many R2R-based products, with a potentially deleterious effect on product performance, yield, and manufacturing cost. Such issues can be effectively addressed by proper process and/or film design changes. But the choice of the particular mitigating steps requires good understanding of the underlying curling mechanism and the potential impact of the problem and possible solutions on process robustness and cost.

## References


1 Schrader, R. W.; Carroll, J. F. Process for Reducing Core-Set Curling Tendency and Core-Set Curl of Polymeric Films. US41417345A, **1979**.

2 Greener, J.; Tsou, A. H.; Ng, K. C.; Chen, W. A. *J. Polym. Sci. B* **1991**, 29, 843–858.

3 Massa, D.; O'Reilly, J.; Perchak, D.; Gillmor, J. In *APS Annual Meeting*; Minneapolis, MN, **2000**.

4 Timoshenko, S. P. *J. Opt. Soc. Am.* **1925**, 11, 233–255.

5 Tsou, A. H.; Greener, J.; Smith, G. D. *Polymer.* **1995**, 36 (5), 949–954.

6 Greener, J.; Tsou, A. H. In *Theoretical and Applied Rheology*; Moldenaers, P., Keunings, R., Eds.; Elsevier: New York, **1992**; pp 799–801.

7 Struik, L. C. E. *Physical Aging in Amorphous Polymers and Other Materials*; Elsevier: New York, **1978**.

8 Struik, L. C. E. *Polymer.* **1987**, 28 (9), 1521–1533.

9 Greener, J.; Gillmor, J. R. *J. Polym. Sci. B* **2001**, 39, 1957–1967.
10 (a) Hutchinson, J. M. *Prog. Polym. Sci.* **1995**, 20 (4), 703–760; (b) Greener, J.; Ng, K. C. Method and Apparatus for Reducing Curl in Wound Rolls of Photographic Film. US5795512, **1998**.
11 Umberger, J. Q. *Photogr. Sci. Eng.* **1957**, 1 (2), 69–73.
12 Chow, T. S. *J. Appl. Phys.* **1976**, 47 (4), 1351–1354.
13 Croll, S. G. *J. Coatings Techol.* **1978**, 50 (638), 33–38.
14 Fu, T. J.; Durning, C. J.; Tong, H. M. *J. Appl. Polym. Sci.* **1991**, 43, 709–721.
15 Lei, H.; Payne, J. A.; McCormick, A. V.; Francis, L. F.; Gerberich, W. W.; Scriven, L. *J. Appl. Polym. Sci.* **2001**, 81 (4), 1000–1013.
16 Lei, H.; Francis, L. F.; Gerberich, W. W.; Scriven, L. E. *AICHE J.* **2002**, 48 (3), 437–451.
17 Chung, H.; Joe, Y.; Han, H. *J. Appl. Polym. Sci.* **1999**, 74 (14), 3287–3298.
18 Igarashi, T.; Kondo, S.; Kurokawa, M. *Polymer.* **1979**, 20 (3), 301–304.
19 Shams Es-Haghi, S.; Cakmak, M. In *Roll-to-Roll Manufacturing: Process Elements and Recent Advances*; Greener, J., Pearson, G. H., Cakmak, M., Eds.; John Wiley & Sons, Inc.: New York, **2017**; Chapter 3.
20 Greener, J. In *Roll-to-Roll Manufacturing: Process Elements and Recent Advances*; Greener, J., Pearson, G. H., Cakmak, M., Eds.; John Wiley & Sons, Inc.: New York, **2017**; Chapter 1.

# 8

# Roll-to-Roll Processing of Glass

*Doug Brackley, Dale Marshall, Gary Merz, and Eric Miller*

*Corning Inc., Corning, NY, USA*

## 8.1 Introduction

Glass is a highly desirable substrate for many applications in multiple form factors. One form factor, thin flexible glass, has numerous potential high volume applications, including handheld electronics, displays, appliances, architecture, and photovoltaic (PV) energy. Developing roll-to-roll (R2R) processes for the thin flexible glass form factor is an area of active investigation for a number of research groups around the world. This chapter provides an overview of the Corning® Willow® Glass and its manufacturing process and summarizes the challenges and keys to success for R2R processing of glass. Its potential uses and advantages as a substrate for future applications are also discussed.

Willow® Glass is produced by a continuous manufacturing process where it is formed from raw materials, processed, and then rolled onto spools for shipping. Beginning with a brief history of the rolled (or spooled) glass efforts at Corning, the chapter goes on to review the properties of glass that provide significant advantages over more traditional R2R substrate materials. These include dimensional stability, optical transmission, surface roughness, and hermeticity. The chapter then identifies those properties that have an impact on the R2R process, such as modulus of elasticity and mechanical reliability. Finally, key process factors that are important elements of a successful continuous R2R glass process are outlined.

## 8.2 History of Rolled Glass at Corning

As early as the 1970s, Corning began experimenting with rolls of continuous glass webs. Corning produced 13 μm thick by 25 mm wide glass through a slot draw process, which was then wound onto large diameter spools. Slot draw is

*Roll-to-Roll Manufacturing: Process Elements and Recent Advances*, First Edition.
Edited by Jehuda Greener, Glen Pearson, and Miko Cakmak.

a down-draw glass forming process in which glass is delivered to an elongated conditioning system and removed through a machined slot in its bottom by a pair of pulling rollers [1]. The glass was wound with a paper interleaf material and was targeted for the manufacture of capacitors.

At the request of a potential customer for an automotive application, an initial rolled glass experiment was conducted with slot drawn glass in the mid-1980s. This glass was 150 μm thick by approximately 16″ wide. As formed, it had relatively thin edge beads (approximately twice the thickness of the center web), allowing the glass to be wound on a spool without removing the edge beads. A thick polyester interleaf material was used to prevent the beads from touching and eliminate surface-to-surface contact of the wound convolutions. A dozen or so rolls of material were produced as customer samples, but the application was never commercialized.

This experiment was then repeated in mid-2009 at the request of another potential customer for a PV application. Glass web lengths of 100–250 ft were successfully wound onto a few large diameter (24″) spools. Learnings from that trial and a subsequent trial a few months later were then used to upgrade the spooling process. In early 2010, a multi-day, extended run produced a large number of spools of glass that were later used for process development of continuous bead removal and edge protection.

As a result of these development efforts, Corning was able to implement a continuous converting process for fusion drawn glass. Also a down-draw process, the molten glass is fed into a trough called an "isopipe," overfilling until the glass flows evenly over both sides. It then rejoins, or fuses, at the bottom, where it is drawn down to form a continuous sheet of flat glass [2]. The fusion draw process has several advantages over the slot draw process. It can produce much wider glass (>1 m), with more pristine surface quality and glass compositions that are amenable to more diverse applications. However, it also presents an additional challenge from a continuous processing perspective, as the sheet has thicker, stiffer edge beads than glass made with the slot draw forming technique. This process was started up in 2012. Process learnings were then used to implement major upgrades in 2014.

## 8.3 Key Attributes of Glass

Several key properties of glass provide significant advantages over more traditional R2R materials used as substrates for the fabrication and performance of high-quality electronic devices [3]. These include:

- Dimensional stability/stiffness: Glass is much more dimensionally stable (stiffer) than polymers with respect to stress and temperature as shown in Figure 8.1.
- Optical transmission: Glass has higher optical transmission over a wider spectrum of wavelengths than most polymers as indicated in Figure 8.2.

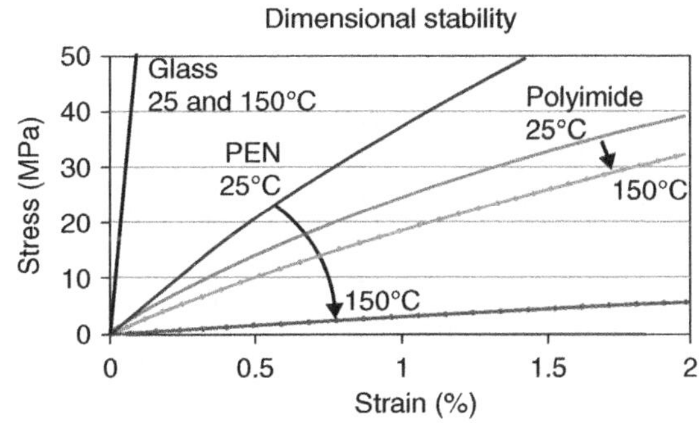


**Figure 8.1** Stress-strain curves for glass and two high-performance polymer films at two temperatures.

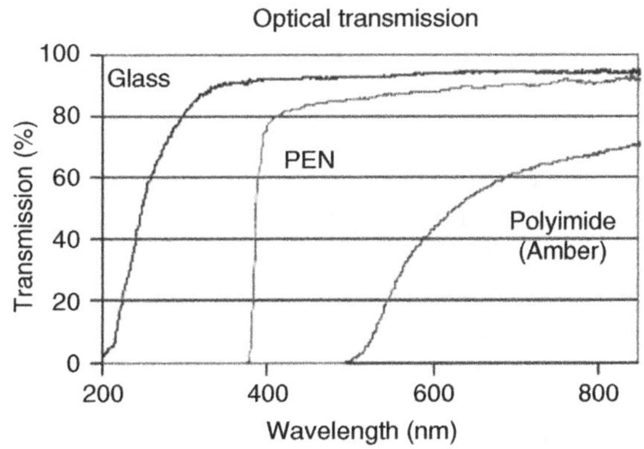


**Figure 8.2** Transmission spectra of glass versus polymer films.

- Surface roughness: The surface roughness of glass is significantly lower than that of polymers as shown in Figure 8.3.
- Hermeticity (barrier properties):
  Glass is a high-performance barrier material. In fact, glass substrates are typically used as the benchmark when measuring the barrier properties of various films. This is important in applications such as organic light-emitting diodes (OLED) and PV where oxygen or moisture permeation has a significant impact on product life.
- Thermal stability:
  The outstanding thermal stability of glass enables high temperature processing, such as sintering and high temperature backplane fabrication, to be performed. By way of comparison, the glass transition temperature ($T_g$) for alumino-borosilicate glass is approximately 700°C, while polyethylene naphthalate (PEN), a high-performance polymer, has a $T_g$ of 120°C.

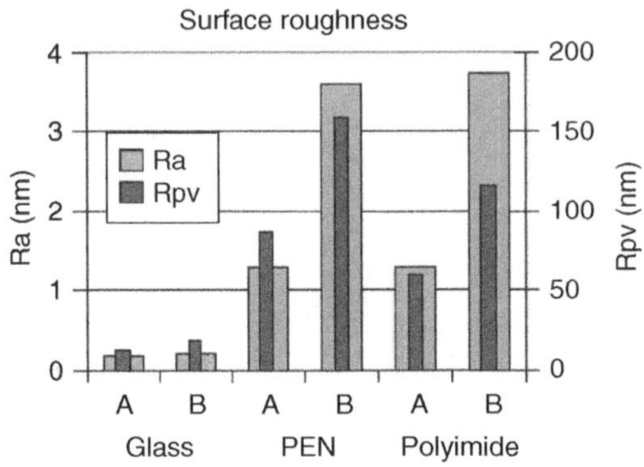


**Figure 8.3** Comparison of surface roughness characteristics of glass and high-performance polymers.

## 8.4 Properties of Glass That Impact R2R Processing

Although stiffer than polymers and papers, glass is sufficiently flexible to be wound onto a spool at thicknesses below approximately 300 μm. In fact, a 50 μm glass film has approximately the same stiffness as a 120 μm PEN film.

The modulus of elasticity of glass is approximately 73 GPa, which is significantly higher than that of film or paper (~2.5 GPa). Thus, at tensions typically applied in web conveyance (1–2 PLI (pounds-force per linear inch width)), glass does not experience as much strain as polymer or paper webs. As discussed in the next section, this has significant implications to web tensioning, handling, and wound roll integrity.

Mechanical reliability is an important aspect of R2R processing. As formed, pristine glass surfaces are extremely strong. Bend strengths of greater than 6 GPa have been measured. It is the subsequent handling that can reduce the strength by a factor of 100. Maintaining the quality of the glass surface and producing strong, defect-free edges on the glass web are critically important, as demonstrated by the data in Figure 8.4. The information presented is for glass types typically used in the display industry. However, glass composition does not have a major impact on these properties [3]. Furthermore, mechanical reliability requires controlling defects and applied stress during web conveyance and handling as indicated in Figures 8.5 and 8.6.

This has significant implications on the type of web conveyance and web steering that can be utilized in the R2R process, as well as how the glass surfaces need to be protected when wound on a spool. These implications will be discussed in the next section.

Similar to more traditional web materials (i.e., polymers, papers, and metals), edge wrap, thickness variation, and camber are important issues that need to be considered and addressed in R2R glass processing.

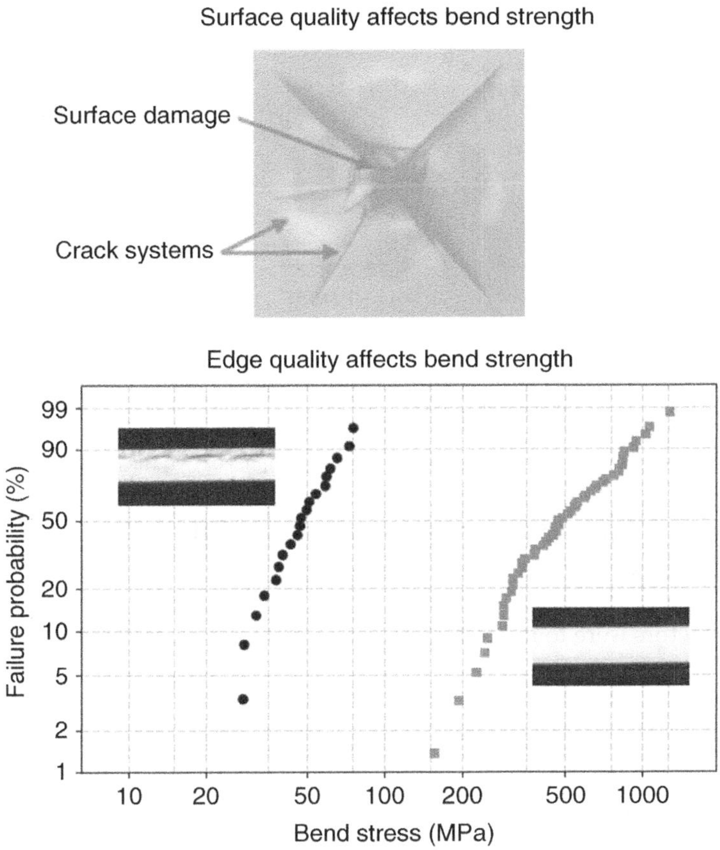


**Figure 8.4** Effect of bend stress and edge quality on failure probability and (top) image of crack induced by surface damage.

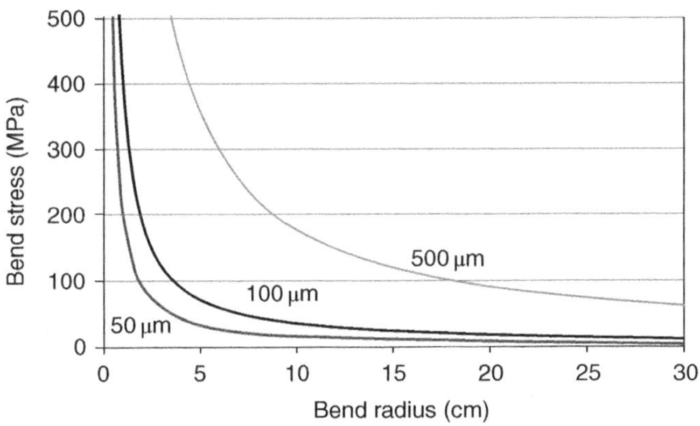


**Figure 8.5** Effect of bend radius on bend stress for glass films of different thicknesses.

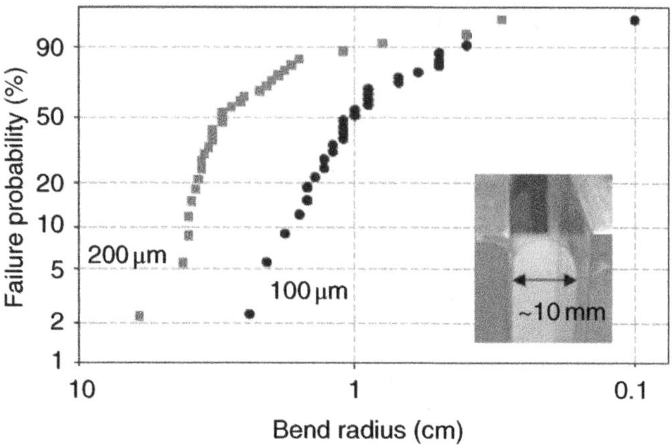


**Figure 8.6** Effect of bend radius on failure probability for glass films of two thicknesses.

## 8.5 Important Considerations for a Successful R2R Glass Process

The following factors were found to be important enablers of a successful R2R glass process:

- *Line speed:*
  Line speeds typically in the range of 3–30 m/min, depending on the application and in-line process requirements, were found to work best with R2R glass webs. However, web conveyance of glass does not appear to be limited to this range.
- *Process isolation and web conveyance:*
  There are several unique features of glass that create challenges for successful conveyance and stabilization of the glass web. First, the glass is extremely sensitive to surface defects, which create stress points that generate cracks and lead to breakage (see Figure 8.4). Therefore, roller conveyance that directly contacts the web surface, as is typically done with polymer, paper, or metal webs, is not always feasible. One of the key benefits of roller conveyance is that it provides lateral tracking forces that stabilize web conveyance. With glass webs, non-contact air conveyance often replaces roller conveyance. In the absence of these lateral tracking forces, glass web conveyance is inherently unstable laterally. Second, maintaining lateral alignment of the glass web as it travels through glass manufacturing equipment may be complicated by misalignment of components. Further, instabilities, perturbations, vibrations, and transient effects that may exist in manufacturing environments or in processing and handling equipment may cause intermittent or extended misalignment of the glass web in the lateral direction. Third, the forming process can produce differential cross web camber (continuous curvature in one

direction) in the glass web. Camber, thickness variation, and possible residual stresses would cause the glass web to shift laterally towards its “long” side rather than conveying in a straight line.

Laying out a system level isolation, tensioning, and appropriate conveyance scheme is an important part of implementing a successful continuous R2R glass process. For applications where contact conveyance is acceptable, traditional methods, such as displacement guiders, roller steering, sidelay, and turn bars, can be employed. For applications where non-contact conveyance is required, unique methods to control lateral position and tension have been developed and successfully implemented.

- *Stress and web path geometry:*
  Overall web path geometry should be considered to minimize conveyance length and avoid producing undesirable stresses in the glass web. Web bend radiuses that are within allowable limits should be used throughout the process web path. This is particularly important for process steps such as converting. In addition, any opportunity to reduce the number of bends or stress cycles the glass experiences should be considered. This will reduce the growth of edge or surface defects in the web.
- *Spooling requirements:*
  There are several unique features of glass that create challenges for successfully implementing a spooling step. First, glass is extremely sensitive to surface defects, which create stress points that generate cracks and lead to breakage. Thus, it is not advisable to have direct surface-to-surface contact of glass to itself, as is typically found in a spooled roll of material. Second, the forming process can introduce differential cross web thickness and/or camber (continuous curvature in one direction caused by differential cooling between the two edge beads). When spooling a roll with differential cross web thickness and/or camber, lateral forces are generated in the spooled roll, resulting in an angled or dished, rather than straight, side wall for the wound roll. Third, glass does not stretch under tension, which results in extremely low interlayer roll pressure. As a result, the resistance of the wound roll to lateral side shifting and telescoping is compromised.

  To overcome these issues, a thickness compliant interleaving material is typically used. The glass web is introduced into the interleaving web path. This interleaving/glass combination is then wound onto the cylindrical core, resulting in a spooled roll of glass where successive wraps are separated from each other by a layer of the interleaving material, as shown in Figure 8.7.

  Furthermore, the interleaving material is used to control the tension, and thus the resultant interlayer pressure, in the wound roll. The interleaving web tension is set low (<0.5 PLI) and is tapered as a function of increasing wound roll diameter. This tapering, combined with active sidelay (shifting the winder in the cross web direction to the correct position to “catch” the web and place it on the core in the right position), allows the production of wound rolls with straight side walls. A typical glass R2R windup station is shown in Figure 8.8.

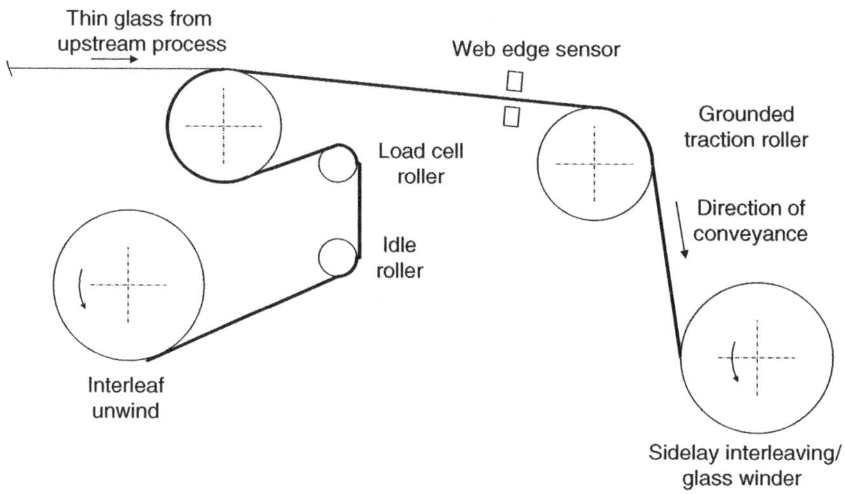


**Figure 8.7** Typical spooling configuration for a glass R2R process.

**Figure 8.8** A typical glass R2R windup station.

- *Handling considerations:*
  To reduce the growth of defects in the glass roll, care should be taken in the transportation of the roll from the manufacturing site to the packaging area. Roll handling equipment similar to what is used in polymer films and paper are sufficient for internal handling.
- *Packaging and shipping considerations:*
  To reduce the growth of defects in the glass roll, care should be taken to mitigate undesired effects of vibration and impact from handling and shipping.

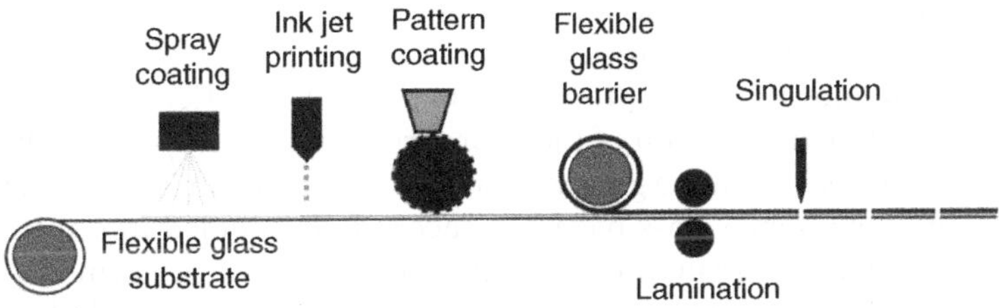


**Figure 8.9** Examples of typical R2R glass film converting steps.

Packaging should be designed to isolate the glass roll from outside disturbances that could degrade the quality of the product. International Safe Transit Association (ISTA) guidelines should be followed in package testing.

Once glass has been successfully spooled into rolls, there are a number of downstream process steps that can be performed in a continuous R2R manner [4] to convert the glass substrate into a fully functionalized film (see Figure 8.9).

Possible converting steps include, but are not limited to:

- Edge coating
- Surface coating, including slot die wet coating and vacuum deposition
- Photolithography and patterning
- Lamination
- Printing (both decorative and functional)
- Defect inspection
- Spooling

Collaborations with external partners, such as the Center for Advanced Microelectronics Manufacturing (CAMM), Frontier Industrial Technology, Industrial Technology Research Institute (ITRI), and MDI Advanced Processing GmbH, have demonstrated capabilities for key building block processes [5].

## 8.6 Summary

This chapter presented an overview of the Corning Willow Glass and its manufacturing process. This is a novel continuous manufacturing process where glass is formed from raw materials, processed, and then rolled onto spools for shipping. Beginning with a brief history of the rolled (or spooled) glass efforts at Corning, the chapter goes on to review the properties of glass that provide significant advantages over more traditional R2R substrate materials. These include dimensional stability, optical transmission, surface roughness, and barrier performance. The chapter then identifies those properties that have an impact on a continuous R2R process, such as modulus of elasticity and mechanical reliability. Finally, key process factors that are important elements of a successful continuous R2R glass process are outlined. Of particular importance are process isolation, selection of the appropriate web conveyance technology, management of stress in the glass web, and utilization of an interleaving material when spooling. Once glass has been successfully spooled into rolls, there are a number of downstream process steps that can also be performed in a continuous R2R manner to convert the glass into a fully functionalized film.

## References

1 Brown, D. Sheet Glass Forming Apparatus. US 2422466, June 17, **1942**.
2 Dockerty, S.M.; Shay, G.C. Downflow Sheet Drawing Method and Apparatus. US 3149949, September 22, **1964**.
3 Garner, S.; Merz, G.; Tosch, J.; Matusick, J.; Li, X.; Marshall, D.; Chase, C.; Steiner, J.; Yepez, D.; Switzer, J.; Moschak, P. Flexible Glass Substrates for Continuous Manufacturing, Flexible Electronics and Displays Conference, Phoenix, February 9, **2011**.
4 Garner, S.; Merz, G.; Tosch, J.; Matusick, J.; Marshall, D.; Chase, C.; Steiner, J.; Yepez, D.; Switzer, J.; Moschak, P. Flexible Glass for Roll-to-Roll Electronic Device Manufacturing, Flexible Electronics Symposium, Binghamton University, August 16, **2010**.
5 Web, J.; Garner, S. Flexible Glass Substrates for Electronic Applications, Flexible & Printed Electronics Conference & Exhibition, Phoenix, February 4–6, **2014**.

# 9

# Novel Hybrid Composite Films by Roll-to-Roll Processing

*Saurabh Batra[1], W. Zhao[2], Baris Yalcin[3], and Miko Cakmak[4]*

[1] *3M Company, St. Paul, MN, USA*
[2] *Department of Polymer Engineering, University of Akron, Akron, OH, USA (Present address: SABIC, Mt Vernon, USA)*
[3] *Department of Polymer Engineering, University of Akron, Akron, OH, USA (Present address: The Coca Cola Company, Atlanta, GA, USA)*
[4] *Department of Materials & Mechanical Engineering, Purdue University, West Lafayette, IN, USA*

## 9.1 Introduction

This chapter describes a novel roll-to-roll (R2R) process for producing hybrid polymer films with unique functional properties. The polymer films generated consist of single, laminated, complete, and/or partially embedded nanofibers. The novel R2R manufacturing line is a hybrid process that combines the well-known electrospinning and solution casting processes [1]. The integration of these two technologies was achieved through an endless belt system acting as the carrier that combined the continuous mass production of fibers by electrospinning with the mass production of films by solution coating. Hybrid functional films that have synergistic properties of films and fibers, thus generated, lead to distinctive properties such as flexibility, conductivity, and so on.

The solution coating process used to create polymer thin films has often been referred to as tape or band casting in the industry. This process involves spreading of a polymer solution using a doctor blade, reverse roll coater, or slot die on a substrate, which also acts like a carrier. The solution is then dried with the application of heat [2–4] leading to films that can range from millimeter to a few microns thick, uniform coatings.

The solution coating process has several advantages specifically for materials that are difficult to melt due to melt instabilities incurred while making films at small thickness ranges. Commercially available solution coating machines have been manufactured for many years and usually include a solid coating surface, a drying chamber with the means to control the airflow over the coated

*Roll-to-Roll Manufacturing: Process Elements and Recent Advances*, First Edition.
Edited by Jehuda Greener, Glen Pearson, and Miko Cakmak.

solution, an air heater to control the heating through convection, underbed heaters to set up the desired temperature profile, and an adjustable drive to control carrier speed to set the residence time for drying [5–7]. The carrier belt itself is made of stainless steel; however, coated paper or polymeric substrate with required surface energy is typically used to get desired wettability and peelability. The stainless steel belt may include the bottom heater to dry the solvent through conductive heating. To control the drying/solidification of the cast fluid, convective heating from the air above and conductive heating below the belt are often used simultaneously. Multiple carriers in a single solution casting machine can also be used.

The other process used to create functional films using this hybrid machine is electrospinning [8]; this method of producing fibers has also been known as electrostatic spinning in the industry. Electrospinning involves tailored solutions that are capable of forming fibers, the requirements of which have been detailed in numerous journals and patents. During the electrospinning process, a DC electric field is applied to the surface of the solution that leads to repulsive forces in the solution, which create a jet of liquid carrying an electrical charge. These electrically charged jets are attracted to the body or an object that is at a suitable electrical potential. As the liquid jet moves toward the object, it starts to elongate [9] and hardens steadily during the motion, thereby forming a fiber. The hardening of the fiber may be caused by cooling of the liquid if it is solid at room temperature or by drying of solvent due to evaporation. The hardening of fibers can also be physically induced such as by thermal, chemical, or UV curing during the flight of the fibers. The fibers produced by electrospinning techniques are collected on a suitably located charged receiver and subsequently removed from the receiver if needed.

In this chapter, we will discuss the use of electrospun/solution-coated fiber processing machine for various demanding applications such as flexible transparent conductive films, reinforced aerogels, and biaxially stretchable hydrogels. Due to extensive literature on electrospinning, covered in many articles and patents, it is not discussed in detail in this chapter. Chapter 2 in this volume provides more information on the solution coating process.

## 9.2 Process Overview

Figure 9.1 shows the schematic of the novel hybrid film manufacturing line. The framework of this line is around a solution coating equipment, which consists of a mirror finished endless stainless steel belt, four underbed heating zones with independent controls, and two zones of air circulation chambers where each zone has its independent air temperature control system along with variable fan speeds. Along with the stainless steel belt that can be used as a substrate to coat solutions, the machine is equipped to handle a removable

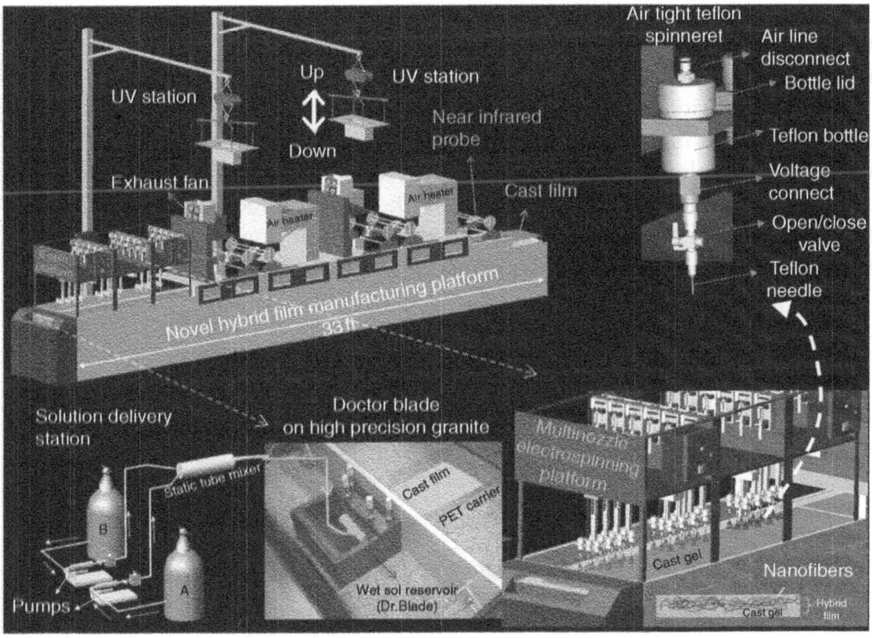


**Figure 9.1** Schematic of the electrospinning/solution casting hybrid machine [10]. © Royal Society of Chemistry.

polymer film carrier system. The polymer film substrate usually runs on top of the stainless steel belt and can be used to coat the polymer solution. The lower surface energy of the polymer carrier makes it easy to remove the coated film from the substrate.

The electrospinning bench consisting of 36 electrospinning nozzles was mounted over the carrier in front of the machine following the coating unit. The bench was divided in three different sections, where each section was mounted on an actuator to independently control the linear motion of the electrospinning nozzles. Each zone in the electrospinning section consisted of 12 independently controlled electrospinning bottles that could move back and forth through the width of the carrier containing the coated solution with a precise speed control. Each individual nozzle has its independent control over pressure on the solution, applied voltage, and type of solution. This hybrid process is also equipped with two UV stations, which can be utilized to process UV-curable materials, and with near-infrared probes that can be used to monitor the molecular structure change in the process.

The back and forth motion of the electrospinning units on the linear actuator leads to a uniform deposition of nanofibers through the width (transverse direction) of the web. Typically the electrospinning process is limited in throughput; however, the use of 36 independent nozzles increased the throughput considerably.

The R2R hybrid process can be operated with two different operation modes to create functional films with embedded nanofibers. In the first mode, the nanofibers are electrospun directly in the cast solution, which is carried using a substrate on the endless belt system. In the second mode, the nanofibers are first electrospun on the substrate, and then the solution is cast on top of the nanofiber mat. These two modes of operation are depicted in Figure 9.2.

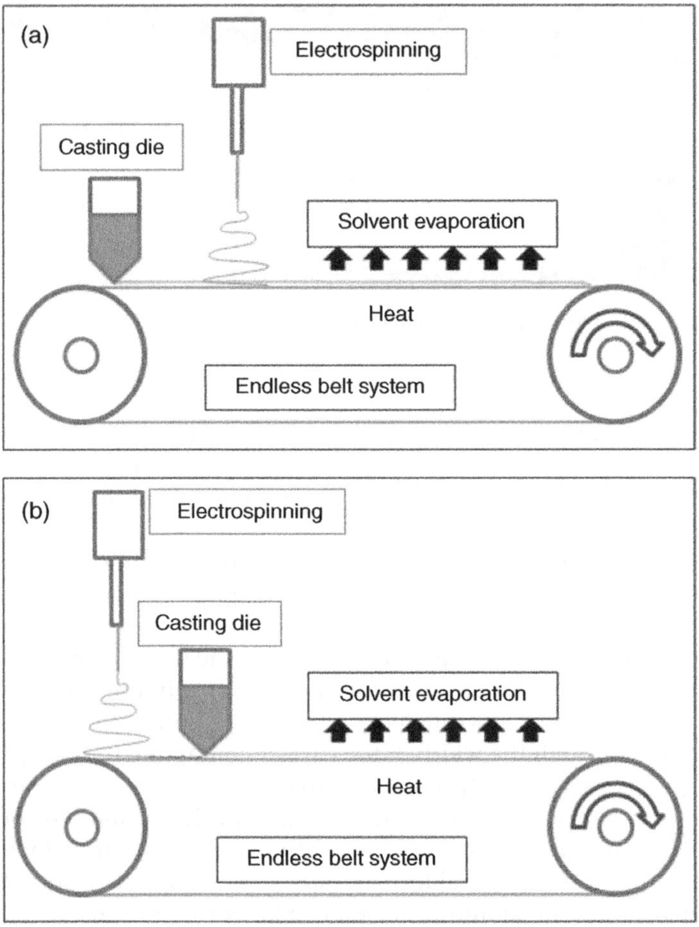


**Figure 9.2** Two operation modes of electrospinning/solution casting R2R machine. (a) Electrospun nanofibers directly into cast solution. (b) Cast solution over electrospun nanofiber mat [11]. © John Wiley & Sons.

## 9.3 Transparent Electrically Conductive Films

Transparent and electrically conducting films are essential components primarily used for various electro-optical devices such as liquid crystal displays (LCDs), plasma displays (PDs), touch panels, organic light-emitting diodes (OLEDs), and photovoltaics. They have also been used in electrostatic discharge (ESD) applications that require resistivity, which are magnitudes higher around $10^{9}$–$10^{12}\,\Omega^{-2}$, as compared to that required for the display industry, usually lower than $10^{3}\,\Omega^{-2}$ [12].

Coating of indium tin oxide (ITO) on polymer films is the current material of choice for transparent conducting films since it exhibits unique properties in combination with good optical transparency (90%) and low electrical resistivity ($100\,\Omega^{-2}$). However, ITO cannot be used for applications that need flexible films as it shatters when bent or rolled repeatedly or if coated on flexible substrates such as polyester films due to the inherent brittle behavior of ITO coatings. These coating lose their conductivity even at small strains, and the loss becomes severe when they are stretched to large strains along with the bendability and rollability requirements [13, 14]. There are devices and processes where flexibility and stretchability are both desired, which include various R2R lines for manufacturing displays or during thermoforming the films into a desired shape. In addition, devices such as wearable electronics, future displays, sensors attached to skin, solar cells, and so on need not only flexibility and stretchability but also subsequent recovery from strain while maintaining low resistivity.

Due to the issues discussed, most of the recent studies have been focused on the preparation of the transparent conductive films by completely eliminating the use of ITO coatings. Carbon nanotubes [15], graphenes [16, 17], nano-metal materials [18–20], and intrinsically conducting and solution-processable polymers [21–27], such as poly(3,4-ethylenedioxythi ophene):polystyrene sulfonate (PEDOT:PSS) [28–30], have been the most commonly employed materials. Films based on both carbon nanotube and graphene rely on the formation of a continuous thin carbon film layer on top of a flexible substrate. The preparation of these films can be achieved using R2R manufacturing equipments [31, 32]; however, those processes are costly and have various limitations, making them commercially unviable. Some of the processes used to coat a thin continuous electrically conductive layer on flexible substrates include chemical vapor deposition (CVD), sputtering, reactive evaporation, spray coating and various sol–gel methods, and so on. Coating on flexible substrates using abovementioned processes will retain the electric conductivity with repeated bending and flexing of the substrate; however, the tolerance of such thin coatings decreases dramatically with increasing level of stretching leading to increase in resistivity.

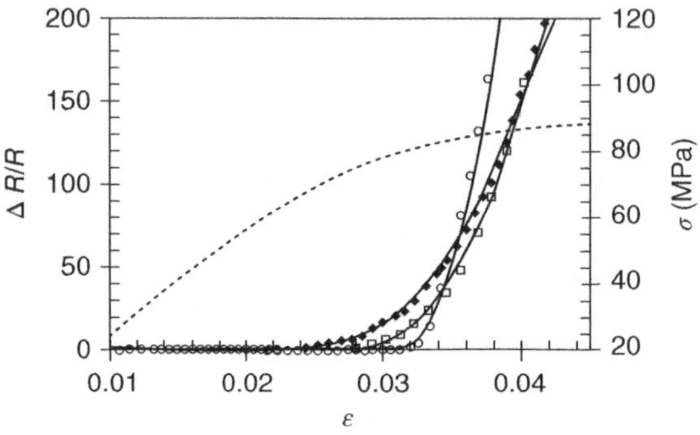


**Figure 9.3** Change in resistance during the stretching of ITO coated PET film [13]. © John Wiley & Sons.

The transparent conductive functional film market has grown considerably especially in the flexible electronics and display markets as we move toward flexible displays and wearable electronics. As discussed previously, most transparent conductive electrodes are made from ITO deposited onto glass or polymer film substrates. Due to the brittle nature of the coating, they can withstand only a small strain of about 2.5% before the thin ITO layer cracks and loses its ability to conduct electricity making it unsuitable as a component for flexible electronic devices (Figure 9.3).

To provide flexible transparent conductive electrodes for use in future flexible electronic applications, the hybrid processing line was designed by Cakmak et al. [10] to produce polymer films with a conductive submicrometer/nanofiber network partially embedded in them. The typical structure of the nanofiber embedded film is shown in Figure 9.4. The submicrometer/nanoscaled conductive fibers were generated by electrospinning using a solution of an intrinsic conductive polymer and a polymer carrier. The electrospinning process can generate nanoscaled fibers and nonwoven mats using a simple device. The electrospun conductive fiber mat was then embedded into a flexible transparent polymer film by casting the solution of film material onto the conductive fiber mat, followed by removing the solvent from the cast solution.

These flexible conductive films shown in Figure 9.4 were produced using the hybrid machine in the operation mode described in Figure 9.2b. Using different electrospinning conditions, it is possible to embed conductive fibers ranging from submicrometer to nanometer scale, which helps to control the properties of the final functional film. Along with the fiber dimensions, it is possible to control the substrate such as polycarbonate (PC), poly(methyl methacrylate) (PMMA), polyimide (PI), and so on.

The effect of substrate and fibers on conductivity, stretchability, and so on was studied by making samples with dimensions of 15 cm in width and up to

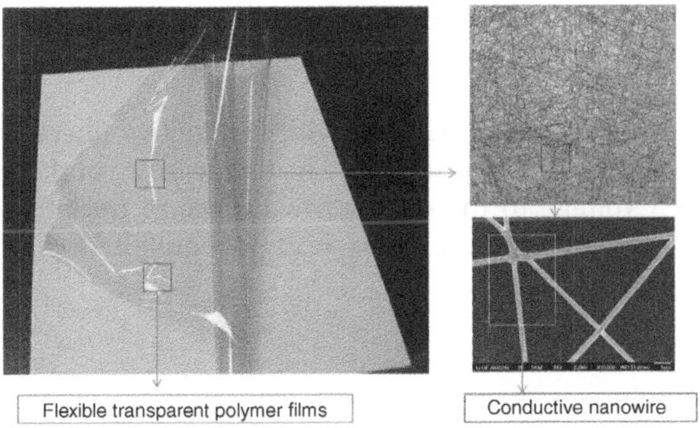


**Figure 9.4** Structure of the transparent conductive films with embedded nanofiber [11]. © John Wiley & Sons.

3 m in length. The conductive fibers were then embedded in the substrates with different weight ratios. These fibers were generated by electrospinning the PEDOT/PEO solution with 0.7 wt % PEO concentration. The films prepared showed high transmittance that could reach well above 80% at a wavelength of 550 nm as can be seen in Figure 9.5a. Along with transparency, these films were flexible and could be bent over a 2 mm diameter rod for over 300 times while maintaining the resistivity below 1 k$\Omega^{-2}$ (as seen in Figure 9.5d). ITO coated films, that were used for comparison, lost their conductivity when they were bent over a rod with a diameter of 6 mm as shown in Figure 9.5c.To further demonstrate flexibility, both the transparent conductive films from the hybrid process, and the ITO coated films were subjected to a folding test. In this test, both the films were folded and then placed under a load of 2.5 kg for 15 min. Flexible transparent films made on the hybrid line maintained low resistivity even after the fold test; however, ITO-coated films lost their conductivity completely after the test. Along with outstanding flexibility, the conductive fibers embedded films also show high stretchability. When these films were stretched to about 1.8 times of their original lengths above their glass transition temperatures, they maintained their conductivity to a certain degree, which has not been seen previously by any other transparent conductive films (Figure 9.5b). Again, in comparison with the conductive fiber embedded polymer films, the ITO-coated films lost their conductivity at very small strains as shown in Figure 9.5c and d.

The applicability of these transparent conductive films for real-life applications was shown using two examples: (i) resistive heaters and (ii) electrodes for flexible displays. In the past, there have been several attempts to replace ITO coatings in transparent heater applications. Various potential systems have been studied as their replacements, including carbon nanotubes that have been a prime topic of academic research. However as transparent heaters, carbon nanotubes suffer from common disadvantages such as transparency. In addition, they are expensive and it is usually difficult to obtain pure conductive carbon nanotubes.

The first example includes PMMA and PC films with conductive fibers as transparent heaters with a very short response time. The temperature of these films increased with applied voltage to the conductive surface of the films. Higher voltages and low resistance of the films lead to equilibration of

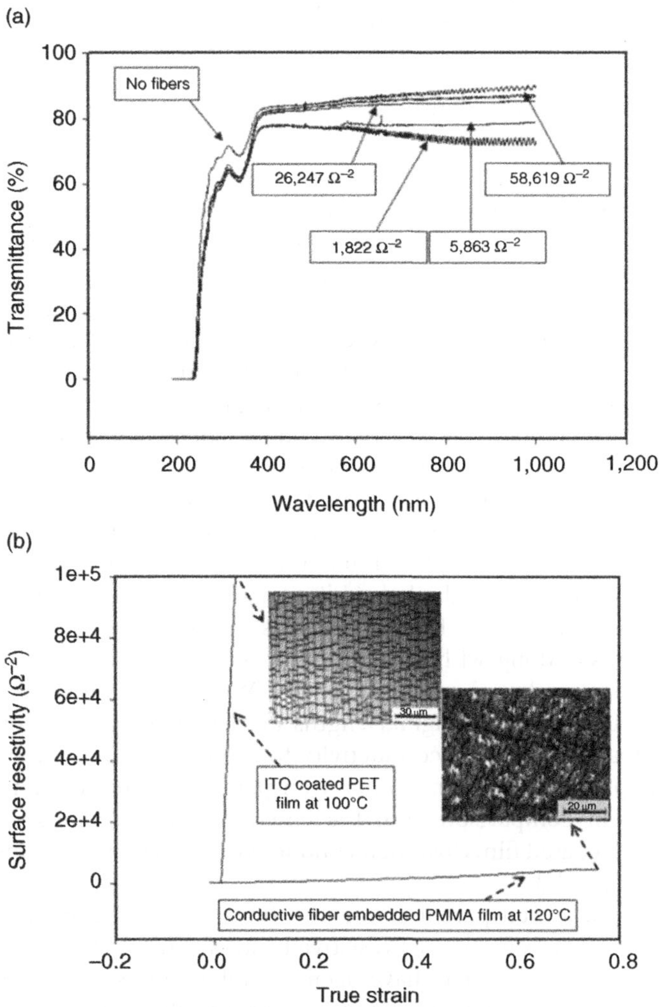


**Figure 9.5** (a) UV-Vis spectra of PMMA conductive films, (b) stretchability comparison of ITO-coated PET film and fiber embedded PMMA film,

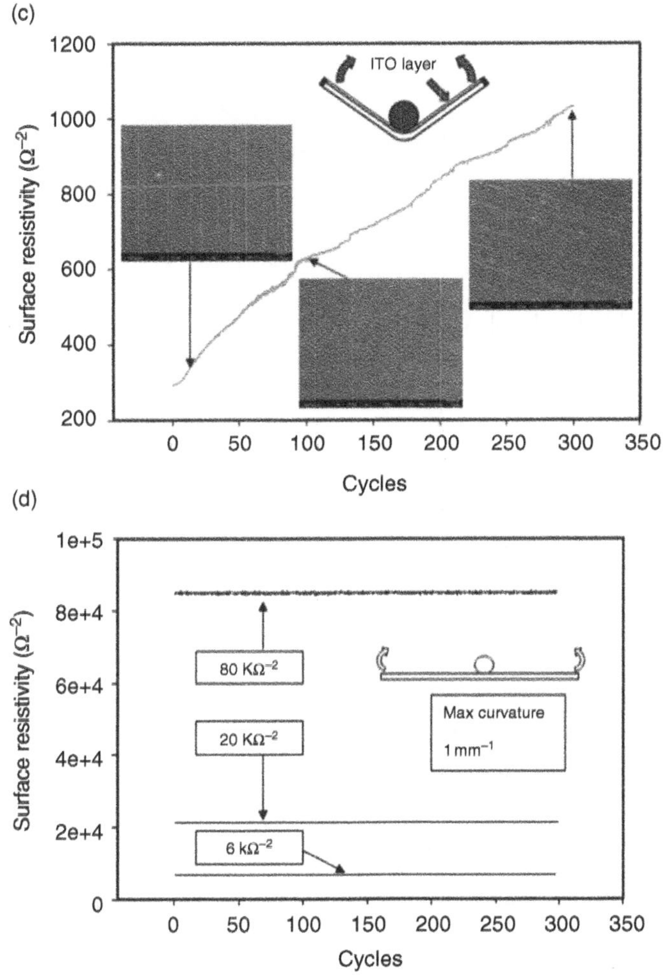


**Figure 9.5 (Cont'd)** (c) crack development in 16.5 nm ITO coating layer while being bent around 6 mm mandrel with ITO coating in compression, and (d) flexing tests of transparent conductive flexible PI films around 2 mm mandrel with the conductive side in tension [11]. © John Wiley & Sons.

temperatures as can be seen Figure 9.6a. The tested films had a short response time, and the temperature dropped as soon as the voltage was turned off. It took less than 20 s for the films to heat up to an equilibrium temperature after the application of voltage. The films that were created during this study were not only transparent but also flexible, making the film suitable for special applications where there is a need for flexibility.

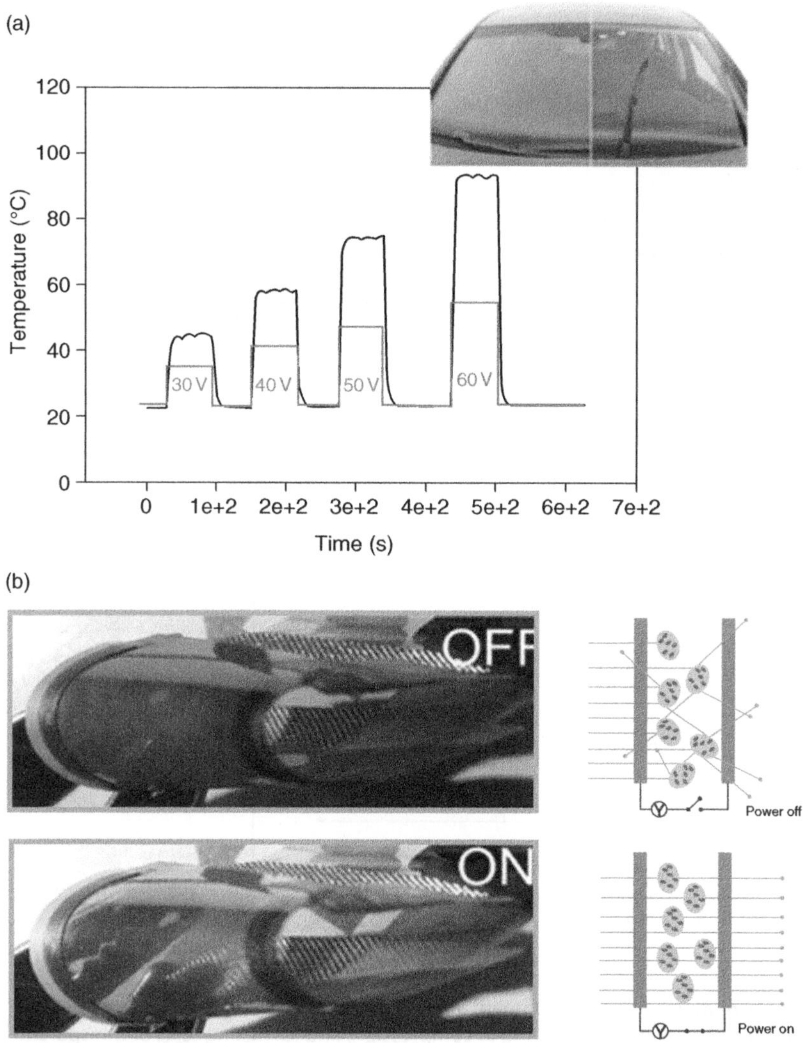


**Figure 9.6** (a) Hybrid functional films of PC and electrospun conductive fibers as heaters and (b) on and off states of the bent PDLC display from PI films [11]. © John Wiley & Sons.

The second example shows the use of hybrid films as flexible electrodes in PDLC devices. Display devices can be manufactured by encapsulating polymer dispersed liquid crystal (PDLC) between two glasses or polymer substrates that are coated with transparent conductive layers. The PDLC displays are widely used as a light valve since they are capable of adjusting and controlling the amount of light that passes through by controlling the voltage. They have also found applications in privacy screens and windows in various green houses. Most PDLC devices use electrodes that are ITO coated because of its optical clarity and good electrical conductivity. However, ITO-coated electrodes cannot be used for flexible devices. There have been some attempts to prepare flexible electrodes with polymers that are intrinsically conductive such as polyaniline (PANI) and poly(3,4-ethylenedioxythiophene) (PEDOT).

The flexible PDLC devices created using the hybrid line were prepared from polyimide films with embedded conductive PEDOT/PEO fibers. All devices prepared using flexible electrodes from the hybrid line showed response times that were close to the PDLC devices prepared using ITO-coated PET film as electrodes. The gray scale of all the flexible PDLC devices could be controlled by adjusting the applied voltage. PDLC devices became transparent at higher voltages, until the saturation voltage above which there was no change in transparency. The ITO-coated PDLC devices, on the other hand, were not flexible and lost their conductivity while bending the device. All polymer PDLC devices made using films from the hybrid line were still functioning after bending the display to 180° as can be seen in Figure 9.6b.

## 9.4 Bendable Aerogels (Xerogel)

Aerogels are unique materials formed by extraction of a gel using a supercritical fluid that leads to porous solids with extremely light open cells [33, 34]. These materials have very high internal surface area, large pore volume, and extremely low density. Aerogels have been sought for various applications that are not limited to flame retardants, thermal insulation, and so on. Their thermal conductivity can be as low as 0.13 W/m·K [35]. Along with thermal properties, due to their porous structure, they also exhibit a low dielectric constant, high impendence to acoustic waves, and various unique properties that are difficult to find in other materials. However even with these superior material properties, their use is pretty limited due to their inherent brittle nature. Therefore, aerogels cannot be used for devices that are flexible or conformable. Thus, there is a need to develop aerogels that are flexible and can conform to various devices as we move toward a flexible future. These flexible aerogels could be used in devices for home care, aerospace, transportation, and so on where there is need for thermal and acoustic insulation.

Cakmak et al. [36] have shown, for the first time using the hybrid process, a way to prepare flexible aerogel films. This was done using PDMS hybrid silica aerogel (xerogel) thin films, which were reinforced by continuous electrospun nanofibers. Using this method, the fragility of the aerogel was reduced by reinforcing the mesoporous structure. These aerogels were reinforced in three dimensions: perpendicularly, by macroscopically electrospinning nanofibers into the silica aerogel matrix; along with microscopically interconnecting interparticle necks within the mesoporous silica network in the planar dimension; and by crosslinking the silica particles with a silanol-terminated PDMS oligomer to modify the brittleness and stiffness of the silica matrix. Nanofiber reinforcement of aerogels was accomplished by electrospinning these fibers into the sol whose gelation kinetics was delayed to make the nanofiber penetrate easily into the low viscosity wet sol film. Controlling the gelation kinetics ensures that nanofibers are either embedded or anchored on the surface of the aerogel, which acts as a reinforcement by creating a continuous nanofiber network leading to flexibility and enhanced mechanical properties.

There are three major aspects to manufacturing nanofiber electrospun aerogel (xerogel) using the hybrid line: firstly the chemical composition of the sol and the gelation kinetics of the sol–gel process, secondly the composition of nanofiber materials, and thirdly the interaction between the sol component and the nanofibers during the gelation process and subsequent aging and washing process. These three factors provide a substantial opportunity to manipulate the morphology, microstructure, and so on and thus control the physical and mechanical properties of the reinforced aerogel films. During this study, various formulations with different kinetics and nanofibers with different material properties were prepared.

The scanning electron microscopy (SEM) image of fiber-reinforced silica aerogels can be seen in Figure 9.7c. The micrograph shows that the continuous nanofibers are embedded in the aerogel film. These fibers are uniformly dispersed creating a seamless boundary between the silica aerogel and the nanofibers. As can be seen from the top view of the fractured sample in Figure 9.7d, there are a number of unbroken nanofibers crisscrossing it, which indicates that the fibers are well embedded in the silica aerogel film. The thermal conductivity of these hybrid films was similar to that of the silica aerogels without the nanofiber reinforcement (13 mW/m·K). Hybrid aerogel films had lower thermal conductivity than that of air (26 mW/m·K) and comparable values to that of glass wool (40 mW/m·K) and polyurethane foam (24–30 mW/m·K).

This was the first reported synthesis method of an *in situ* electrospun nanofiber-reinforced aerogel/xerogel film with controlled silica sol gelation kinetics. BET surface area analysis shows that the mesoporous PDMS-modified hybrid aerogels, which were reinforced by using PU/PEO nanofibers, have higher surface area with a large pore volume. As discussed previously there is a seamless connection between the fibers and silica mixtures, the well-integrated fiber network not only increases the mechanical properties but also makes them bendable as can be seen in Figure 9.7c. These aerogels still not inherently flexible possess all of the useful attributes of pristine aerogels, that is, high porosity, low thermal conductivity, and so on. These properties, combined with the bendable nature of the entire composite film facilitated by the nanofibers, render these films useful for applications that require flexing, wrapping, and rolling [36].

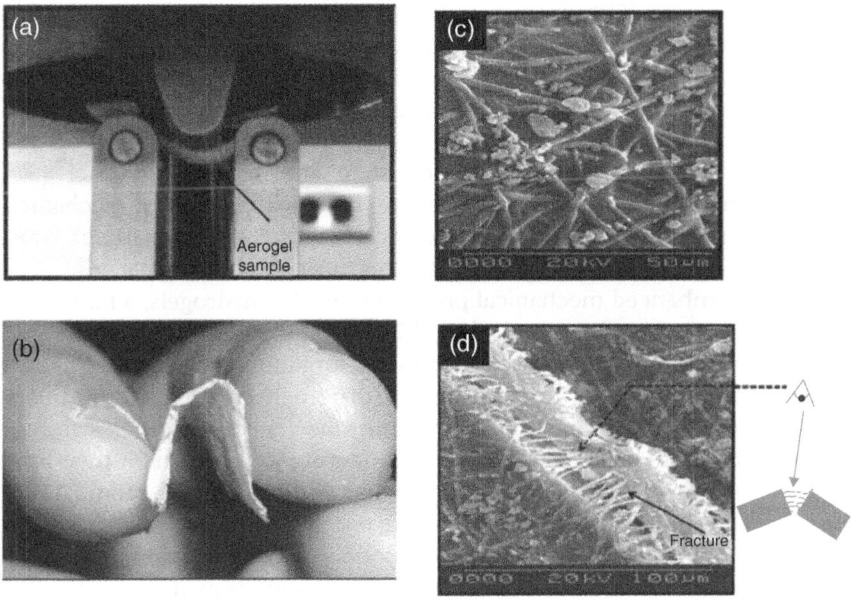


Figure 9.7 (a) 80 mil thick aerogel–nanofiber hybrid sheet flexing during the bending test, (b) 0.46 mm thick aerogel–nanofiber hybrid film bent almost by 180°, (c) SEM micrograph of a silica aerogel film reinforced with fibers, and (d) SEM micrograph of a fractured silica aerogel film reinforced with fibers [36]. © American Chemical Society.

## 9.5 Flexible Hydrogels

Unlike aerogels, as discussed in the previous section, which consists of mostly air due to the large number of pores, hydrogels are crosslinked polymer chains with interpenetrating networks that can encapsulate large quantities of water in between the polymer chains. Hydrogels are known for their excellent biomaterial characteristics [37] that include biocompatibility, nontoxicity, noncarcinogenicity [38], and biodegradability. These unique properties can lead the way to various applications such as contact lenses [39], wound dressing [40], separation membranes [41], drug delivery [42], artificial skin [43, 44], and in replacement of articular cartilage [45]. Most hydrogels currently being used in these applications are completely made of polyvinyl alcohol (PVA) or have PVA as one of the major components. PVA is a good candidate for hydrogels; however, it has poor stability in water due to its high hydrophilicity [46] and is usually crosslinked before use through different methods, which include chemical crosslinker [47], gamma radiation [48], electron beam, or freeze–thaw [49]. The use of chemical crosslinker can limit the functionality of these hydrogels and their utility in various biomaterial applications. Simpler methods have been reported [50] where crosslinked PVA gels can be obtained using solution casting of PVA and drying at room temperature without any chemical agents as the hydrogen bonds and microcrystals formed during drying can themselves act as physical crosslinking points. The improved physical crosslinking can lead to enhanced film stability, but these films are still mechanically weak during the swollen state after absorbing water [51]. Therefore, there is a need to create hydrogel films with enhanced mechanical properties for applications such as articular cartilages [52] as they are subjected to repeated large forces over millions of times during their lifetime [45].

To achieve enhanced mechanical properties in PVA hydrogels, a number of reinforcing agents have been looked at which include cellulose nanofibers [53], rigid fillers such as organoclay, for example, organophilic montmorillonite [54], and sodium alginate for transdermal drug delivery [55]. These fillers, however, only lead to limited enhancement in mechanical properties and are often hindered in their performance. Nanofibers are well known for their large aspect ratio and high surface-to-volume ratio and are a good candidate for interfacial adhesion [56]. Most hydrogels are tested for their uniaxial, mechanical, and stretching properties making a nanofiber nonwoven mat [57] promising for increased reinforcement in these fragile materials as they create a *soft skeleton* within the soft hydrogels.

In the following text, we discuss the results for nanofiber-reinforced hydrogels created on the hybrid line. A soft polyurethane (PU) nanofiber network was chosen to reinforce the PVA hydrogels as these fibers have low modulus at low strains but exhibit high strain hardening at large deformation, which leads to excellent compatibility with the soft hydrogels. Along with mechanical reinforcement, PU exhibits hydrolytic stability, low $T_g$, transparency, and toughness [58]. The hydroxyl group of PVA film and urethane group of PU nanofibers has strong hydrogen bonds leading to good interfacial linkages between the nanofibers and PVA. PU nanofibers have also been used as biomaterials [59] in biosensors [60], wound dressing [61], and drug delivery devices [62].

The hybrid process was used to manufacture samples where PU fibers were electrospun on the PVA cast films to enhance biaxial stretching properties of PVA hydrogels. Two sets of nanofibers with different diameters were chosen to reinforce the hydrogels; the average diameters of these nanofibers were 332 and 194 nm. The nanofibers had a broad diameter distribution, but the surface was smooth and they were randomly oriented in the plane to achieve isotropy in the reinforced hydrogel films, thereby increasing the biaxial stretching properties [63].

The first set of composite films produced on the hybrid process line consisted of PU larger size nanofibers electrospun on top of the PVA cast film before solidification. This was then followed by a drying step that entraps the embedded nanofibers in the PVA matrix. The morphology at the interface between the matrix and the fibers was studied by looking at the cross section of the composite films through SEM, as shown in Figure 9.8. The PU

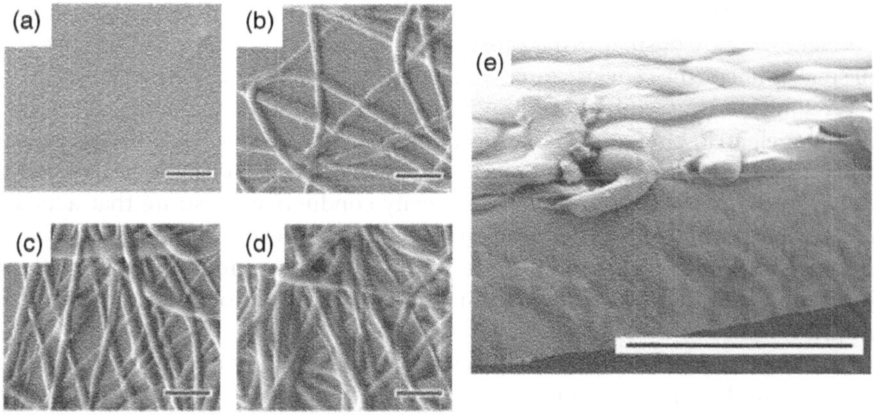


**Figure 9.8** Morphology of reinforced PVA films by larger nanofibers with different fiber contents before stretching: (a) neat, (b) 2%, (c) 3.7%, and (d) 5.4%. (e) Cross-sectional morphology. The scale bar is 5 μm [63]. © John Wiley & Sons.

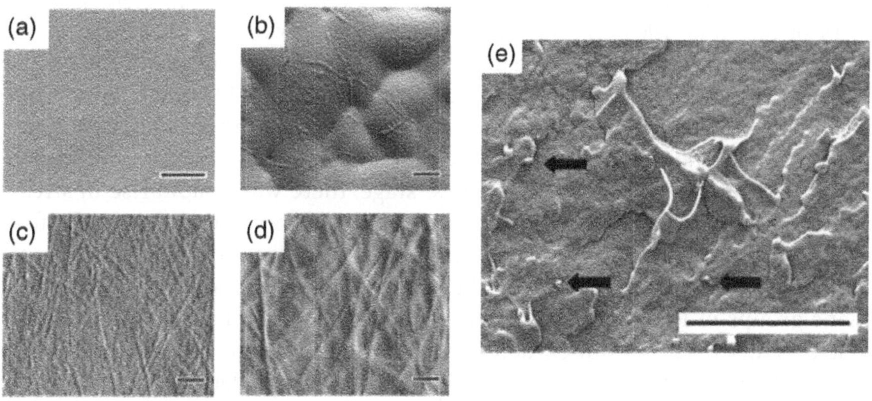


**Figure 9.9** Morphology of reinforced PVA films by smaller size nanofibers with different fiber contents before stretching: (a) neat, (b) 2%, (c) 3.7%, and (d) 5.4%. (e) Cross-sectional morphology. The scale bar is 5 μm [63]. © John Wiley & Sons.

nanofibers had good interfacial adhesion with the PVA, and the surface of the neat PVA film was smooth as most of the fibers were embedded in the film. However, with increasing the fiber volume in the film, more fibers are deposited on top of the film as can be seen in Figure 9.8b–d. As shown in Figure 9.8e, the majority of fibers resided on the surface of the films, while only a few got partially embedded in the matrix.

The second set of samples was made using the smaller diameter nanofibers, and, as can be seen from Figure 9.9b–e, the fibers were able to fully embed in the PVA films. The following factors led to the increased fiber embedding efficiency in case of smaller nanofibers: Firstly due to the small pore size in case of smaller fibers, the capillary forces are much higher that lead to a rapid uptake of PVA solution thus embedding the fibers in the solution. Secondly, the large residual electric charge per volume of smaller nanofibers leads to increased force that draws the nanofibers in the matrix in the established electric field between the nanofibers and the electrically conductive substrate that acts as electrical ground. As can be seen from the cross-sectional SEM view of the fractured sample (Figure 9.9), there is no apparent nanofiber delamination indicating that there is strong adhesion between the nanofibers and the PVA matrix.

The fiber-reinforced films were then put in a humidity chamber to achieve different swelling ratios before they were tested for their mechanical properties. A schematic of the process is shown in Figure 9.10a. Mechanical properties of these films were tested using a bubble testing device as shown in the schematic in Figure 9.10b. Before testing the films, they were swollen by treating it in the humidity chamber. Figure 9.11a and b shows the true stress–true strain behavior of the 6 μm thick films that were reinforced by nanofibers with average diameters of 330 and 190 nm. The films, which were swollen with a high water content, did not fail (break) even at the maximum stress available in the biaxial bubble stretching device; therefore, the curves in Figure 9.11a and b do not represent failure properties of the films. It was observed that there were two stages in the stress–strain curves: during the first stage, the stress increased linearly with the strain. After the end of the first stage, there was a rapid rise in stress, which was caused due to strain hardening. It was observed that during the first stage of the curve that the slope increased with increasing fiber content. Similarly, strain hardening during the second stage also increased as the urethane groups in PU nanofiber surfaces formed hydrogen bonds with the PVA matrix. Thus, these nanofibers acted as soft but increasingly hardening skeleton segments, which can carry large stresses. Films embedded with smaller size nanofibers had a higher initial increase in slope. This can be attributed to more effective reinforcing of the smaller nanofibers in the matrix due to their large interfacial area that led to better interface adhesion with the matrix.

The surface morphology of the stretched samples was studied using SEM to understand the interfacial adhesion of the fiber matrix. The surface morphology of hydrogel films embedded with large nanofibers after stretching is shown in Figure 9.12. As can be seen in the images, even after large stretch ratios, both the nanofibers and the matrix were still intact. However, there was some observable debonding between the nanofibers and the matrix as seen in Figure 9.12b–d. This can be attributed to interference of absorbed water that disengages the hydrogen bonds between the PU nanofibers and the PVA matrix leading to separation of fibers from the matrix during stretching. The smaller size nanofibers by contrast did not show any debonding between the

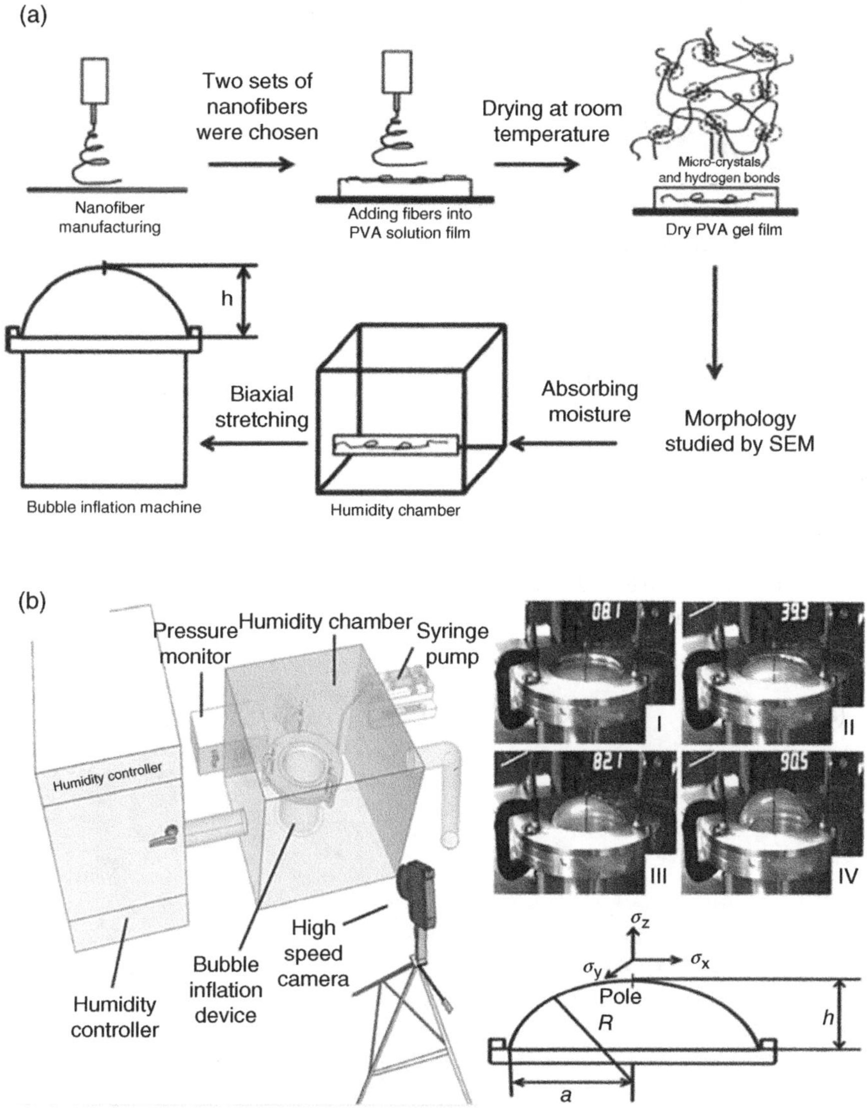


**Figure 9.10** (a) Schematic of the experimental procedure and (b) apparatus for measuring biaxial mechanical properties [63]. © John Wiley & Sons.

nanofibers and the matrix as can be seen from the micrographs in Figure 9.13. This can be attributed to the good interfacial adhesion between the fibers and the matrix due to the smaller fiber size and large surface area along with better embedding of the smaller nanofibers [63].

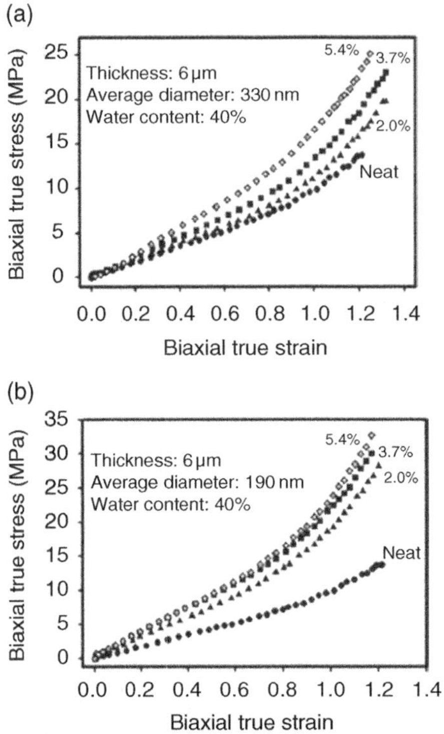


**Figure 9.11** Fiber effect on mechanical properties of 6 µm PVA films with 40% water content: (a) reinforced by 330 nm nanofibers and (b) reinforced by 190 nm nanofibers [63]. © John Wiley & Sons.

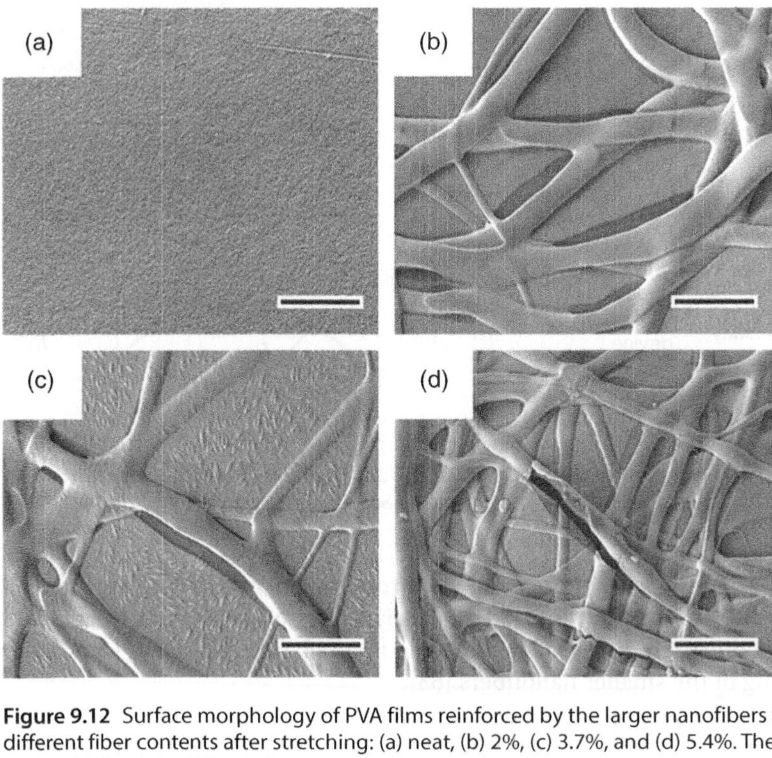


**Figure 9.12** Surface morphology of PVA films reinforced by the larger nanofibers with different fiber contents after stretching: (a) neat, (b) 2%, (c) 3.7%, and (d) 5.4%. The scale bar is 5 µm [63]. © John Wiley & Sons.

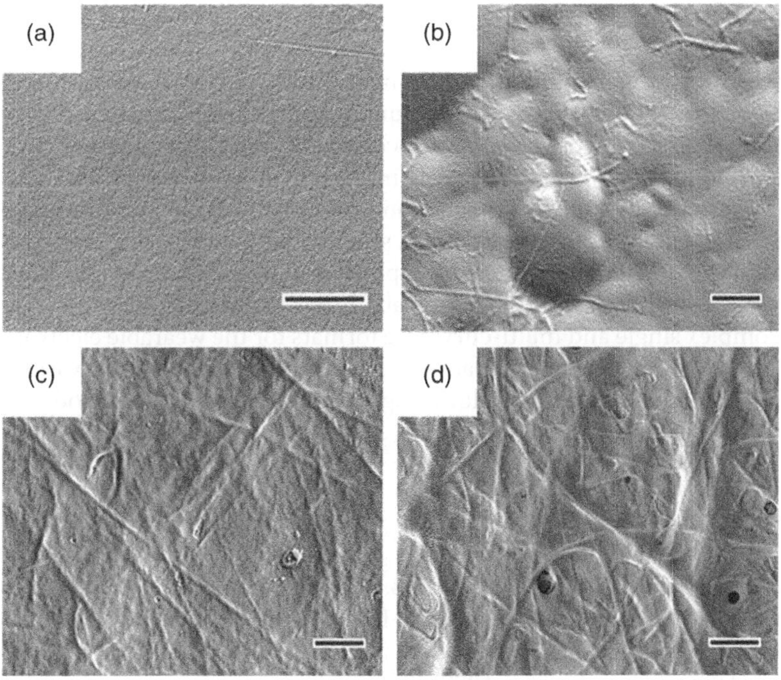


**Figure 9.13** Surface morphology of PVA films reinforced by the smaller nanofibers with different fiber contents after stretching: (a) neat, (b) 2%, (c) 3.7%, and (d) 5.4%. The scale bar is 5 μm [63]. © John Wiley & Sons.

The results discussed previously clearly illustrate that one of the biggest advantages of nanofibers is their ability to reinforce thin films in swollen state. It was also observed that hydrogel films with smaller nanofibers had a stronger reinforcement effect that was consistent with results from aerogel-reinforced fibers [64]. This is because at a constant fiber content and with reducing fiber diameter, the number of fibers and interfacial surface area of fibers increases for a specified volume. Higher nominal loads and larger displacements are needed for interfacial debonding of smaller fiber diameter-reinforced composites. Nanofiber-reinforced PVA thin films were successfully made using the electrospinning/solution casting hybrid process and reinforcement of the films is directly related to the size of the nanofibers. The largest effect of nanofiber presence was observed in thinner PVA films in a highly water-swollen state. The presence of nanofibers in PVA hydrogels increased Young's modulus and decreased the onset of strain hardening leading to earlier strain hardening and stabilization of the thickness uniformity in these soft films. The results of this research have important ramifications in a wide range of hydrogels or swollen films for applications including biomedical devices, membranes, and liquid-swollen battery separators.

## 9.6 Conclusion

A novel electrospinning/solution casting hybrid R2R manufacturing process was employed to generate films with unique properties, which could not be otherwise prepared via conventional processes. Polymer films with surface electrical conductivity were produced using a R2R manufacturing line that exhibited resistance to applied deformation while retaining electrical conductivity, as well as exhibiting good flexibility, stretchability, and twistability. This novel manufacturing technology can be applied to allow preparation of various electronics systems that are more "stretchable," thus facilitating thermoformability of complex single and multi-curvature formats for the wearable electronics platforms of the future. Bendable and reinforced aerogels for various thermal insulation and biological device applications could also be produced using this hybrid R2R process.

## References

1 Cakmak, M.; Reneker Darrell, H.; Yalcin, B. Hybrid Manufacturing Platform to Produce Multifunctional Polymeric Films. US 8889054 B2 November 18, **2014**.
2 Powers, G. W.; Collier, J. R. Experimental Modeling of Solvent-Casting Thin Polymer Films. *Polym. Eng. Sci.* **1990**, 30, 118–123.
3 Park, W. R. R. *Plastic Film Technology*; New York: Van Nostrand Reinhold Co., **1969**.
4 Sittig, M. *Plastic Films from Petroleum Raw Materials*; Park Ridge: Noyes Development, **1967**.
5 Hansen, C. M. Polymer Coatings. Concepts of Solvent Evaporation Phenomena. *Ind. Eng. Chem. Prod. Res. Dev.* **1970**, 9, 282–286.
6 Rajagopalan, D. Impact of Viscoelasticity on Gage Variation in Film Casting. *J. Rheol.* **1999**, 43, 73–83.
7 Powers, G. *The Effect of Temperature on the Solvent Removal from Solution-Cast Thin Polymer Films*; Ohio: Ohio University, **1988**.
8 Reneker, D. H.; Yarin, A. L. Electrospinning Jets and Polymer Nanofibers. *Polymer (Guildf)* **2008**, 49, 2387–2425.
9 Ramakrishna, Z. M. S.; Fujihara, K.; Teo, W.-E., Lim, T.-C. *An Introduction to Electrospinning and Nanofibers*; Singapore: World Science Publishing Company, **2005**.
10 Zhao, W.; Nugay, I. I.; Yalcin, B.; Cakmak, M. Flexible, Stretchable, Transparent and Electrically Conductive Polymer Films via a Hybrid Electrospinning and Solution Casting Process: In-Plane Anisotropic Conductivity for Electro-Optical Applications. *Displays* **2016**, 45, 48–57.
11 Zhao, W. *Flexible Transparent Electrically Conductive Polymer Films for Future Electronics*; Akron: University of Akron, **2011**.

**12** Crawford, G. P., Ed. *Flexible Flat Panel Displays*; Chichester: John Wiley & Sons, Ltd, **2005**.

**13** Cairns, D. R.; Witte, R. P.; Sparacin, D. K.; Sachsman, S. M.; Paine, D. C.; Crawford, G. P.; Newton, R. R. Strain-Dependent Electrical Resistance of Tin-Doped Indium Oxide on Polymer Substrates. *Appl. Phys. Lett.* **2000**, 76, 1425.

**14** Zhao, W.; Cakmak, M. Temporal Signatures of Resistivity in Bending of Indium Tin Oxide-Coated Flexible Transparent Conductive Films for Flexible Electronics: Influence of Coating Thickness and Bending Radius. *J. Soc. Inf. Disp.* **2014**, 22, 260–266.

**15** Wu, Z. Transparent, Conductive Carbon Nanotube Films. *Science* **2004**, 305, 1273–1276.

**16** Li, X.; Zhu, Y.; Cai, W.; Borysiak, M.; Han, B.; Chen, D.; Piner, R. D.; Colombo, L.; Ruoff, R. S. Transfer of Large-Area Graphene Films for High-Performance Transparent Conductive Electrodes. *Nano Lett.* **2009**, 9, 4359–4363.

**17** Kim, K. S.; Zhao, Y.; Jang, H.; Lee, S. Y.; Kim, J. M.; Kim, K. S.; Ahn, J.-H.; Kim, P.; Choi, J.-Y.; Hong, B. H. Large-Scale Pattern Growth of Graphene Films for Stretchable Transparent Electrodes. *Nature* **2009**, 457, 706–710.

**18** Hu, L.; Kim, H. S.; Lee, J.-Y.; Peumans, P.; Cui, Y. Scalable Coating and Properties of Transparent, Flexible, Silver Nanowire Electrodes. *ACS Nano* **2010**, 4, 2955–2963.

**19** Wu, H.; Kong, D.; Ruan, Z.; Hsu, P.-C.; Wang, S.; Yu, Z.; Carney, T. J.; Hu, L.; Fan, S.; Cui, Y. A Transparent Electrode Based on a Metal Nanotrough Network. *Nat. Nanotechnol.* **2013**, 8, 421–425.

**20** Lee, J.; Lee, P.; Lee, H.; Lee, D.; Lee, S. S.; Ko, S. H. Very Long Ag Nanowire Synthesis and Its Application in a Highly Transparent, Conductive and Flexible Metal Electrode Touch Panel. *Nanoscale* **2012**, 4, 6408.

**21** Shirakawa, H. The Discovery of Polyacetylene Film: The Dawning of an Era of Conducting Polymers (Nobel Lecture). *Angew. Chem. Int. Ed.* **2001**, 40, 2574–2580.

**22** MacDiarmid, A. G. Synthetic Metals: A Novel Role for Organic Polymers. *Synth. Met.* **2001**, 125, 11–22.

**23** Heeger, A. J. Semiconducting and Metallic Polymers: The Fourth Generation of Polymeric Materials (Nobel Lecture). *Angew. Chem. Int. Ed.* **2001**, 40, 2591–2611.

**24** Sun, Y.; Ruckenstein, E. Surface Conductive Films via Polymerization on a Liquid Surface. *Synth. Met.* **1996**, 82, 35–40.

**25** Kim, J.; Kim, E.; Won, Y.; Lee, H.; Suh, K. The Preparation and Characteristics of Conductive poly(3,4-Ethylenedioxythiophene) Thin Film by Vapor-Phase Polymerization. *Synth. Met.* **2003**, 139, 485–489.

**26** Jang, J.; Oh, J. H. Fabrication of a Highly Transparent Conductive Thin Film from Polypyrrole/Poly(methyl Methacrylate) Core/Shell Nanospheres. *Adv. Funct. Mater.* **2005**, 15, 494–502.

27 Ouyang, J.; Yang, Y. Conducting Polymer as Transparent Electric Glue. *Adv. Mater.* **2006**, 18, 2141–2144.

28 Wang, P.-C.; Liu, L.-H.; Alemu Mengistie, D.; Li, K.-H.; Wen, B.-J.; Liu, T.-S.; Chu, C.-W. Transparent Electrodes Based on Conducting Polymers for Display Applications. *Displays* **2013**, 34, 301–314.

29 Ouyang, J. "Secondary Doping" Methods to Significantly Enhance the Conductivity of PEDOT:PSS for Its Application as Transparent Electrode of Optoelectronic Devices. *Displays* **2013**, 34, 423–436.

30 Groenendaal, L.; Jonas, F.; Freitag, D.; Pielartzik, H.; Reynolds, J. R. Poly(3,4-Ethylenedioxythiophene) and Its Derivatives: Past, Present, and Future. *Adv. Mater.* **2000**, 12, 481–494.

31 Kobayashi, T.; Bando, M.; Kimura, N.; Shimizu, K.; Kadono, K.; Umezu, N.; Miyahara, K.; Hayazaki, S.; Nagai, S.; Mizuguchi, Y.; et al. Production of a 100-M-Long High-Quality Graphene Transparent Conductive Film by Roll-to-Roll Chemical Vapor Deposition and Transfer Process. *Appl. Phys. Lett.* **2013**, 102, 23112.

32 Bae, S.; Kim, H.; Lee, Y.; Xu, X.; Park, J.-S.; Zheng, Y.; Balakrishnan, J.; Lei, T.; Ri Kim, H.; Song, Y. I.; et al. Roll-to-Roll Production of 30-Inch Graphene Films for Transparent Electrodes. *Nat. Nanotechnol.* **2010**, 5, 574–578.

33 Pierre, A. C.; Pajonk, G. M. Chemistry of Aerogels and Their Applications. *Chem. Rev.* **2002**, 102, 4243–4265.

34 Aguado-Serrano, J.; Rojas-Cervantes, M. L. Titania Aerogels: Influence of Synthesis Parameters on Textural, Crystalline, and Surface Acid Properties. *Microporous Mesoporous Mater.* **2006**, 88, 205–213.

35 Lu, X.; Arduini-Schuster, M. C.; Kuhn, J.; Nilsson, O.; Fricke, J.; Pekala, R. W. Thermal Conductivity of Monolithic Organic Aerogels. *Science* **1992**, 255, 971–972.

36 Li, L.; Yalcin, B.; Nguyen, B. N.; Meador, M. A. B.; Cakmak, M. Flexible Nanofiber-Reinforced Aerogel (Xerogel) Synthesis, Manufacture, and Characterization. *ACS Appl. Mater. Interfaces* **2009**, 1, 2491–2501.

37 Wise, D. L. *Handbook of Pharmaceutical Controlled Release Technology*; Boca Raton: CRC Press, **2000**.

38 DeMerlis, C.; Schoneker, D. Review of the Oral Toxicity of Polyvinyl Alcohol (PVA). *Food Chem. Toxicol.* **2003**, 41, 319–326.

39 Hassan, C. M.; Peppas, N. A. Structure and Applications of Poly(vinyl Alcohol) Hydrogels Produced by Conventional Crosslinking or by Freezing/Thawing Methods. In Abe, A.; Albertsson, A.-C.; Cantow, H.-J.; Dušek, K.; Edwards, S.; Höcker, H.; Joanny, J. F.; Kausch, H.-H.; Kobayashi, T.; Lee, K.-S.; McGrath, J. E.; Monnerie, L.; Stupp, S. I.; Suter, U. W.; Thomas, E. L.; Wegner, G.; Young, R. J., Eds., *Biopolymers PVA Hydrogels, Anionic Polymerisation Nanocomposites*; Berlin, Heidelberg: Springer, **2000**; pp. 37–65.

40 Yoshii, F.; Zhanshan, Y.; Isobe, K.; Shinozaki, K.; Makuuchi, K. Electron Beam Crosslinked PEO and PEO/PVA Hydrogels for Wound Dressing. *Radiat. Phys. Chem.* **1999**, 55, 133–138.

41 Upadhyay, D.; Bhat, N. Pervaporation Studies of Gaseous Plasma Treated PVA Membrane. *J. Membr. Sci.* **2004**, 239, 255–263.

42 Morimoto, K.; Nagayasu, A.; Fukanoki, S.; Morisaka, K.; Hyon, S. H.; Ikada, Y. Evaluation of Polyvinyl Alcohol Hydrogel as a Sustained-Release Vehicle for Rectal Administration of Indomethacin. *Pharm. Res.* **1989**, 6, 338–341.

43 Peppas, N. A.; Huang, Y.; Torres-Lugo, M.; Ward, J. H.; Zhang, J. Physicochemical Foundations and Structural Design of Hydrogels in Medicine and Biology. *Annu. Rev. Biomed. Eng.* **2000**, 2, 9–29.

44 Sreenivasan, K. Synthesis and Evaluation of Multiply Templated Molecularly Imprinted Polyaniline. *J. Mater. Sci.* **2007**, 42, 7575–7578.

45 Stammen, J. A.; Williams, S.; Ku, D. N.; Guldberg, R. E. Mechanical Properties of a Novel PVA Hydrogel in Shear and Unconfined Compression. *Biomaterials* **2001**, 22, 799–806.

46 Patachia, S.; Valente, A. J. M.; Papanceaa, A.; Lobo, V. M. M. *Poly (Vinyl Alcohol)[PVA]-Based Polymer Membranes*; New York: Nova Science, **2009**.

47 Suzuki, M.; Yoshida, T.; Kobayashi, S.; Koyama, T.; Kimura, M.; Hanabusa, K.; Shirai, H. Proton Conduction in New Polymer Hydrogel Films Consisting of Crosslinking Partially Phosphorylated Poly(vinyl Alcohol)s. *Phys. Chem. Chem. Phys.* **1999**, 1, 2749–2753.

48 Wang, B.; Kodama, M.; Mukataka, S.; Kokufuta, E. On the Intermolecular Crosslinking of PVA Chains in an Aqueous Solution by $\gamma$-Ray Irradiation. *Polym. Gels Networks* **1998**, 6, 71–81.

49 Hassan, C. M.; Peppas, N. A. Structure and Morphology of Freeze/Thawed PVA Hydrogels. *Macromolecules* **2000**, 33, 2472–2479.

50 Otsuka, E.; Suzuki, A. A Simple Method to Obtain a Swollen PVA Gel Crosslinked by Hydrogen Bonds. *J. Appl. Polym. Sci.* **2009**, 114, 10–16.

51 Finch, C. A. *Polyvinyl Alcohol: Properties and Applications*; 362nd ed.; London: John Wiley & Sons, Ltd, **1973**.

52 Corkhill, P. H.; Trevett, A. S.; Tighe, B. J. The Potential of Hydrogels as Synthetic Articular Cartilage. *Proc. Inst. Mech. Eng. H* **1990**, 204, 147–155.

53 Tang, C.; Liu, H. Cellulose Nanofiber Reinforced Poly(vinyl Alcohol) Composite Film with High Visible Light Transmittance. *Compos. A Appl. Sci. Manuf.* **2008**, 39, 1638–1643.

54 Kokabi, M.; Sirousazar, M.; Hassan, Z. M. PVA–clay Nanocomposite Hydrogels for Wound Dressing. *Eur. Polym. J.* **2007**, 43, 773–781.

55 Kulkarni, R. V.; Sreedhar, V.; Mutalik, S.; Setty, C. M.; Sa, B. Interpenetrating Network Hydrogel Membranes of Sodium Alginate and Poly(vinyl Alcohol) for Controlled Release of Prazosin Hydrochloride Through Skin. *Int. J. Biol. Macromol.* **2010**, 47, 520–527.

56 Zucchelli, A.; Focarete, M. L.; Gualandi, C.; Ramakrishna, S. Electrospun Nanofibers for Enhancing Structural Performance of Composite Materials. *Polym. Adv. Technol.* **2011**, 22, 339–349.

**57** Ramakrishna, S.; Fujihara, K.; Teo, W.-E.; Lim, T.-C.; Ma, Z. *An Introduction to Electrospinning and Nanofibers*; Singapore: World Science Publishing Company, **2005**.

**58** Pedicini, A.; Farris, R. J. Mechanical Behavior of Electrospun Polyurethane. *Polymer (Guildf)* **2003**, 44, 6857–6862.

**59** Sheikh, F. A.; Kanjwal, M. A.; Saran, S.; Chung, W.-J.; Kim, H. Polyurethane Nanofibers Containing Copper Nanoparticles as Future Materials. *Appl. Surf. Sci.* **2011**, 257, 3020–3026.

**60** Han, J. H.; Taylor, J. D.; Kim, D. S.; Kim, Y. S.; Kim, Y. T.; Cha, G. S.; Nam, H. Glucose Biosensor with a Hydrophilic Polyurethane (HPU) Blended with Polyvinyl Alcohol/vinyl Butyral Copolymer (PVAB) Outer Membrane. *Sensors Actuators B Chem.* **2007**, 123, 384–390.

**61** Heo, D. N.; Ko, W.-K.; Bae, M. S.; Lee, J. B.; Lee, D.-W.; Byun, W.; Lee, C. H.; Kim, E.-C.; Jung, B.-Y.; Kwon, I. K. Enhanced Bone Regeneration with a Gold Nanoparticle–Hydrogel Complex. *J. Mater. Chem. B* **2014**, 2, 1584.

**62** Kenawy, E.-R.; Abdel-Hay, F. I.; El-Newehy, M. H.; Wnek, G. E. Processing of Polymer Nanofibers Through Electrospinning as Drug Delivery Systems. *Mater. Chem. Phys.* **2009**, 113, 296–302.

**63** Guo, Y.; Drum, J.; Qu, C.; Cakmak, M. Mechanical Reinforcement of Hydrogels by Nanofiber Network Undergoing Biaxial Deformation. *Polym. Compos.* **2016**, 37, 709–717.

**64** Peterson, W.; Hossain, N. M. A.; Woo, K.; Jenkins, C. Strength of Flexible Composites: Damage Tolerance in Compliant Unidirectional Fiber Reinforced Composites. In *52nd AIAA/ASME/ASCE/AHS/ASC Structures, Structural Dynamics and Materials Conference*; Reston, VA: American Institute of Aeronautics and Astronautics, **2011**.

# 10

# Roll-to-Roll Manufacturing of Flexible Displays

*E. Montbach and D. Davis*

*Kent Displays, Inc., Kent, OH, USA*

## 10.1 Introduction: Thin and Flexible Substrates

Flexible electronic displays made using roll-to-roll (R2R) manufacturing on thin, bendable, plastic substrates have been discussed for well over a decade [1]. Initially the vision for flexible displays was a "rollable" handheld device; however, due to the global acceptance of smartphones and tablets, the current need for flexible substrates is to improve display ruggedness, reduce bulk [2], and allow for organic shapes. Manufacturers are motivated to develop thinner displays for use in wearable electronics due to the perceived added value by the consumer, novel designs that differentiate from competitors, and potential to charge a higher price at introduction for unique products. Furthermore, the display industry is experiencing saturation and needs innovations to break stagnation through product differentiation and to create new markets [3]. Displays of the future will be installed on objects of various shapes and will be ubiquitous because the demand for information will be everywhere so these displays will be flexible, or at least conformable, and have low price [4]. As handheld computing on smartphones are migrated to wearables like watches, bracelets, and so on, demand for thinner, conformable, and even flexible displays will grow.

In display product categories beyond wearables and handheld, such as laptops, monitors, and TVs, consumers are also demanding thinner flat panel displays (FPDs) to reduce bulk and to take up less space as shown in Table 10.1. This challenges the FPD manufacturer to combine functionalities and produce the most compact designs on the thinnest possible substrates rather than using the display substrate as a rigid carrier for electrical circuits, optical coatings, and functional layers. Since thin substrates will be less mechanically rigid, this

*Roll-to-Roll Manufacturing: Process Elements and Recent Advances*, First Edition.
Edited by Jehuda Greener, Glen Pearson, and Miko Cakmak.

**Table 10.1** Customer-desired attributes for flat panel display product categories.

| Product category | Thin | Lightweight | Rugged | Novel designs |
|---|---|---|---|---|
| Wearable | √ | √ | √ | √ |
| Handheld | √ | √ | √ | √ |
| Laptop | √ | √ | √ | |
| Monitor | √ | | | |
| TV | √ | √ | | √ |

means that these thin substrates will be easy to bend, flex, and wind into rolls. R2R manufacturing focuses on processing flexible, thin substrates in high volume at low cost, so R2R manufacturing readily comes to mind for producing next-generation display technologies.

### 10.1.1 Thinner Display Architectures

Information displayed on a thin screen rather than a bulky cathode ray tube (CRT) has been the driving force behind FPD pioneering work such as Aiken's tube-based approach to a flat screen in the late 1950s and Brody's development of the active matrix in the late 1970s. In fact, the name flat panel display itself typically refers to an information display that is less than 10 cm thick. Current FPDs can be broken into three main types: transmissive, emissive, and reflective. Each FPD type will be discussed briefly.

Active matrix liquid crystal displays (AMLCDs), the leading transmissive type FPD used in TVs, monitors, laptop computers, and smartphones, combine thin-film transistors (TFTs) and other circuits in a two-dimensional array of pixels on a glass substrate, which apply voltages across a liquid crystal-filled cavity to a transparent electrode on a second glass substrate. Polarizers and other light management films on either side of the AMLCD make each pixel function as a light control valve passing a variable amount of light, coming from a light source behind the display (known as a backlight) through the display to the viewer on the front side. Information is displayed through the combination of bright, dark, and varying level grayscale pixels. This is an efficient use of depth, taking only a few millimeters in a smartphone compared with over 45 cm for a CRT. AMLCDs can also have the backlight replaced with a light guide plate that is side lit by light-emitting diodes (LEDs) to reduce thickness.

The thinnest instance of an FPD would be to replace the transmissive structure of AMLCDs, filters, films, and backlight with a patterned emissive layer such as an organic light-emitting diode (OLED) material on a thin, active matrix substrate to produce an array of multiple colored, light-emitting pixels producing an active matrix organic light-emitting diode (AMOLED). The thinner overall structure of the AMOLED offers a high degree of flexibility; there is no backlight to flex and flow of the liquid crystal layer during flexing is avoided.

Another path to a thinner, simpler design is to use reflective electro-optic materials that do not require backlights or emit light themselves but instead are better suited to indoor lighting and outdoor daytime applications. Reflective display technologies seen most frequently on flexible substrates include cholesteric liquid crystal (ChLC) and electrophoretic (EPD) and microfluidic displays. The ultimate application of R2R manufacturing methods in FPD production is to produce patterns for electronic circuits and light-emitting or -reflecting layers on thin flexible substrates using printing techniques in high volume at low cost.

### 10.1.2 Challenges in Migrating to Roll-to-Roll

R2R manufacturing has existed for well over 100 years, with some of the first manufacturing lines for products like newspapers and photographic film. The R2R manufacturing sector has grown in the number of applications to include packaging films for food and nonfood items, construction films such as optical films for windows and vapor barriers, and even high value electronics, such as flexible solar cells, batteries, radio frequency identification (RFID) tags, among others. R2R covers a variety of manufacturing techniques that continuously process flexible substrates (e.g., plastic film, paper, metal foil) where the thickness of the substrate is very small compared with its width and length so the material can be wound into a roll and processed in mass on long webs. In contrast to other manufacturing methods, like sheet-to-sheet processing, R2R offers significant advantages in the manufacture of products at high production rates in large volumes, resulting in low cost since material handling between processes is eliminated due to the substrate providing the means of conveyance. In addition, the continuous nature of R2R results in a short transfer time between sequential processes. There is great interest in applying R2R manufacturing techniques to the production of FPDs, with the potential for dramatically reducing manufacturing costs and enabling new product formats like flexible displays [5] by using thin, bendable substrates.

There is a large FPD manufacturing infrastructure with annual volume in excess of 200 million square meter per year with excess capacity. This sheet-to-sheet based industry is very mature at 20 years old and uses high levels of process automation such as robotic handling of large glass sheets between process modules of cluster tools. Current LCD production is based on photolithography semiconductor processes, which are subtractive, where a layer is applied to a substrate then systematically removed to create a pattern that defines a circuit or color filter. An example of a photolithography process is shown in Figure 10.1, where the core processes for production of FPDs are repeated multiple times with multiple materials to produce patterned layers.

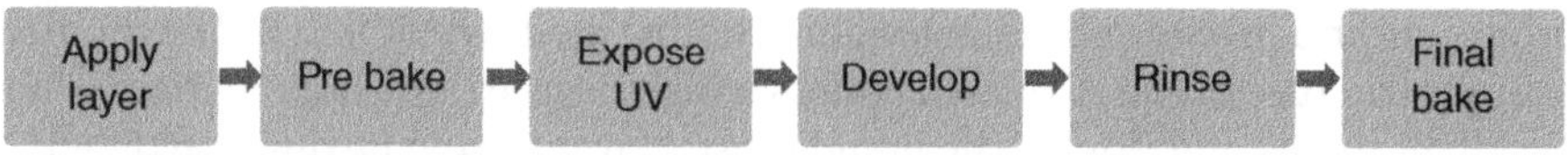


**Figure 10.1** Generic flowchart for LCD/semiconductor photolithography patterning process.

In contrast, most R2R processes are additive, for example, printing the material only at locations where needed to produce a pattern or multiple layer coatings as in photographic film production. This leads to a shorter number of steps to make the final product. R2R manufacturing is a natural fit with additive patterning processes. The usual way patterned images are made in R2R manufacturing is ink-jet printing, screen printing, and offset printing, among others; however, when the resolution of additive patterning processes (printing) is compared with subtractive patterning processes (photolithography), it is difficult to achieve high resolution patterned images using additive patterning. The resolution limit for additive printing processes is due to the fundamental materials science behind the various printing techniques, which requires control of the material spreading for the additively printed material, mechanical control of the jets or screens, and control of how the substrate surface receives the printed material. In contrast, the resolution limit for subtractive patterning processes can be clearly defined only by how the mask imparts an image (collimation of the light source) and how the developed materials separate from the undeveloped materials. The resolution limit for additive printing processes makes it difficult for FPD production to depend solely on traditional continuous manufacturing processes and frequently results in manufacturing solutions that are a hybrid of both R2R manufacturing and sheet-to-sheet processes.

R2R manufacturing is a high volume manufacturing process capable of producing large areas of materials comparable with the sheet-to-sheet processes used in FPD manufacturing. LCD Fab area production rates are compared with narrow and wide web R2R coating lines in Figure 10.2. A state-of-the-art generation 10 LCD line with 180 s tact time per step processing 3 m × 3 m glass sheets can only process about a tenth of the area of 1.5 m wide R2R coating line running at 25 m/min.

The current state of the art in flexible display manufacturing is to use a hybrid process combining sheet-to-sheet and R2R manufacturing methods. The large, global semiconductor and display manufacturing infrastructure is the reason for this hybrid approach. In this method, the flexible substrate is bonded to rigid substrate for processing and then removed, resulting in its own challenges like bond/debond, bowing and thermal mismatch problems [2]. Transitioning display manufacturing completely from current sheet to sheet to R2R depends on the materials science problem on how to get same performance from new materials processed by additive patterning processes as from the current materials used in subtractive processes.

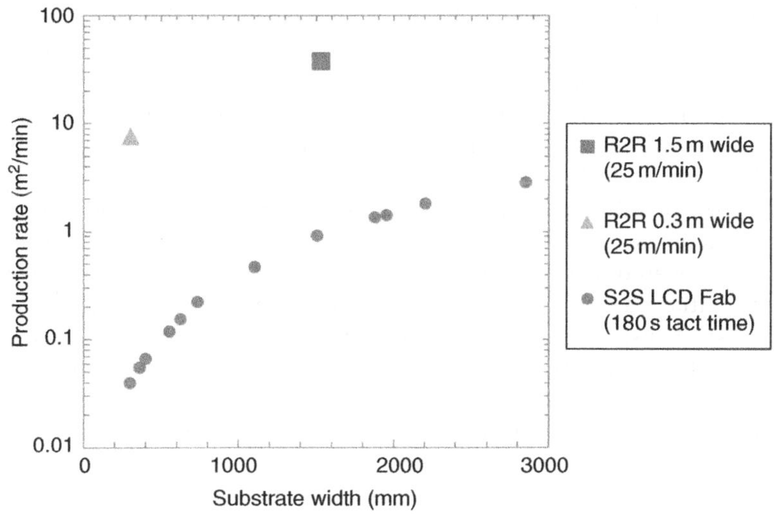


**Figure 10.2** LCD Fab Production Rate (up to Gen10) compared with roll-to-roll (0.3 and 1.5 m wide) [6]. *Source:* Based on Oikawa [6].

R2R manufacturing offers an intuitive way to make a flexible display since it is the ideal manufacturing method for processing flexible substrates. FPDs made by R2R manufacturing have unique attributes compared with FPDs made by sheet-to-sheet production methods. Flexible displays made R2R use novel device technologies or serve a unique application either for product differentiation to serve new markets or specifically designed with R2R manufacturing in mind. As discussed later in this chapter, new and disruptive technologies like OLED lighting and eWriters have been the first FPD-related products made R2R; however, polymer-dispersed liquid crystal (PDLC) switchable windows have been manufactured R2R for over a decade. All of these devices have seen success using R2R process technology due to their low resolution requirements and simple structure while also being designed explicitly for R2R production and thus do not require flexible substrates to be laminated to rigid substrates and then later removed. Flexible displays represent not only a potential paradigm shift in the manufacturing technique but also a potentially different final product form factor, which may be another cause for the slow adoption of flexible displays to the marketplace. As flexible displays become more pervasive, device designers will begin to identify new and different applications and form factors, creating a market pull for flexible displays that will drive the demand for flexible displays from not only the design perspective but also the general public as they begin to expect the flexible display form factor. This increase in demand will also drive development to happen at a fast rate and will reduce the technical gaps between R2R manufacturing and display manufacturing as the benefits of the R2R process will be alluring for both financial and manufacturing reasons. Several of the flexible display technologies discussed here exist even without a strong marketplace pull for flexible displays simply because those technologies are not compatible with rigid glass substrates, but instead require flexible substrates to function as designed.

The adoption of R2R manufacturing for flexible displays will be mainly determined on a case-by-case basis by the individual display manufacturer, which needs to overcome several initial barriers to enter R2R flexible display production. Initially a display manufacturer must commit to the large equipment purchase of custom specialized R2R manufacturing lines that will likely have higher cost than traditional R2R manufacturing lines due to the need for vision systems for alignment of various patterned layers, unique web paths that do not come in contact with the side of the substrate that will have high resolution patterned traces, and some form of patterning whether additive or subtractive, along with increased cleanliness. Purchasing new R2R production equipment would not be expected to be more costly than current glass fabs; however, this production equipment would be unproven and need development, making this a high risk venture. In addition, the potential R2R flexible display manufacturer must develop an entirely new supply chain, utilizing in some cases materials that have not been previously used in displays. This requires extensive specification developments and risks taken by all companies in the supply chain to develop materials that may not be viable in other markets. Finally, certain applications, such as cell phones, will need the ability to quickly and easily scale up to large volumes if the product is a success so initial R2R production facilities must be built large enough to provide those volumes. In addition, R2R display manufacturers will likely want some customer commitment of initial large volume orders with larger display area to develop an entirely new manufacturing stream [7], resulting in the proverbial chicken and egg type situation. This simply highlights that R2R manufactured displays are somewhat of a risk when initial production investments are considered.

To enable flexible display R2R manufacturing, several materials science advancements still remain. Continuous web manufacturing requires the use of flexible substrates. As mentioned previously in the literature [8], flexible substrates have a number of issues associated with being used in typical sheet-to-sheet FPD manufacturing, such as optical properties, thermal properties, and surface smoothness, among others. This is why the technologies discussed in this chapter have found new manufacturing processes or display requirements to allow for the flexible substrates of today. In addition, there has been considerable work by the flexible substrate manufacturers to improve flexible substrates to meet industry requirements. Other materials that require further development for flexible display R2R manufacturing are conducting layers, electro-optic layers, barrier layers, and TFTs, among others, which are specific to certain display types.

The past several years has seen considerable developments in conductor materials for flexible substrates. This is due to the need of conductors on flexible substrates for the large and growing touchscreen industry. Initially, flexible substrates with conductors needed to flex in the final product as users engaged the touchscreen, in terms of resistive type touchscreens; however, more recent touchscreens do not require substrate flexure (projected capacitive). The demand from several industries including FPD, touchscreen, battery, solar cells, and others has resulted in several new flexible substrate conductors, including carbon nanotubes, carbon nanobuds, fine trace-patterned copper, poly(3,4ethylenedioxythiophene)polystyrene sulfonate (PEDOT/PSS), and even more rugged conducting oxide layers, among others. All of these conductor types have seen developments in base materials along with improved coating layers surrounding the conductors to allow for better adhesion to the substrate and better interaction with the electro-optic layers.

The various electro-optic layers for displays are always undergoing considerable research and developments. Whether the electro-optic layer is transmissive, reflective, or emissive, in part, defines how many subsequent layers and subcomponents are required to build a functioning display. The typical transmissive LCD requires several layers above and below the display, making flexibility more elusive for the final product. However, displays can be built R2R but be utilized in final rigid assembly or a conformed curvature, where the display is flexed only one time. This chapter mainly covers several reflective and emissive electro-optic-type displays and does not review the issues associated with typical transmissive AMLCDs. However, as further advances occur in the reflective and emissive display types, those advances will surely migrate to transmissive-type displays if it is advantageous to the market for either a final flexible use model or R2R manufacturing, both likely decided by the economics and need of the final product.

Barrier layers and TFTs are not needed by all flexible displays, but some do require these enabling technologies. Barrier layers are seeing continual improvements and developments in alternative industries, such as medical and food packaging. As will be seen in this chapter, there are some wearable devices, smartwatches, that utilize flexible AMOLED displays; however, those displays are manufactured in a bond/debond sheet-to-sheet manufacturing process. The main technology gaps for TFTs manufactured R2R include thermal properties of flexible substrates, compatibility of high resolution photolithography patterning with R2R processes, cleanliness of R2R processes, and careful handling of substrates through the R2R process to prevent scratches and broken traces.

Beyond the materials science, advancements required there are several fundamental processes that disconnect between R2R manufacturing and traditional sheet-to-sheet FPD manufacturing that results in difficulties transferring FPD manufacturing to R2R:

- Careful handling
  - Display quality and throughput require that thin conductive traces and pristine coating surfaces have minimal handling defects, such as scratches, for uniform FPD appearance. Typical scratches from R2R manufacturing are on the order of millimeters in length and tend to repeat. Even for TVs that have a pixel size of 0.66 mm × 1 mm, this is a yield-limiting issue.
- Patterning processes
  - Trace and space widths must be in the tens of micron range or less for conductor traces going to individual pixels and within the TFT structure. In addition, color filters require small space widths and typically use subtractive patterning processes.
- Cleanliness
  - FPD quality and throughput require that particulate be minimal due to the thinness of the electro-optic layers, and R2R manufacturing static enhances particulate issues. One dust particle can make a thin conductive trace electrically open, resulting in a dead pixel.

In contrast, some seemingly similar high-tech products are being made R2R, which would seem to imply that displays should be more fundamentally compatible with the process. Some examples include flexible solar cells, batteries, and RFID tags. In particular, each of those technologies enjoys the ability to have reduced trace requirements (wider conductive traces) along with the ability to use an additive manufacturing process to create the printed electronics portion of the product. In addition, flexible solar cells, batteries, and RFID tags do not have the same requirements as LCDs such as a narrow spacing between two substrates for the electro-optic material or use multiple substrates layered with a critical thickness between them, making cleanliness less of a concern. All of this indicates that those who wish to manufacture R2R must choose a product that does not require processes incompatible with the current R2R manufacturing capabilities; however, improvements and developments are swiftly happening. Not all flexible display designs are compatible with R2R manufacturing due to material sets or processes, but this manufacturing method will always be of interest due to its unique ability to simply process flexible substrates.

### 10.1.3 General Description of Roll-to-Roll Manufacture of Flexible Flat Panel Displays

Flexible FPDs made by R2R manufacturing include six general steps.

#### 10.1.3.1 Flexible Substrate

A flexible substrate is selected for the correct mix of optical, mechanical, thermal, and cosmetic properties, among others. Display products typically have stringent specifications around incoming substrates and require substrates not to deform or change under demanding processing steps. The fundamental form of the base flexible substrate will greatly govern the form of the final flexible display; therefore, choosing a substrate that can provide the properties needed for the final product and survive process steps is key.

#### 10.1.3.2 Application of Electrode

Depending on the FPD technology, different electrode types will be required on the flexible substrate. Examples of electrodes include metallic transparent conductors like indium tin oxide (ITO) or conducting polymers (CP) like PEDOT:PSS. The application method depends on the electrode type and could include vacuum deposition followed by patterning or solution coating and drying. Active matrix FPDs also require metallic/oxide based or organic TFTs produced on the flexible substrate.

#### 10.1.3.3 Electro-optic Layer

The electro-optic layer is the key element of FPD that converts electrical signals to optical outputs. Various electro-optic materials can be used including liquid crystals, OLEDs, or electrochromic materials, among others.

#### 10.1.3.4 Cover Layer/Encapsulation

The type of electro-optic layer determines the use and type of cover layer or encapsulation. For example, a liquid crystal display requires a second substrate with a conductor on it, while for OLEDs, a top conductor is coated over the OLED material and then a barrier layer is applied by vacuum deposition over both the OLED material and conductor.

#### 10.1.3.5 Singulation

In singulation, the displays are cut from the web or made singular from a continuous web of displays. The singulation process can be performed by die cutting with a metal tool or by scanning a laser across the web.

#### 10.1.3.6 Integration and Test

In this step the display is inspected for defects, connected to any ancillary drive electronics, and installed in the final assembly and is a critical step to ensuring good production. In this step additional films may also be incorporated into the display module, which improves performance like brightness enhancement films, polarizers, and the like.

## 10.2 Roll-to-Roll Display Technologies

This chapter will review some FPD technologies that are or could be made using R2R manufacturing methods.

### 10.2.1 Cholesteric Liquid Crystal Displays

The cholesteric liquid crystal display (ChLCD) technology has inherent reflective color and grayscale and requires no power to maintain a static image [9, 10]. The simple structure and inherent properties of ChLCDs make a reflective, flexible, low power display possible, which can also be simple to manufacture when compared with competing technologies. ChLC material is a natural reflector as the inherent chiral structure of ChLC leads to a Bragg-type reflection of the incident light. Reflective ChLCDs have selective reflection properties. For normal incidence, the reflection of light occurs within a certain wavelength bandwidth:

$$\Delta\lambda = \left(n_{\parallel} - n_{\perp}\right)p, \tag{10.1}$$

where $n_{\parallel}, n_{\perp}$ are the components of refractive index locally parallel and, perpendicular to the liquid crystal director, $p$ is the cholesteric pitch. An example of ChLC reflection spectra is shown in the Boogie Board section, Figure 10.5 (left). The pitch is defined through the concentration $c$ of a chiral dopant and helical twisting power (HTP) of the chiral dopant as follows:

$$p = \frac{1}{c \cdot \mathrm{HTP}}. \tag{10.2}$$

ChLCDs are bistable when alignment is designed appropriately and can be switched between two stable states: planar, the reflective state, and focal conic, the slightly scattering transmissive state. Being bistable, ChLCDs require power only when being switched from one image to another. More details of the fundamental ChLC technology can be found in the literature [11]. As will be seen, several creative techniques for switching ChLCDs to create and image have been developed.

#### 10.2.1.1 Industrial Technology Research Institute of Taiwan

The Display Technology Center at the Industrial Technology Research Institute (ITRI) in Taiwan uses multidisciplinary teams of scientists and engineers with expertise in optoelectronics, mechanical engineering, chemistry, and materials science to be a world leading center in R2R display manufacturing research.

ITRI, working with Kodak, has demonstrated a flexible, banner size ePaper display [12, 13] made using R2R processes. This display uses a commercially available ITO-coated polyethylene terephthalate (PET) film that is laser patterned, then followed by slot die coating layers of a ChLC gelatin mixture followed by a colored absorption layer, and finally screen printing a silver electrode pattern. Prototype FPDs of greater than 3 m of length have been demonstrated with these R2R processes as shown in Figure 10.3. The display is based on a one-substrate architecture starting with 125 μm PET and all subsequent coatings totaling about 30 μm producing a total thickness of only 155 μm.

**Figure 10.3** ITRI's 3-meter long prototype FPD made using roll-to-roll processes [13]. © John Wiley & Sons.

The display performance reported is a reflectance of 27%, contrast ratio of 10 : 1 using 15 wt % gelatin in the liquid crystal with 8 grayscale levels in a 300 dpi image. This flexible display uses a ChLC and is bistable with two states, namely, focal conic and planar. For this design with the reflective backplane, the content information is displayed in the focal conic state. The drive voltage is high (80 V), which is led to the development of a novel thermal addressing process using a low-cost thermal print head producing high resolution images with almost no limitation to panel length. Heating a ChLC with the appropriate alignment and then cooling it down places it in the focal conic state. The potential application of this display includes e-banners, e-signage, or even small format e-toys or e-cards that could easily be rewritten with new data.

This work has been extended by using liquid crystals reflecting different wavelength to produce a black and white display [14], which is 25 cm × 100 cm, 300 dpi addressed with refined thermal print head and better grayscale and smoother Ag ink. Another work [15] involves adding photo tunable material to the ChLC layer and exposing this layer to a UV light through a mask to produce adjacent red and green pixels in a single layer over a blue absorbing layer. The result demonstrated is a 75 dpi, 16 grayscale A4 size display showing multiple colors.

Workers at ITRI have also developed different drive schemes including a multiline electrical driving approach [16], thermal addressing, laser addressing for high resolution, and thermal driving. An interesting "rolling writer" that uses probes to contact the display while it is continuously moved by driven rollers past the drive electronics has been used to address a 10 cm × 3 m long, no ITO pattern, and 30 dpi display.

#### 10.2.1.2 Kent Displays, Inc.

In the flat panel industry, one of the first flexible displays to be R2R manufactured and commercially available in mass volume are the nontraditional Boogie Board® eWriter liquid crystal films [17] in the unique and exciting writing tablet application (Figure 10.4). The Boogie Board is a novel device that is a pressure-sensitive writing device with electronic erase marketed as a replacement for paper for note-taking, doodling, and sketches. An image is written on the display when the user applies pressure to the top surface with either a stylus or a finger tip. It is flexible, made with rugged plastic substrates, has unique handwriting experience, and is low cost. The pressure creates a bright line on a dark background, which can be erased with a simple voltage pulse. The Boogie Board eWriter liquid crystal film has a good writing response with sharp written lines. The Boogie Board fills the need for a lightweight, relatively inexpensive electronic device that offers a natural writing experience with the feel of paper for handwritten and drawn images on a reusable medium that is also environmentally conscious in a technologically driven world.

The appeal of the Boogie Board eWriter is the natural paper-like writing experience. The writing is analog with no pixels to distort the image. The Boogie Board has seen a tremendous adoption due to the similarity to paper,

**Figure 10.4** Left: A photograph of a flexible eWriter liquid crystal film. Right: A photograph of a Boogie Board product that allows electronic erasing of handwritten notes.

ease of use, and simple yet fun operation. It quickly became popular and successful in the consumer market in simple applications such as note-taking, lists, and literally anywhere a scratch paper was used. Now, the product space is entering rapid expansion into other areas such as education, healthcare, business/office use, government, and so on [18]. It has also already been adopted in mobile accessories such as smartphone cases. In addition, it has entered mass retail in toy applications (see later sections). The demand for Boogie Board eWriters has been organically grown over the last 6 years of commercial availability with millions of units sold into the market.

eWriters are a perfect fit for R2R manufacturing due to the simple, yet elegant, structure of two flexible substrates, a ChLC/polymer dispersion and light-absorbing backcoat. R2R processes allow for low-cost and high volume manufacturing due to reduced substrate and materials handling and the continuous web transport of displays at a rapid rate. Web processing displays take advantage of the inherent display flexibility during the manufacturing process, unlike the batch processes that attach the flexible substrates to rigid carriers.

#### *10.2.1.2.1 Fundamental Science*

The ChLC material in the eWriter fundamentally functions the same as other ChLC materials described earlier. The only difference is that the eWriter liquid crystal film uses pressure to write lines to the planar texture and requires a small voltage pulse to switch or erase to the focal conic texture. An example of typical reflection spectra for the Boogie Board display is shown in Figure 10.5 (top). A schematic showing the transitions between the different textures of the eWriter liquid crystal film is also shown in Figure 10.5 (bottom), where the planar texture is created by pressure or flow of the ChLC and the focal conic texture is created by applying a small voltage pulse to the display.

When the written image is erased, the written planar image is electrically switched to the focal conic texture, and during the high voltage pulse, the cholesteric helix is untwisted to the metastable homeotropic texture that relaxes to a metastable transient planar texture and then to a stable planar texture. During the following low voltage pulse, the switched planar texture is electrically switched to the focal conic texture [11].

The Boogie Board eWriter liquid crystal film consists of two flexible substrates, each with a PEDOT/PSS acid conductor layer, also referred to as CP, a light-cured dispersion of ChLC and polymer between the substrates and spacers to maintain the cell gap. A cross section of a typical Boogie Board eWriter liquid crystal film is shown in Figure 10.6. The back of the Boogie Board eWriter liquid crystal film is typically flood coated with an absorbing ink or coating, but as has been shown in previous publications, patterned ink images can be coated on various display layers and different color backcoats can be used [19]. The conductors for this display are unpatterned, and the display can be simply cut into the desired shape for the final product. The eWriter

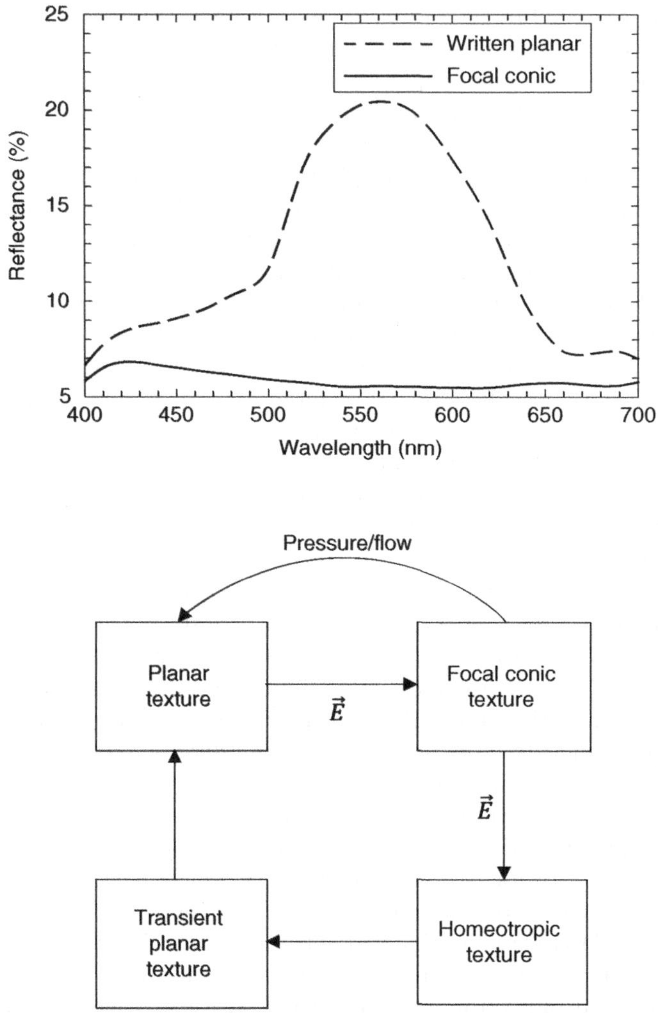


**Figure 10.5** Top: Reflection spectra for a typical Boogie Board eWriter liquid crystal film. The written planar texture has a peak wavelength in the yellow green, approximately 570 nm. The focal conic texture appears as the color of the backcoat, in this case the backcoat is black. Bottom: A schematic showing the texture transitions for the Boogie Board eWriter liquid crystal film.

electro-optic layer contains ChLC materials and polymer structure to create good line acuity [20]. Figure 10.7 shows SEM micrographs of the polymer pillar structure confirming that the ChLC is not encapsulated by the polymer, but is allowed to flow around the polymer making the display rewritable without destruction. The pressure induces flow in the ChLC, resulting in a written bright line on a dark background.

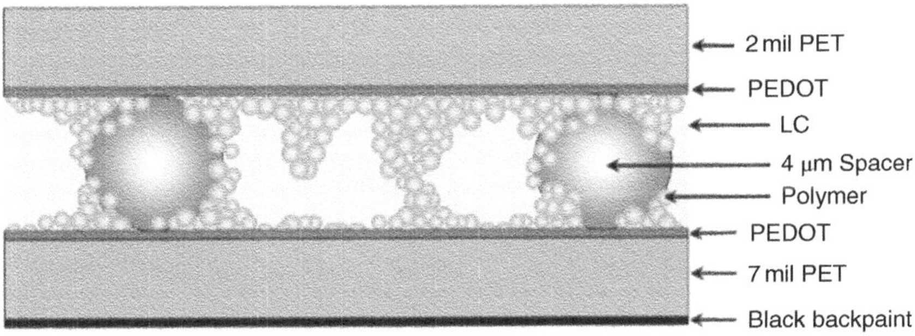


**Figure 10.6** Cross-sectional view of the Boogie Board eWriter liquid crystal film. The polymer forms pillars that support the substrates but do not encapsulate the ChLC.

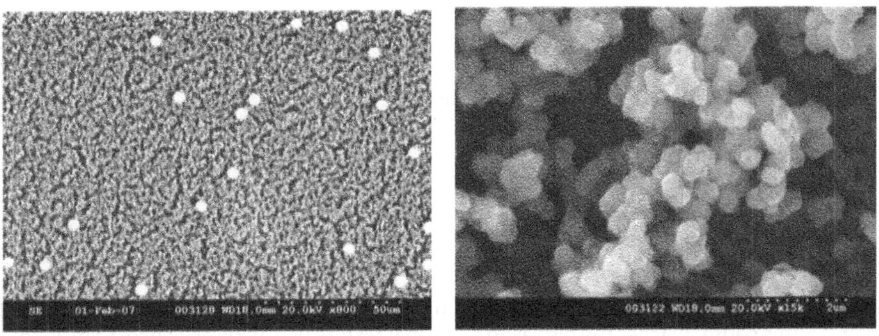

**Figure 10.7** SEM micrographs of the polymer structure formed within the Boogie Board eWriter liquid crystal film. The spacers show up as solid white circles in the low magnification image on the left. In the higher magnification image on the right, a channel structure for the liquid crystal is evident (the black regions).

The Boogie Board eWriter liquid crystal film has a focal conic state that appears dark until the user presses on the top substrate with a point stylus or finger causing the ChLC to flow to the bright planar texture, creating a bright line on a dark background. To erase the image, a small voltage is applied to the eWriter liquid crystal film, as shown in Figure 10.8, creating a uniformly black blank page to write on. The initial pulse applied to the display is $V_P$, 3.3 Hz with a 50 ms pause switches the display to the planar texture ($V_P$ can be anywhere from 20 to 36 V depending on the product). The following two pulses are $V_{FC}$, 3.3 Hz with a 50 ms pause fully erases the eWriter liquid crystal film to a focal conic texture ($V_{FC}$ can have an amplitude of 9.5–18 V). This waveform was chosen to switch the display because it fully erases the eWriter liquid crystal film for the entire operating range of the product, +10 to +40°C. Shorter pulses are possible to fully erase the eWriter liquid crystal film; however, those require a different waveform for different temperatures.

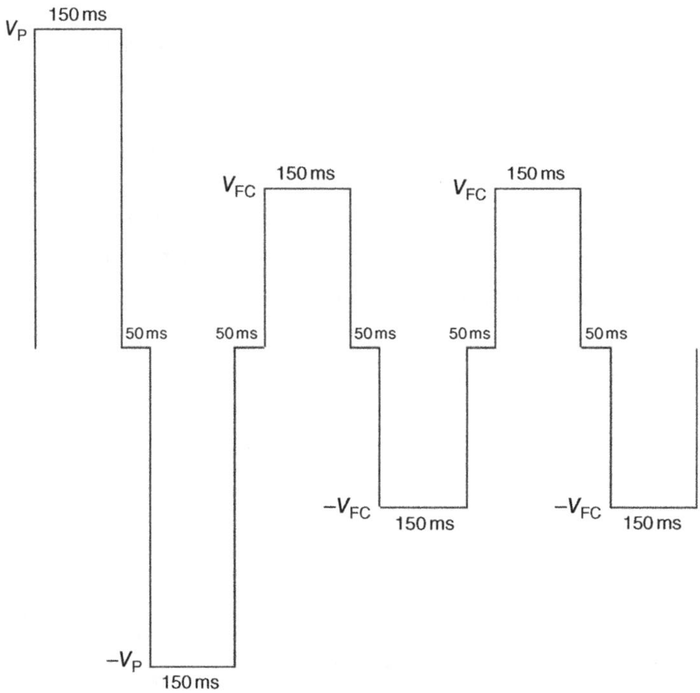


**Figure 10.8** Boogie Board switching waveform to focal conic texture. Initial pulse at $V_P$, 3.3 Hz with a 50 ms pause switches the display to the planar texture ($V_P$ can be anywhere from 20 to 36 V, depending on the product). The following two pulses at $V_{FC}$, 3.3 Hz with a 50 ms pause fully erases the display to a focal conic texture ($V_{FC}$ can have an amplitude of 9.5–18 V).

The Boogie Board eWriter liquid crystal film is very touch sensitive, and different line widths can be created on the display by the user varying their writing pressure and speed. The thickness and reflectivity of the written line increase with pressure. In addition, the Boogie Board eWriter liquid crystal film linewidth and pressure sensitivity can be tuned by varying materials during the manufacturing process. For instance, in the Boogie Board product family, there are variations in linewidth, depending on the market need for the particular device. It is possible to vary the linewidth of the Boogie Board products through processes and materials, as shown in Figure 10.9.

#### *10.2.1.2.2 Roll-to-Roll Manufacturing*

eWriters are made on two R2R manufacturing lines at the company headquarters in Kent, Ohio. The volume R2R manufacturing of Boogie Board eWriter liquid crystal films is a simple yet elegant process that begins with unwinding rolls of PET substrates coated with CP followed by the necessary process steps to effectively convert the coated film into finished flexible displays at the end of

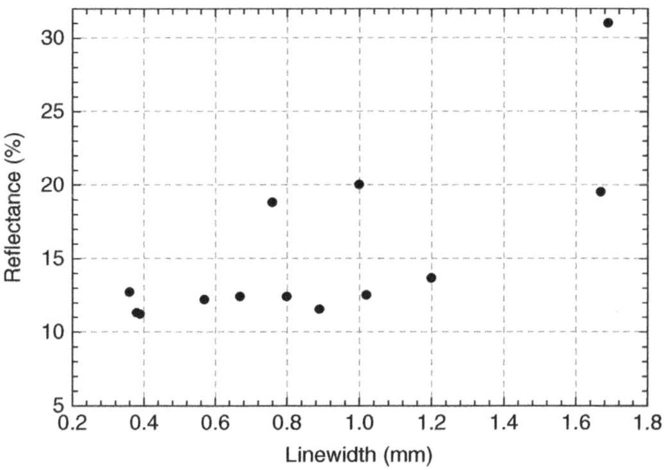


**Figure 10.9** Boogie Board eWriter liquid crystal film pressure sensitivity for different displays at the same pressure and speed. This measurement is made with a written line speed of 75 mm/s and a 150 gF applied to the stylus.

the manufacturing line. Both production lines are inside a clean room. Volume manufacturing has been ongoing for over 6 years.

Neither production line produces waste water or chemicals requiring posttreatment, and both utilize a closed loop system for process cooling. Environmentally friendly UV-curable materials and processes are used throughout the manufacturing process. UV curing lowers air pollution, as well as the energy required to treat exhausts by dramatically reducing solvent vapor emissions compared with more conventional thermal curing processes [21]. An example of one of the R2R manufacturing lines is shown in Figure 10.10.

After UV curing, the display is laser singulated (or cut) from the processed web [22]. Laser singulation permits the manufacture of displays of nonconventional shapes, for example, see Figure 10.11. By carefully controlling the cutting parameters, the laser is able to simultaneously cut through the flood-coated conductor and substrates while welding the top and bottom substrates together without electrically shorting the conductors together from the top and bottom substrates. The laser welds add sufficient strength to the edges of the display such that it does not delaminate and can be considered sealed.

#### *10.2.1.2.3 Alternative Color Films and ChLC*

R2R manufactured eWriter liquid crystal films can also be manufactured using semitransparent back substrates instead of the typical opaque-coated black film. This allows the eWriter to be used in both the reflective and the transparent modes.

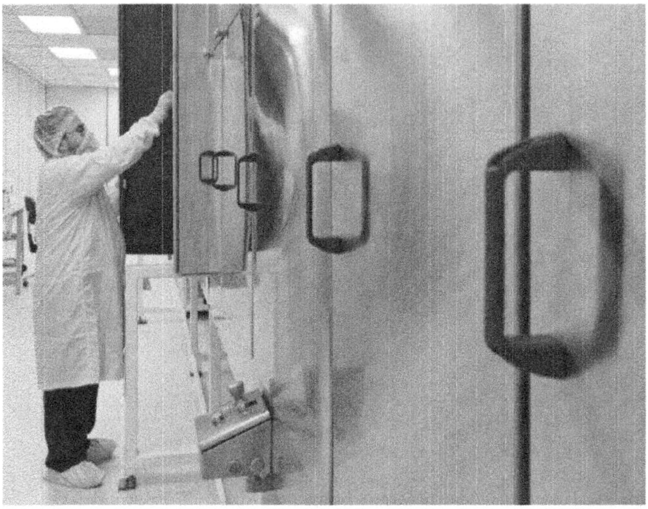

**Figure 10.10** Photograph of a roll-to-roll production line for eWriter liquid crystal films. The line is located in a clean room.

**Figure 10.11** Photograph of eWriter display with a non-conventional shape.

The semitransparent eWriter liquid crystal films shown in Figure 10.12 has high contrast in both reflection and transmission mode and has high clarity in transparent mode [23]. This allows one to view objects behind the semitransparent Boogie Board while in use. With these properties, the semitransparent eWriter liquid crystal films can be used as a standalone device and also as electronic tracing paper.

The semitransparent eWriter liquid crystal films rely on a colored lower substrate or semitransparent colored coating applied in an R2R process onto the

**Figure 10.12** Boogie Boards with semitransparent eWriter liquid crystal films. Both devices include a semitransparent display that allows for use in either the reflective mode or the transparent mode.

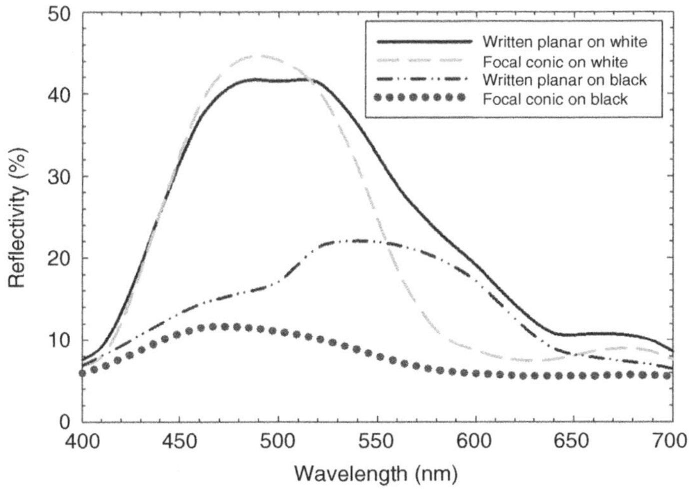


**Figure 10.13** The reflection spectra of the Play n'Trace eWriter, as measured on a white and black background.

lower substrate in the ChLCD to function in transparent mode. The exact color of the final display with semitransparent blue backcoat is a combination of the additive reflective colors of the ChLC layers and the subtractive reflective colors of the semitransparent layer along with the color of whatever is behind the device as shown in the reflection spectra of Figure 10.13.

Another type of eWriter liquid crystal film can be manufactured using multiple color ChLC materials in one eWriter liquid crystal film (Figure 10.14). In the multicolor eWriter liquid crystal film, the writing surface is the color of

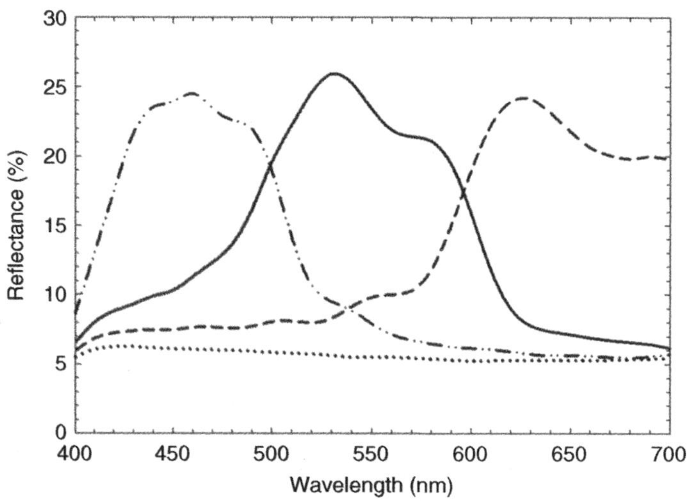


**Figure 10.14** Boogie Board with multicolor eWriter liquid crystal film. Bottom: The reflection spectra of multicolor eWriter liquid crystal film.

the absorbing background until written on, at which time the color of the writing will depend on the color reflected by the ChLC layer and the absorbing background at that particular writing location [24]. The multicolor eWriter liquid crystal film has the pressure sensitivity greatly increased to allow pressure from a rubber stamp or roller to create images while maintaining palm rejection [25].

To create an eWriter liquid crystal film whose writing color changes with location requires changing the peak Bragg reflection wavelength of the ChLC at various locations across the display. This is achieved by varying the amount of chiral dopant in the ChLC to change the reflected color. The multicolor eWriter liquid crystal film is still made with the same low-cost, environmentally friendly R2R high volume manufacturing processes as all other Boogie Board eWriter liquid crystal films.

## 10.2.2 Active Matrix Organic Light-Emitting Diode Displays

### 10.2.2.1 Background

#### *10.2.2.1.1 Introduction*

AMOLED consists of a repeating dot pattern of red, green, and blue light-emitting organic materials arranged in subpixels as a single layer on top of an array of TFTs. The subpixels are energized to emit a pattern that creates an image as seen by the viewer and updates in real time creating full-color video.

#### *10.2.2.1.2 Market Outlook*

The AMOLED display became mainstream in 2015 with the introduction of smartphones and watches containing rigid (e.g., Samsung Galaxy S6) and flexible (e.g., Galaxy S6 Edge, Apple Watch) OLED displays [26, 27]. Samsung, the main producer of mobile OLED displays, introduced the Galaxy S6 with its OLED on glass display and Galaxy S6 Edge with OLED on plastic substrate with one curved edge. Samsung continues to develop new products containing flexible OLEDs with the introduction of the Galaxy S7 Edge [28] with two curved sides in the early 2016. The OLED display in the Galaxy S7 Edge made on plastic substrates with over 500 dpi and 1440 × 2560 resolution demonstrates the state of the art in flexible display performance with basically the same performance as the glass-based OLED display in the Galaxy S7 [29]. It is estimated that the market for plastic OLED displays will nearly double from 2015 to 2016 growing from 1.5 to 2.9 billion USD [30]. Korean-based AMOLED manufacturers have announced extensive investments to expand production capacity [31–33], and interestingly, all of these investments are reported to be sheet-based processes to leverage existing manufacturing infrastructure rather than R2R methods.

#### *10.2.2.1.3 Functionality*

An OLED is a type of LED using an organic compound as the emissive layer sandwiched between two conductors that emit light when an electric current passes through it. The intensity of the light is controlled by the amount of electric current, and the color of the emitted light depends on the chemical characteristics of the OLED material. At least one electrode must be transparent to allow light to escape from the structure. Since OLEDs are fundamentally a thin layer coated on a substrate, they are planar emitters and are well suited for use in FPDs. An OLED display can achieve better black levels (by not emitting light) and can be thinner and lighter than a liquid crystal display (since it does not require a second substrate or ancillary films). The interested reader is referred to other resources for in-depth discussions on operation and material science of OLEDs [34, 35].

There are generally two main architectures for flexible OLEDS, namely, using red-, green-, and blue-emitting OLED materials, which are viewed directly or using red-, green-, and blue-patterned color filters over white light-emitting OLEDs. These two main AMOLED architectures and corresponding pros and cons are shown in Figure 10.15.

#### 10.2.2.2 Challenges in AMOLED Manufacturing

There are four key challenges in manufacturing AMOLED FPDs, namely, developing a good barrier layer, patterning OLED material (when separate RGB colors used), separating flexible substrate with TFT from a carrier substrate, and reliability after flexing of the display.

##### *10.2.2.2.1 Barrier Layer Structure and Performance Testing*

To protect AMOLEDs on polyimide film substrates from water vapor and oxygen penetration, barrier layers such as $SiN_X/SiO_X$ dyads are applied to the film substrate by plasma-enhanced chemical vapor deposition (PECVD) [37]. Water vapor ingress causes dark spots where the OLED stops emitting light and can also corrode metal traces. One issue in evaluating a barrier layer system is measuring the low levels of water vapor transmission rates (WVTRs) through barrier-coated substrates, on the order of $10^{-6}$ g/m$^2$/day, which is required for OLED device reliability and a lifetime of several years. Typically accelerated reliability testing at high temperature and high humidity (e.g., 85°C at 85%RH) is used to evaluate the diffusion of water vapor through the barrier coatings coupled with an analytical test to detect the presence of water. Various analytical techniques have been used including atmospheric pressure ionization mass spectrometry, cavity ring-down spectroscopy, and differential pressure system, as well as indirect methods like calcium corrosion [38, 39]. Water diffusion through barrier layers can also be measured using the concentration–depth profile of deuterium by secondary ion mass spectrometry (SIMS) coupled with sputter profiling. This technique was used to develop two new methods based on detecting water concentration using changes in electrical capacitance and changes in mechanical stress in the barrier film [40]. The diffusion coefficient of water through the barrier film was determined by the three techniques and found within 20% error. The results at extreme temperatures were extrapolated to typical environment of 38°C and 90% relative humidity corresponding to steady state WVTR of $1.5 \times 10^{-7}$ g/m$^2$/day for the 3 μm barrier coating. The one-monolayer lifetime under these conditions is calculated to be 13 years.

| | RGB (SMS) : Samsung display | WOLED : LG display |
|---|---|---|
| Pixel structure | 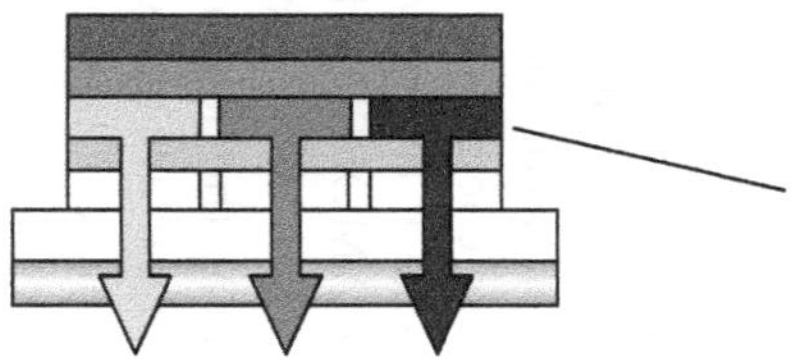  | 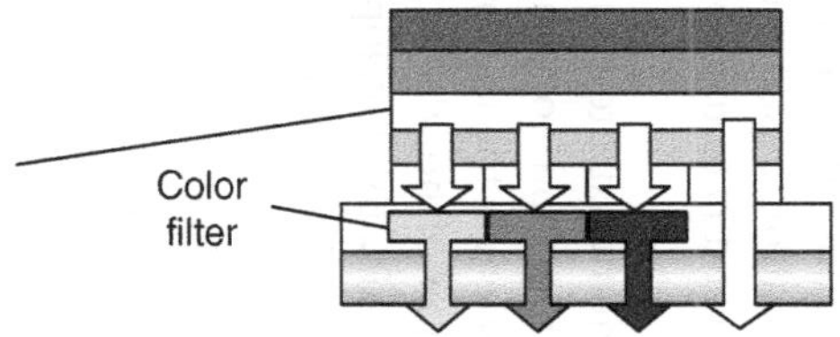  |
| Process method | 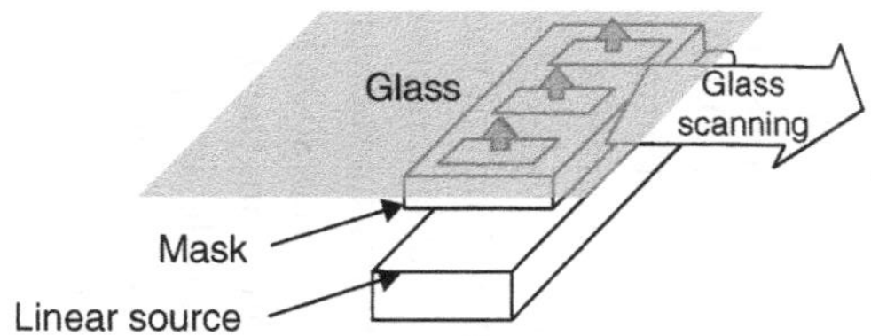  | 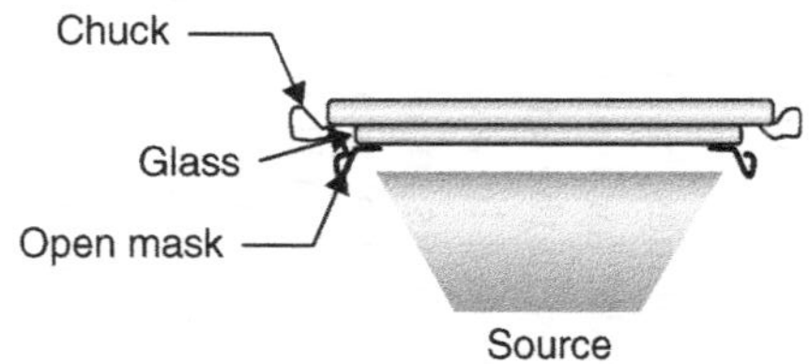  |
| Merits | ▪ Low power and long lifetime<br>▪ Relatively good viewing angle | ▪ Easy process with > gen. 8 glass<br>▪ High resolution |
| Demerits | ▪ Relatively difficult process<br>▪ Color mixing issue<br>▪ Thickness uniformity | ▪ High power consumption<br>▪ Viewing angle issue<br>▪ Short lifetime |

**Figure 10.15** Comparison of direct and indirect view AMOLEDs [36]. *Source:* From Kwon [36].

##### 10.2.2.2.2 OLED Material Patterning

The second challenge in AMOLED FPD manufacturing is the production of patterned OLED material arranged in individual pixels of three colors red, green, and blue. There are several approaches to produce the pattern including shadow mask deposition methods [41, 42] and solution-based method using ink-jet [43] or nozzles [44]; however, the current trend is to use solution coating methods.

Two solution coating techniques have been demonstrated in AMOLED manufacturing, namely, slot die for blanket coatings and continuous nozzle printing for patterned coatings. In the latter, a nozzle jets the material onto the substrate and a pattern is created by a combination of nozzle movements in one axis and substrate movements in the other axis. By developing custom metrology software based on optical profilometry [45], modeling the coating flow process and formulating the OLED inks have produced ±5% uniformity in a 40 nm functional coating over topology demonstrating an AMOLED panels made with solution coating methods on Gen 4 size glass substrates with the same or better short-range luminance uniformity as commercial AMLCDs. Recently, ink-jet printing OLED materials have been introduced for producing the red, green, and blue pixels processing large AMOLED panels for TV mass production [46].

##### 10.2.2.2.3 Separating TFTs from a Carrier Substrate

For AMOLEDs manufactured on a glass carrier substrate, ultimately the flexible substrate containing the TFT and OLED layers must be separated from the rigid glass carrier substrate. A laser ablation process is used, called laser lift-off. In this process typically an excimer laser with a wavelength of 308 nm is passed through the glass carrier, which is indexed using pulse energies of $275\,J/m^2$ to evaporate a thin layer (about 200 nm) of the polyimide at the polyimide glass carrier interface [47]. After the ablation process, the AMOLED substrates are separated from the glass carriers under very low strain of 0.2% as 0.8% can damage the device. Recent work has shown that during the delamination of a polyimide, AMOLED substrate from the glass carrier surface charges of 800 V can be produced on the polyimide and glass [37]. The resulting current flow penetrates the polyimide to the TFT layer damaging the devices. The electrostatic charge from separation can be eliminated by using a metallic interlayer that dissipates the charge.

##### 10.2.2.2.4 Bending Reliability of AMOLEDs

Flexible AMOLED displays are an FPD technology that offers the potential to be a freeform display that can be rolled, bent, and folded due its thinness and ruggedness. This will lead to new industrial designs and use models for FPDs including extreme curvature for immersive viewing and even rollable displays for ease of transport and storage. For this extreme flexibility the flexible OLED displays will have to be produced without cover glass, currently used for protection, once their ruggedness is improved. Core to this capability is the OLED display's ability to pass basic rolling and bending reliability testing, for example, bending over a mandrel of a certain diameter. As the displays become more robust and rugged these tests become more severe, ultimately demonstrating that the flexible OLED will no longer need a cover glass.

Foldable AMOLED display designs have recently been demonstrated to survive folding over a 2 mm radius of curvature 100,000 times [48] as well as designs, which can be folded inward and outward as a trifold display [49]. Finite element models of the folding processes have been used to optimize layer thicknesses and placement such as moving the TFT and OLED layer to the neutral plane so that stresses for this layer were minimized [50].

#### 10.2.2.3 OLED Manufacturing Examples

The following are three examples of OLED displays made on flexible substrates.

##### *10.2.2.3.1 OLED Lighting Panels*

Konica Minolta has been leading the commercialization of phosphorescent OLEDs made using R2R manufacturing for lighting applications [51, 52]. They make the OLED panel by solution coating active layers directly on barrier films with WVTR less than $10^{-5}$ g/(m$^2$·day). To capture a large market, they report that R2R manufacturing is needed to lower costs, reduce manufacturing complexity and OLED material waste, and simplify processing by eliminating the glass carrier. Konica Minolta uses multiple R2R lines (example shown in Figure 10.16) all located in clean rooms and comprising the steps of patterning and cleaning, wet coating and drying, vacuum deposition of electrode, encapsulation, and packaging [54]. The solution-based, custom-formulated OLED materials are slot die coated in four layers on plastic barrier films. To increase yield they optimized

**Figure 10.16** Konica Minolta's roll-to-roll OLED manufacturing line [53]. © John Wiley & Sons.

die design and improved drying to reduce coating defects. By studying the morphology of the coated OLED layers using small angle X-ray scattering for particle size investigation and wide angle X-ray diffraction for molecular orientation investigation, with these improvements they have achieved similar performance from solution and evaporated OLED layers conditions as those seen in shadow mask deposition methods. In 2014, Konica Minolta built an R2R manufacturing plant [53] in Japan capable of producing 1 million OLED panels per month sized 150 × 60 mm that are 0.35 mm thick and weighs 5 g. OLED lighting is being commercialized by Acuity Brands selling OLED lighting fixtures in the United States through DIY home improvement stores.

#### *10.2.2.3.2 G Flex Smartphone*

LG introduced the G Flex smartphone in 2013, which contains a 5.98″ AMOLED display [55, 56] shown in Figure 10.17 on a flexible substrate made by bonding the substrate to a glass sheet and then processing on a conventional LCD manufacturing line. Critical issues in this process are the mechanical deformation of TFTs during release of the flexible substrate from the glass, coating the polyimide film on the carrier glass, and the use of low temperature polycrystalline silicon for the TFT structure.

To protect the OLED layer from water vapor, a multilayer organic/inorganic barrier layer was developed with a WVTR of approximately $10^{-6}$ g/m$^2$/day, which protects the display through high temperature and high humidity testing of 85°C/85% RH testing. The flexible substrate is a polyimide that is slot die coated on a glass carrier then baked. The polyimide was developed to be

**Figure 10.17** LG's flexible OLED flat panel display (left) featured in the G Flex smartphone (right) [57]. © John Wiley & Sons.

optically isotropic, with good light transmission and low CTE mismatch with glass. The process temperature was kept at less than 150°C due to the polyimide's thermal stability and to allow low temperature polycrystalline silicon TFTs to be used. The polyimide is released from the glass using laser irradiation, which generates hydrogen gas in a release layer that was applied between the polyimide and glass. The gas effectively separates the polyimide from the glass. The release process was developed to minimize change in the TFT performance before and after removal of the flexible polyimide substrate from the glass carrier.

The AMOLED panel bending radius capability of 15 mm is limited by smartphone case. By removing brittle inorganic layers, a high reliability flexible AMOLED display was produced. New materials, thinner films integrating touch, and barrier films will be the focus of future work.

##### 10.2.2.3.3 18″ AMOLED FPD

LG has scaled their AMOLED production from 6″ diagonal to 18″ and anticipate producing 55″ diagonal AMOLEDs in 2017 by continuing to scale up their processes [57].

The major challenges reported in producing this large top emission AMOLED display include developing a multilayer barrier coating, highly uniform TFT characteristics over a large area, flexible encapsulation, laser lift process for large areas, and bending reliability. Figure 10.18 shows a cross section and condensed process flow chart for the fabrication of this OLED FPD.

A multilayer barrier layer on the polyimide substrate using a $SiN_X$ tie layer instead of $SiO_2$ to improve adhesion was demonstrated with a WVTR approximately $10^{-5}$ g/m$^2$/day. Amorphous indium–gallium–zinc–oxide (a-IGZO) TFTs were successfully produced on the polyimide base film due to their high electrical uniformity and low parasitic capacitance, leading to improved electrical uniformity over the entire panel. A flexible encapsulation was developed to seal the edge of the display. The laser release process was refined for large areas including changing beam shape, uniformity, laser power, and stage control accuracy. Attention was paid to contamination control as a particle can block the laser light, resulting in locations where the polyimide substrate is firmly attached to the glass carrier, leading to damage of the TFTs during the release process. The AMOLED displays were subjected to bending tests consisting of repeating bending over a 30 mm radius for 100,000 cycles without dark spots or line defects observed. The excellent flexibility of this FPD is shown in Figure 10.19.

The OLED displays shown in Figures 10.17 and 10.19 were made using sheet-to-sheet processes, and there is continued research into producing OLED displays using R2R manufacturing processes.

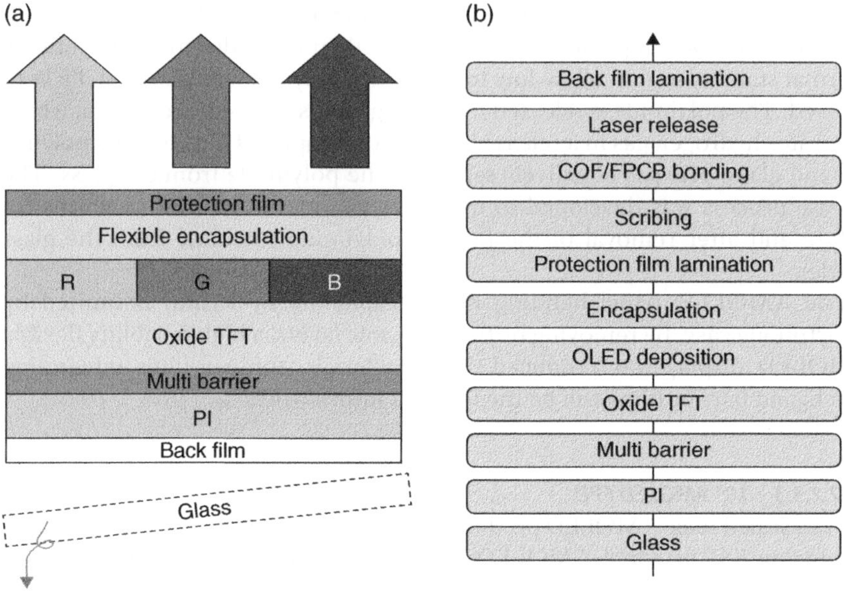


**Figure 10.18** Cross section (a) and process flowchart (b) for 18-inch flexible AMOLED display on a flexible substrate [58]. © John Wiley & Sons.

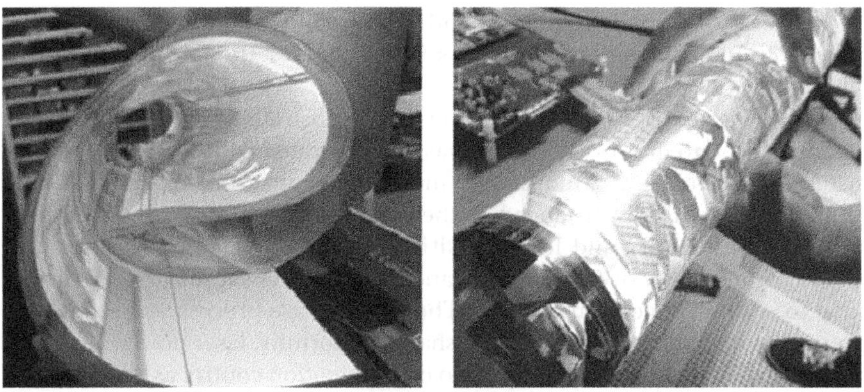

**Figure 10.19** Flexibility of 18″ AMOLED FPD showing that it can be rolled [58]. © John Wiley & Sons.

## 10.2.3 Electrophoretic Displays

### 10.2.3.1 Function

Electrophoretic displays are flexible, reflective, sunlight readable, active matrix addressable, and possibly full color and are an easy fit for R2R manufacturing. An electrophoretic ink, typically consisting of multiple types and colors of charged particles, is suspended in an insulating fluid and driven electrophotetically from the top to the bottom of the display, allowing for the desired color to be written to each pixel [59]. Typically the white particles consist of titanium dioxide and are approximately 1 μm in diameter along with transparent colored particles (either dye or pigment) dispersed in a hydrocarbon oil [58]. The particles have the surface modified with surfactants, charging agents, and polymers to cause the particles to take on an electric charge.

#### 10.2.3.2 Structure

Some electrophoretic displays are built using tiny microcapsules filled with the electrophoretic ink, shown in Figure 10.20. The microcapsules allow the display to be made on flexible substrates and the entire structure to be produced by an R2R process and wound into a roll of flexible electrophoretic material and then shipped to the final display assembly. The final display can be constructed by cutting out portions of the flexible microencapsulated electrophoretic material and laminating to a backplane using an adhesive with controlled resistivity properties.

Another technique for encapsulating the electrophoretic ink is to place it in an embossed microcup structure, as shown in Figure 10.21. Once again, an optically clear adhesive is used to bond the electrophoretic material to a backplane or electrode structure. The grids of the microcup structure limit the motion of the particles in the lateral direction so that solvent and charged particles are not displaced and collect around pixel edges or more frequently used pixels.

Multicolored and full-color displays have been demonstrated by including several different colored particles in one microcup, as shown in Figure 10.22. Each color is created by bringing the desired color or colors required to mix to the front of the display. The different particles are oppositely charged and respond to various threshold voltages, allowing the display to separate the desired colors. The full-colored electrophoretic display structure is very appealing as it should allow for full reflectivity from each color as the other color particles are hidden underneath the desired color and eliminate the need for color filters and side-by-side RGB pixel smearing, both of which reduce individual color's reflectivity.

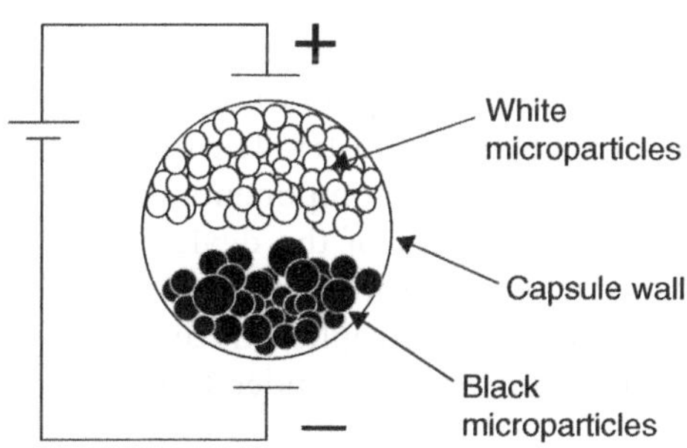


**Figure 10.20** Cross section of encapsulated electrophoretic display [58]. © Nature Publishing Group.

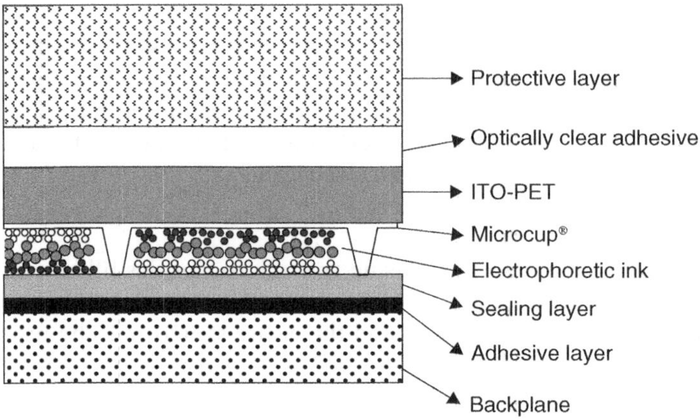


**Figure 10.21** Cross section of microcup-type electrophoretic display [60]. © John Wiley & Sons.

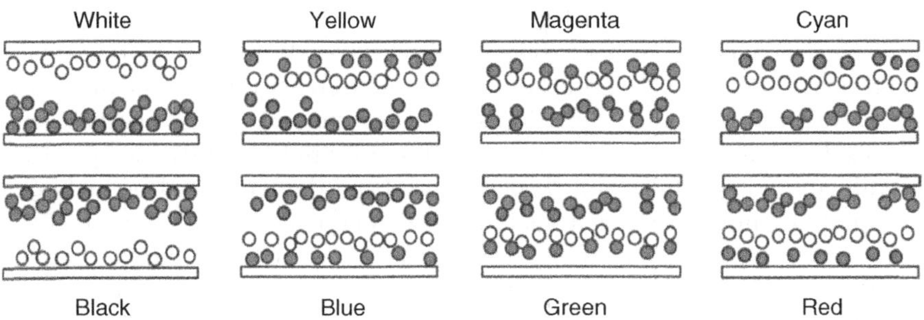


**Figure 10.22** Cross section of various colored configurations for full-color electrophoretic display [61]. © John Wiley & Sons.

#### 10.2.3.3 Manufacturing

The microcup process flowchart is shown in Figure 10.23. The film is coated with an embossable material, embossed, cured, and then filled and sealed. The sealing process can include a top film laminated to the embossed film or a phase-separated material that separates to the top of the opening of the microcup structure and then cured. All of these process steps are very amenable to R2R manufacturing and utilize processes similar to those found in the R2R manufacturing industry. The difficulty lies in defining the optimum materials for producing the components and also allow for good long-term reproducibility and reliability of the device. In addition, high yield and throughput are desired so fast curing materials would be required for cost-effective mass production. The electrophoretic display electro-optic material is a great fit for R2R manufacturing and is indeed manufactured that way at present; however, the

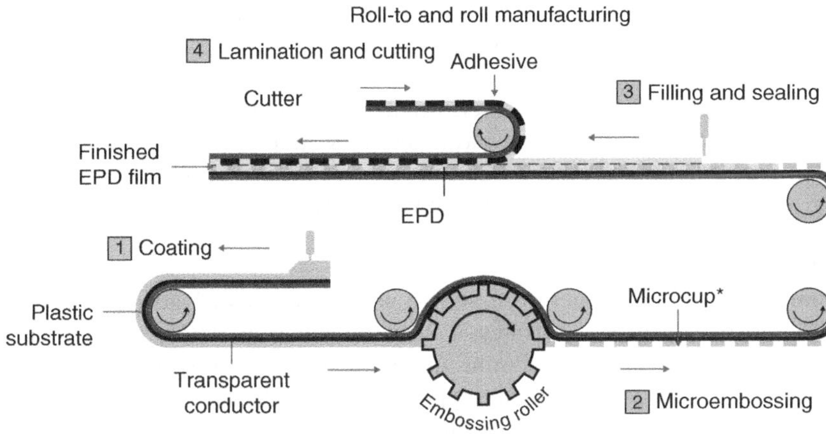


**Figure 10.23** Flowchart of Microcup® imaging film manufacturing [60]. © John Wiley & Sons.

final display building process frequently uses TFT backplanes that are manufactured in a batch process. The final assembly of electrophoretic displays includes laminating the electro-optic layer to the batch-processed TFT structure followed by final testing or interconnect.

### 10.2.4 Microfluidic Displays

#### 10.2.4.1 Function

Microfluidic displays are a good fit for reflective ePaper as they demonstrate reflectivity similar to that of paper, are multicolored, and allow for grayscale. In addition, this technology could be manufactured by an R2R process as the corresponding film structure is amenable to such an operation. As a relatively new technology that has just entered the flexible display arena, there is development to be done and even pilot production is yet to be demonstrated. Microfluidic displays are promising because the device uses a subtractive color that allows for the display to be much brighter than additive color systems and the possibility of a multicolored reflective display, where each color can have high reflectivity without the need for color filters or RGB pixels. This is accomplished by multiple fluid colors being swapped between upper and lower chambers of the device.

Microfluidic displays function by electrically transporting an immiscible black ink and a clear oil fluid to the desired pixel locations using electromechanical pressure overcoming their initial inherent geometrical Laplace pressure imparted by the thin-layer and porous structures of the display [62]. The display has bistability since the field can be removed and the image remains [60]. Upon a voltage in the opposite chamber, the fluid redistributes and is erased. There are upper and lower channels separated by a white porous film, allowing the fluids to travel such that they are viewable (above center white film) or hidden (below center white film). The published reflectivity of the white state is measured at a maximum value of 76% and an estimated dark state reflectivity of less than 3%. The switching speed for a 150 PPI device was measured at less than 15 ms for on and off, thus making it capable of video rates [60].

#### 10.2.4.2 Structure

The microfluidic film structure is shown schematically in Figure 10.24. The display consists of three substrates: a clear top substrate with patterned ITO electrodes, an opaque white aluminum center film, and a lower film coated with a metal electrode, creating an upper and lower channel with micropores bridging the two channels. The opaque white center film has micropores of two sizes to allow for transport of the black ink (pore diameter ~32 μm) and the clear oil (pore diameter ~6 μm), spacers to allow for splitting a volume of fluid from within, and a surface microtexture to allow for optimum white scattering of light. The opaque white center film is hydrophobic to allow for improved fluid transport from upper to lower channels. The density of pores affects the switching speed—more is faster; however, more pores also result in lower reflectivity due to the visible dot where each black oil pore is located so this requires a balance to create the optimum display function. The pore size has been designed to accept a manufacturing variation of 10% or more. The structure of this display is improved for ease of manufacturing such that the porous film is not required to be aligned with the patterned pixels due to the random placement of the pores and spacers.

#### 10.2.4.3 Technology

All the novel features of this display technology directly stem from the unique and sophisticated fluid mechanics of the porous white film. Several iterative cycles of theory and experiment identified that fundamentally, the film must satisfy several key principles. First, it must be able to transport both ink and oil. Second, it must ensure a portion of the ink is always in a position such that it can receive electromechanical force from an electrode. Third, it must work without the need for alignment to electrodes. Finally, it must allow the merged ink in either channel to be locally split for rapid removal from that channel. The first three requirements are solved by ink and oil pore design, whereas the final requirement is solved by the spacer design (Figure 10.24a, angled view diagram) [63].

The microfluidic displays discussed were made using sheet-to-sheet processes, there is continued research into producing microfluidic displays using R2R manufacturing processes, and in addition, microfluidic displays seem to be designed specifically for continuous web manufacturing.

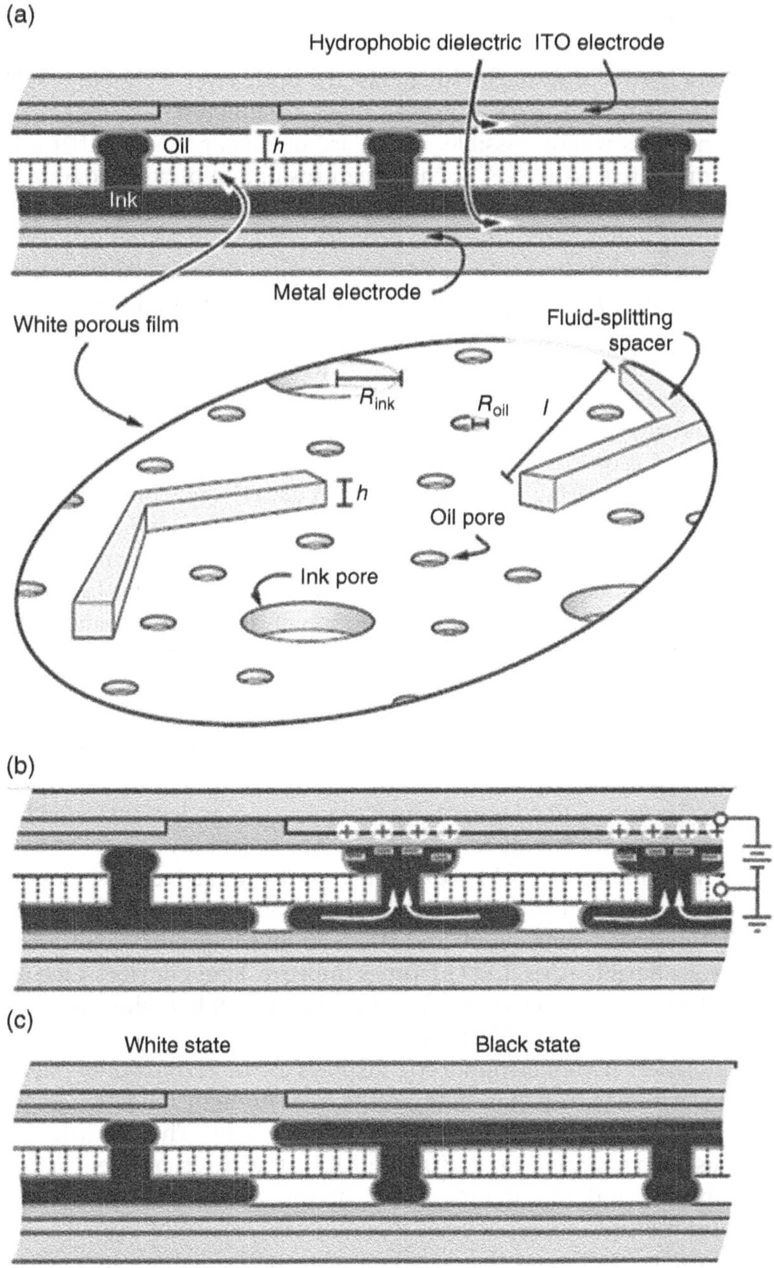


**Figure 10.24** Cross-sectional views of microfluidic display (a) with exploded view of spacer design and pores, (b) voltages and charge flow, and (c) ink and oil locations in white and black states [62]. © Nature Publishing Group.

## 10.3 Conclusions

Flexible FPDs are here to stay. Each type of electro-optic layer technology has different manufacturing processes; however, there are inherent advantages to utilizing R2R methods where possible. There are obvious processes that are not currently compatible with R2R, such as high resolution photolithography patterning for TFT and other components that in some cases require specialized manufacturing equipment where the process is fully conducted in a vacuum and enclosed for cleanliness. In this chapter, several flexible displays were presented that are currently being manufactured at least in part following R2R methodologies. There are many opportunities for improvements and technical developments in this field. In the coming years there will likely be an increase of flexible displays in the market, and this will drive new and exciting technical developments and advances in R2R manufacturing processes.

The Boogie Board is one of the first commercially available flexible displays made with R2R manufacturing and has created an entirely new product category of electronic writing tablets. The convergence of innovations in many technical fields was required to achieve Boogie Board display R2R manufacturing, including high volume manufacturing, materials science and display technology, web process engineering, product development, and commercialization of consumer products. The R2R manufacturing processes were designed to produce Boogie Board displays with high throughput and volume at low cost and while using conventional materials. The R2R processes developed at Kent Displays are mature and allow these flexible displays to be built in a significantly automated process.

In addition to eWriters, ChLCDs made by R2R processes as demonstrated by ITRI can be possibly applied in digital signage, banners, and toys.

FPDs using AMOLED on flexible substrates are being actively developed using a mix of R2R process technologies and conventional LCD manufacturing processes. The increased ruggedness and thinness and improved color have drawn in users. As flexible AMOLED lifetimes improve, larger devices such as TVs and monitors will also move toward flexible AMOLEDs due to their improved ruggedness and less bulky structure.

Almost all eReaders use electrophoretic displays as the electro-optic layer due to the similarities with paper, being reflective, sunlight readable, good black text on white background, and thin and lightweight. The electrophoretic display has been manufactured using multiple structures, both encapsulated microcapsules and microcups. Some of the newer electrophoretic designs allow for multiple color particles that allow for full-color displays. As consumer interest in eReaders with full-color images and video rate increases, these new electrophoretic display developments will be key in ensuring this reflective display technology continues to dominate the eReader market space. Electrophoretic displays with electro-optic layers manufactured R2R have demonstrated large volume capabilities through their eReader market presence, enabling future market increase and potential new product applications.

Microfluidic displays utilize a new electro-optic layer type for the flexible display arena. The possibility of multiple colors with a high reflectivity for each color makes this technology attractive for ePaper type applications. The display structure is amenable to R2R manufacturing, although apparently this is yet to be demonstrated in pilot lines. The substrate structure seems to have been developed specifically with R2R manufacturing option in mind, which is what is required today to manufacture displays cost-effectively.

Although several materials sets and manufacturing processes have been identified for flexible display manufacturing and have resulted in products currently in the market, these flexible displays are not a fit for all display needs. In addition, not all displays types are compatible with these unique and somewhat limited set of materials and processes. However, the further development of materials and processes will continue as long as there is perceived value in flexible displays replacing current rigid displays for either final display ruggedness, light weight, reduced thickness, flexibility, and unique form factor or lower manufacturing cost with high throughput.

## References

1 Davis, D.J. The Future is Plastics! *Inf. Disp.*, **2003**, 19(10), 4.
2 MacDonald, W.A. Latest Advances in Substrates for Flexible Electronics. In *Large Area and Flexible Electronics*; Caironi, M. and Noh, Y., Eds.; John Wiley & Sons, Inc.: New York; **2015**; p 293.
3 Tseng, F.; Cheng, A.; Peng, Y.N. Assessing Market Penetration: Combining Scenario Analysis, Delphi, and the Technological Substitution Model: The Case of the OLED TV Market. *Technol. Forecast. Soc. Chang.*, **2009** 76(7), 897–909.
4 Song, C., Ryu, G. Flexible Displays. In *Large Area and Flexible Electronics*; Caironi, M. and Noh, Y., Eds.; John Wiley & Sons, Inc.: New York; **2015**; p 411.
5 Hast, J.; Ihme, S.; Makinen, J.-T.; Keranen, K.; Tuomikoski, M.; Ronka, K.; Kopola, H.; Freeform and Flexible Electronics Manufacturing Using R2R Printing and Hybrid Integration Techniques, 44th European Solid State Device Research Conference, Venezia Lido (Venice), Italy, September 22–26, **2014**.
6 Oikawa, K. Based on graph; Development of OLED Lighting Applications Using Phosphorescent Emission System, Printed Electronics Europe, Berlin, Germany, April 3–4, **2012**.
7 Carlon, K. The only thing stopping flexible displays from taking over is that no one is making them, Android Authority [Online] April 29, **2016**, http://www.androidauthority.com/when-will-flexible-displays-arrive-688227/ (Accessed on September 23, 2017).

8 Crawford, G. P. (Ed.) Flexible Flat Panel Display Technology. In *Flexible Flat Panel Displays*; John Wiley & Sons Ltd: Chichester; **2005**.

9 Yang, D.-K.; Doane, J. W. Cholesteric Liquid Crystal/Polymer Gel Dispersion: Reflective Display Application. *SID Symp. Dig.*, **1992**, 23, 759.

10 Yang, D.-K.; West, J. L.; Chien, L.-C.; Doane, J. W. Control of reflectivity and bistability in displays using cholesteric liquid crystals. *J. Appl. Phys.*, **1994**, 76 (2), 1331.

11 Doane, J. W.; Khan, A. Cholesteric Liquid Crystals for Flexible Displays. In *Flexible Flat Panel Displays*; Crawford, G. P., Ed.; John Wiley & Sons Ltd: Chichester, **2005**.

12 Chen, J.; Shiu, J.-W.; Chiu, W.-W.; Tsai, C.-C.; Huang, C.-Y. Invited Paper: Roll-to-Roll Flexible Display for e-Paper Applications. *SID Symp. Dig.*, **2011**, 42, 107–110.

13 Shiu, J.-W.; Lee, K.-C.; Liang, C.-C.; Liao, Y.-C.; Tsai, C.-C.; Chen, A. Rugged Display: Recent Results of Flexible Cholesteric Liquid-Crystal Displays. *J. Soc. Inf. Disp.*, **2009**, 17(10), 811–820.

14 Liu, P.-W.; Tsai, C.-C.; Lee, C.-H.; Huang, W.-H.; Chen, C.-W.; Wang, C.-Y.; Lee, K.-C. Large Area Black/White Bistable Cholesteric Liquid Crystal Display and the Thermal-Addressing System. *SID Symp. Dig.*, **2010**, 41, 1663–1666.

15 Chen, H.-Y.; Su, J.-Y.; Wu, C.-Y.; Chen, C.-W.; Chin, C.-L.; Lin, H.-K.; Liang, C.-C. High Resolution and Multi-Color R2R Flexible e-Papers. *SID Symp. Dig.*, **2011**, 42, 157–159.

16 Yeh, P.-C.; Chen, H.-Y.; Liang, C.-C.; Wei-Sun, C.; Ju-Yuen Su, J.-Y.; A Multi-Line Driving Scheme for R2R PM Ch-LCD. *SID Symp. Dig.*, **2011**, 42, 1542–1544.

17 Kent Displays Inc. Press Release: Kent Displays Forms Improv Electronics, Launches Boogie Board™ LCD Writing Tablet, **2010**, http://www.prnewswire.com/news-releases/kent-displays-forms-improv-electronics-launches-boogie-boardtm-lcd-writing-tablet-82290782.html (Accessed on March 20, 2017).

18 Khan, A.; Montbach, E. Flexible eWriter Technology and Applications. *SID Symp. Dig.*, **2015**, 46, 242–245.

19 Montbach, E.; Krinock, J.; Chiu, J.; Braganza, C.; Schlemmer, M.; Pishnyak, O.; Marhefka, D.; Khan, A.; Doane, J. W. Patterned Image Flexible Reflex Displays. *SID Symp. Dig.*, **2012**, 43, 175–178.

20 Schneider, T.; Montbach, E.; Davis, D.; Franklin, S.; McDaniel, D.; Lightfoot, M.; Venkataraman, N.; Nicholson, F.; Khan, A.; Doane, J. W. UV Cured Flexible Cholesteric Liquid Crystal Displays, *Radtech UV&EB Curing Technology, Technical Conference Proceedings*, Chicago, IL, May 4–7, **2008**.

21 Green, A.; Montbach, E.; Miller, N.; Davis, D.; Khan, A.; Schneider, T.; Doane, J. W. Energy Efficient Flexible Reflex™ Displays. *Proc. Intl. Disp. Res. Conf.*, **2008**, 28, 55.

22 Schneider, T. Method of simultaneous singulation and edge sealing of plastic displays. US 2A1, June 1, **2007**.

**23** Braganza, C.; Montbach, E.; Khan, A.; Doane, J. W. Flexible Semitransparent eWriter Displays. *SID Symp. Dig.*, **2015**, 46, 334.

**24** Braganza, C.; Echeverri, M.; Kammer, K.; Dyshko, A.; Montbach, E.; Khan, A.; Doane, J. W. eWriter Revealing Multiple Colors. *SID Symp. Dig.*, **2016**, 47, 1539–1542.

**25** Li, J.; Krieger, C.; Urbanski, A. N.; Skinner, C.; Guyler, K.; Glasscock, S. D. Coated art material system. US 2A1 (**2009**).

**26** Samsung Press Release. Relive the Moment – Coldplay's 'A Head Full of Dreams Tour' Live in VR with Samsung and Live Nation. https://www.samsungmobilepress.com/2015/03/02/Beautifully-Crafted-from-Metal-and-Glass,-Samsung-Galaxy-S6-and-Galaxy-S6-edge-Define-Whats-Next-in-Mobility (Accessed on October 31, 2016).

**27** Apple Press Release. Apple Watch Available in Nine Countries on April 24. http://www.apple.com/pr/library/2015/03/09Apple-Watch-Available-in-Nine-Countries-on-April-24.html (Accessed on October 31, 2016).

**28** Samsung Press Release. Samsung Introduces Galaxy S7 and Galaxy S7 edge, Center of the New Mobile World. http://news.samsung.com/us/2016/02/21/samsung-introduces-galaxy-s7-galaxy-s7-edge-center-new-mobile-world/ (Accessed on October 31, 2016).

**29** Soneira, R. M. Galaxy S7 OLED Display Technology Shoot-Out. http://www.displaymate.com/Galaxy_S7_ShootOut_1.htm (Accessed on October 31, 2016).

**30** Guillaume Chansin. Plastic OLED Display Capacity Ramps Up & Glass-Based OLEDs Downgraded. http://www.idtechex.com/research/articles/plastic-oled-display-capacity-ramps-up-and-glass-based-oleds-downgraded-00008222.asp (Accessed on October 31, 2016).

**31** LG Display Press Release. LG Display to Invest in 6th Generation OLED Panel Line for Flexible Displays. http://www.lgdisplay.com/eng/prcenter/newsView?articleMgtNo=4924 (Accessed on October 31, 2016).

**32** LG Display Press Release. LG Display Makes Major Investment in Next-generation OLED Panels. http://www.lgdisplay.com/eng/prcenter/newsView?articleMgtNo=4958 (Accessed on October 31, 2016).

**33** Maeil Business News Korea. Pulse. http://pulsenews.co.kr/view.php?year=2016&no=131912 (Accessed on October 31, 2016).

**34** Tsujimura, T. *OLED Display Fundamentals and Applications*, 1st ed.; John Wiley & Sons, Inc.: New York; **2012**.

**35** Gaspar, D.J., Polikarpov, E., (Eds.) *LED Fundamentals: Materials, Devices, and Processing of Organic Light-Emitting Diodes*, 1st ed.; CRC Press: Bocan Raton; **2015**.

**36** Kwon, J. H.; Frontline Technology: RGB Color Patterning for AMOLED TVs. http://informationdisplay.org/IDArchive/2013/MarchApril/FrontlineTechnologyRGBColorPatterning.aspx (Accessed on October 31, 2016).

37 Lin, Y.-L.; Ke, T.-Y.; Liu, C.-J.; Huang, C.-S.; Lin, P.-Y.; Tsai, C.-H.; Tu, C.-H.; Wang, P.-F.; Lu, H.-H.; Lee, M.-T.; Hwu, K.-L.; Chuang, C.-S.; Lin, Y.-H. A Delamination Method and Ultra-High-Reliable Gas Barrier Film for Flexible OLED Displays. *SID Symp. Dig. Tech. Pap.*, **2014**, 45, 114–117.

38 Paetzold, R.; Winnacker, A.; Henseler, D.; Cesari, V.; Heuser, K.; Permeation Rate Measurements by Electrical Analysis of Calcium Corrosion. *Rev. Sci. Instrum.*, **2003**, 74, 5147.

39 Choi, J. H.; Kim, Y.M.; Park, Y.W.; Huh, J.W.; Ju, B.K.; Kim, I.S.; Hwang, H. N. Evaluation of Gas Permeation Barrier Properties Using Electrical Measurements of Calcium Degradation. *Rev. Sci. Instrum.*, **2007**, 78, 064701.

40 Visweswaran, B.; Mohan, S. H.; Mandlik, P.; Silvernail, J.; Ma, R.; Sturm, J.; Wagner, S.; Predicting the Lifetime of Flexible Permeation Barrier Layers for OLED Displays. *SID Symp. Dig. Tech. Pap.*, **2014**, 45, 111–113.

41 Advantech. http://www.advantechus.com/docs/Micro-Patterned-Deposition-Shadow-Mask.pdf (Accessed on May 16, 2016).

42 Mai, T.A.; Bernold Richerzhagen, B. Manufacturing of 4th Generation OLED Masks with the Laser MicroJet® Technology. *SID Symp. Dig. Tech. Pap.*, **2007**, 38, 1596–1598.

43 Madigan, C. F.; Hauf, C. R.; Barkley, L. D.; Harjee, N.; Vronsky, E.; Van Slyke, S. A. Advancements in Inkjet Printing for OLED Mass Production. *SID Symp. Dig. Tech. Pap.*, **2014**, 45(1), 399–402.

44 Chesterfield, R.; Johnson, A.; Lang, C.; Stainer, M.; Ziebarth, J.; Solution-Coating Technology for AMOLED Displays. *Inf. Disp.* **2011**, 1, 24–29.

45 Parker, I.; Johnson, A.; Frischknecht, K.; Stainer, M. Measurement Methods for Quality Control of Coating Uniformity in Solution Processed OLED Displays. *SID Symp. Dig. Tech. Pap.*, **2015**, 46, 207–211.

46 Kateeva Press Release. Latest on Kateeva and Our Technology. http://kateeva.com/press-full/kateeva-introduces-the-yieldjet-flex-a-manufacturing-equipment-solution-to-mass-produce-innovative-oled-displays-tor-curved-and-flexible-consumer-electronics-products/ (Accessed on May 16, 2016).

47 Delmdahl, R.; Pätzel, R.; Brune, J.; Large-Area Laser-Lift-Off Processing in Microelectronics. *Phys. Procedia*, **2013**, 41, 241–248.

48 Komatsu, R.; Nakazato, R.; Sasaki, T.; Suzuki, A.; Senda, N.; Kawata, T.; Ikeda, H.; Eguchi, S.; Hirakata, Y.; Yamazaki, S.; Shiraishi, T.; Yasumoto, S.; Bower, C.; Cotton, D.; Matthews, A.; Andrew, P.; Gheorghiu, C.; Bergquist, J. Repeatedly Foldable Book-Type AMOLED Display. *SID Symp. Dig. Tech. Pap.*, **2014**, 45, 326–329.

49 Jimbo, Y.; Aoyama, T.; Ohno, N.; Eguchi, S.; Kawashima, S.; Ikeda, H.; Hirakata, Y.; Yamazaki, S.; Nakada, M.; Sato, M.; Yasumoto, S.; Bower, C.; Cotton, D.; Matthews, A.; Andrew, P.; Gheorghiu, C.; Bergquist, J. Tri-Fold Flexible AMOLED with High Barrier Passivation Layers. *SID Symp. Dig. Tech. Pap.*, **2014**, 48, 322–325.

50 Yan, J.-Y.; Ho, J.-C.; Chen, J. Foldable AMOLED Display Development: Progress and Challenges. *Inf. Disp.*, **2015**, 1(15), 12–16.

**51** Minakata, T.; Tanamura, M.; Mitamura, Y.; Imashiro, M.; Horiguchi, A.; Sugimoto, A.; Yamashita, M.; Ujiiye, K.; Sunahiro, S.; Yada, Y.; Ibaraki, N.; Hiroshi Tomiyasu, H. Fully R2R-Processed Flexible OLEDs for Lighting. *SID Symp. Dig. Tech. Pap.*, **2015**, 46, 219–222.

**52** Tsujimura, T.; Fukawa, J.; Endoh, K.; Suzuki, Y.; Hirabayashi, K.; Mori, T. Flexible OLED Using Plastic Barrier Film and Its Roll-to-Roll Manufacturing. *SID Symp. Dig. Tech. Pap.*, **2014**, 45, 104–107.

**53** Konica Minolta Press Release, Konica Minolta Constructs Plant for World's First Mass Production of Plastic Substrate Flexible OLED Lighting Panels. http://www.konicaminolta.com/about/releases/2014/0318_01_01.html (Accessed on July 24, 2015).

**54** Oikawa, K. Development of OLED Lighting Applications Using Phosphorescent Emission System, Printed Electronics Europe, Berlin, Germany, April 3–4, **2012**.

**55** Hong, S.; Yoo, J.; Jeon, C.; Kang, C.; Lee, J.; Ryu, J.; Ahn, B.; Yeo, S. Frontline Technology: Technologies for Flexible AMOLEDs. *Inf. Disp.*, **2015**, 31(1), 6–11.

**56** Hong, S.; Jeon, C.; Song, S.; Kim, J.; Lee, J.; Kim, D.; Jeong, S.; Nam, H.; Lee, J.; Yang, W.; Park, S.; Tak, Y.; Ryu, J.; Kim, C.; Ahn, B.; Yeo, S. Development of Commercial Flexible AMOLEDs. *SID Symp. Dig. Tech. Pap.*, **2014**, 45, 334–337.

**57** Yoon, J.; Kwon, H.; Lee, M.; Yu, Y.-Y.; Cheong, N.; Min, S.; Choi, J.; Im, H.; Lee, K.; Jo, J.; Kim, H.; Choi, H.; Lee, Y.; Yoo, C.; Kuk, S.; Cho, M.; Kwon, S.; Park, W.; Yoon, S.; Kang, I.; Yeo, S. World 1st Large Size 18-inch Flexible OLED Display and the Key Technologies. *SID Symp. Dig. Tech. Pap.*, **2015**, 46, 962–965.

**58** Comiskey, B.; Albert, J. D.; Yoshizawa, H.; Jacobson, J. An Electrophoretic Ink for All-Printed Reflective Electronic Displays. *Nature*, **1998**, 394(6690), 253–255.

**59** Herz, E. Electrophoretic Display Technology: The beginnings, the Improvements, and a Future in Flexible Electronics, Technical Report, Materials Science and Engineering, Cornell University, Ithaca, NY, **2006**.

**60** Wang, M.; Lin, C.; Du, H.; Zang, H.; McCreary, M. Electrophoretic Display Platform Comprising B, W, R Particles. *SID Symp. Dig.*, **2014**, 45, p857.

**61** Telfer, S. J.; McCreary, M. D. A Full-Color Electrophoretic Display, *SID Symp. Dig.*, **2016**, 47, p574.

**62** Hagedon, M.; Yang, S.; Russel, A.; Heikenfeld, J. Bright e-Paper by Transport of Ink Through a White Electrofluidic Imaging Film. *Nat. Commun.*, **2012**, 3, 1173.

**63** Hagedon, M.; Heikenfeld, J. A Hybrid of Microreplication and Mask-Less Photolithography for Creating Dual Porosity and Textured Surface Membranes. *J. Micromech. Microeng.*, **2013**, 23, 117005.

# 11

# Flexible Solar Cells

*Y. Galagan*

*Holst Centre, Solliance, Eindhoven, The Netherlands*

## 11.1 Introduction to Photovoltaic Technologies

The history of modern solar cells began in the 1950s at the Bell Laboratories, when the first solar cells capable of converting sun energy into power were first produced. From that time photovoltaic (PV) technologies evolved very rapidly. In the last few years, PVs have evolved from small-scale applications to becoming a mainstream electricity source and one of the most promising sources of renewable energy. Recent reports [1, 2] suggest that PVs will account for 35% of the additional electricity generation capacity installed globally by 2040.

Solar cells are usually divided into three main categories called generations. First-generation PV cells were made from crystalline silicon. Monocrystalline silicon solar cells were originally developed for space applications; however, starting at the early seventies, this type of solar cell was further developed for various terrestrial applications. Multicrystalline silicon solar cells are made out of materials obtained by casting. These types of solar cells are slightly less efficient compared with single crystalline cells, but the manufacturing of multicrystalline silicon is easier, and the production requires less energy. Currently 85–90% of all manufactured solar cells belong to the first-generation class. These solar cells are somewhat expensive to produce because of the high processing costs needed to create silicon cells with sufficient purity and long-range order. The efficiency of these cells is approximately 25% on a lab scale [3–6].

Second-generation PVs are thin-film solar cells that include amorphous silicon, cadmium telluride (CdTe), and copper indium gallium selenide (CIGS) cells. Amorphous silicon cells have lower efficiency than crystalline silicon approaching a level of 10.2% on the lab scale [3, 6], while thin-film microcrystalline silicon cells have an efficiency of 11.8% [3, 7], and multijunction

*Roll-to-Roll Manufacturing: Process Elements and Recent Advances*, First Edition.
Edited by Jehuda Greener, Glen Pearson, and Miko Cakmak.

amorphous/microcrystalline devices reach 13.6% [3, 6, 7]. The advantage of amorphous silicon solar cells compared with conventional crystalline cells is that their production takes relatively less energy and materials. CdTe solar cells can be produced by a relatively simple process. However, the main drawback of this type of solar cell technology is possible environmental consequences associated with cadmium uses. The efficiency record for CdTe solar cells on a lab scale is 21% [3, 6]. Copper indium diselenide (CIS) and CIGS technologies are the most promising thin-film solar cells in the second-generation class. Currently this technology approaches the level of 21% efficiency on a lab scale [3, 6]. The production technology, however, is significantly more complex than in the case of CdTe solar cells.

Third-generation PVs include a number of emerging technologies. Most of these technologies have not been commercialized as yet and are still in the research and development phase. They cover both expensive high-performance experimental multijunction solar cells that hold the world record in solar cell performance and low-cost emerging technologies such as organic, dye-sensitized, and perovskite solar cells (PSCs). Efficiency of 38.8% has been measured for a 1 $cm^2$ five-junction tandem device fabricated by Spectrolab and measured by the National Renewable Energy Laboratory [3]. Emerging single-junction organic solar cells hold a record of only 11.5% [3, 6]. The biggest advantages of organic photovoltaics (OPV) are a low-cost and large-scale production capability that is based on roll-to-roll (R2R) solution processing. Although the efficiency of this type of PV technology is relatively low, due to their low-cost potential, organic solar cells are promising candidates for further commercialization. PSCs have recently undergone rapid development [6]. After only 5 years of development, this PV technology already achieved 22.1% of power conversion efficiency (PCE). This chapter will discuss R2R manufacturing of organic and PSCs, as these emerging PV technologies can be fabricated using well-known R2R printing and coating processes that are widely used in the industry.

## 11.2 R2R Processing

R2R is a manufacturing platform involving continuous processing of materials on flexible substrates. R2R processing is a high throughput operation and therefore provides significant savings in energy usage and productivity compared with conventional batch manufacturing processes. Today R2R processing is applied for a number of applications including the emerging field of organic electronics and flexible PVs. Typical substrates used in R2R processing are paper, plastic, or metal foil. Metal foil is a durable substrate because it has high temperature tolerance, while processing on plastic substrates is typically limited to 100–150°C.

Many conventional deposition methods such as evaporation, sputtering, and chemical vapor deposition (CVD) can also be implemented in R2R processing [8–11]. However, solution-based deposition methods are more attractive because of potential low-cost processing. Processing from solution has also low potential environmental impact, which is very important for many PV applications. The general benefits of R2R manufacturing of high production volume and high yield contribute to low manufacturing cost.

### 11.2.1 Substrates for R2R Processing

The manufacturing of PV devices starts from the selection of a substrate. The main requirements for substrates for R2R manufacturing are flexibility and durability. Typical flexible substrates are plastic or metal foil. The choice of substrate depends on the solar cells' architecture, the type of electrode, the manufacturing process, and the final requirements for the device. Processing of thin-film PVs is strongly dependent on the substrate properties. Most plastic substrates cannot withstand very high temperatures, so the maximum processing temperature is limited. The ideal substrates for PV application should possess a unique combination of properties, which can provide low-cost manufacturing and long lifetime for the PV device. In addition to substrates, solar cells must be protected from environmental factors such as water and oxygen to ensure successful functioning. Such protection can be provided by a proper barrier layer. In fact, the substrates and barrier layers provide a physical "envelope" protecting the PV device. The right choice of substrate and barrier layer(s) is the key for successful commercialization of the thin-film PV device.

The substrates for OPV devices must satisfy numerous requirements, such as optical quality and transparency to let light reach the photoactive layer (PAL), substrate smoothness in the nanometer range to provide a surface that will promote high-quality deposition of subsequent layers and prevent the penetration of potential substrate spikes or irregularities into the device layers, the ability to support processing at high enough temperatures, good dimensional stability, good resistance to any chemicals used during the processing, and low water absorption.

Metal foil substrates [12–15] offer the advantages of higher process temperature capability, dimensional stability, and excellent barrier properties for oxygen and water. The disadvantages and limitations of the metal foil substrates include nontransparency and surface roughness. However, nontransparent metal foils can be successfully used in some of the device architectures, where illumination of the devices occurs from the top side. The smoothness of the metal substrates can be increased by polishing [13] or by applying additional planarization layers. The planarization layer can be organic, inorganic, or a combination. With high content of organic components, thicker layer of planarization material can be applied without forming cracks, and a very smooth surface can be achieved. This requirement is very important for thin-film devices, because the electro-active layers are very thin and the chance to have short circuit in the devices is much higher with rougher surfaces.

The conductivity of metal substrates can be considered as both an advantage and disadvantage. Sometimes the metal substrate can serve as a back contact [16], but very often the conductive property of the metal substrate is not used. In such case, conducting substrates are completely insulated from the PV device by an additional planarization and isolation layer.

Stainless steel is commonly used in research because of its high resistance to corrosion and to chemicals and its long record of application in amorphous silicon solar cells. Stainless steel substrates can tolerate processing temperatures up to 1000°C. These substrates are dimensionally stable and have excellent barrier properties against water and oxygen. Generally, stainless steel substrates are more durable than plastic substrates.

Polymer foil substrates are highly flexible and therefore are R2R compatible. Transparent plastic substrates have the advantage of being compatible with many solar cell architectures. The most common plastic substrates for the fabrication of OPV devices are polyethylene terephthalate (PET) [17–19], polyethylene naphthalate (PEN) [17, 19–21], polycarbonate (PC) [19, 22–24], polyethersulfone (PES) [19, 22, 25], and polyimide (PI) [19, 22, 25–27]. All of these polymer substrates, except polyimide, have good light transmission in the visible range. Polyimide (e.g., Kapton® from DuPont) absorbs in the visible region (has a yellow color), which makes it unsuitable for application in some device architectures. The advantage of polyimide is its high process temperature capability (>350°C), while other plastic substrates have limited process temperature capabilities.

Another disadvantage of plastic substrates is the lack of dimensional stability during processing at elevated temperatures. In addition, typical water and oxygen permeation rates of flexible plastic substrates are $1–10\,g/m^2/day$ and $1–10\,cm^3/m^2/day$, respectively, which are relatively high with respect to those typically required for organic electronic devices, $10^{-3}–10^{-6}\,g/m^2/day$ and $10^{-3}–10^{-5}\,cm^3/m^2/day$. Plastic substrates therefore require additional barrier layers. Barrier coatings can reduce the absorption and permeability of atmospheric gasses, raise the resistance to process chemicals, improve the adhesion of the device layers, and reduce the surface roughness.

As many different types of plastic substrate have been proposed, the final step concerns the selection of the most suitable one for a particular PV device application. For example, it has been shown [28] that PET has a comparatively low price and can be fabricated with a low average surface roughness, while PEN provides a better thermal stability. Currently, the polymer film producers work on improving the surface smoothness and thermal stability of PEN films as well as of thermally stabilized PET films [19]. However, both PEN and PET show insufficient moisture permeability. PET has an increased brittleness after a damp-heat test, while PEN changes its optical properties. Also, the UV stability of both materials is insufficient for the encapsulation of solar cells. After irradiation with UV light, PEN changes its vapor transmission rate, while PET loses its mechanical stability. Because of these issues, poly(ethylene-tetrafluoroethylene) (ETFE) films are also considered as substrates for barrier film. This material provides much better moisture and UV stability (no UV absorption measured at 300 nm wavelength) but is much more expensive than PET, and a pretreatment is needed to provide a good barrier layer adhesion [19].

### 11.2.2 Solution-Based R2R Methods

The main benefits of solution-based R2R processing are high yield and low manufacturing cost. R2R manufacturing has been applied in the commercialization of a wide range of industrial products. Many different printing and coating processes became available for R2R manufacturing over the last few years. In order to become a viable manufacturing option, the R2R technology must be a low-cost and high yield process. Low cost of R2R solution-based processing is a result of short processing time, low volume of materials, and low-energy consumption. Due to high demand for low-cost solution-based manufacturing, a wide range of different deposition methods have evolved. Each deposition method has its own requirements with respect to machinery, substrates, web speed, web tension, properties of deposited materials (e.g., concentration, viscosity, surface tension, etc.), thickness of substrate, and deposited layer. A brief overview of deposition methods employed in the manufacturing of thin-film PVs is given in the succeeding text.

Slot Die Coating: In the slot die coating process, the coating fluid is squeezed out by gravity or under pressure through a slot and onto the substrate (Figure 11.1a). Slot die coating has several main advantages. First, it allows the potential to coat at very high speed. This coating method has very good control of layer thickness by regulating the pumping rate to line speed ratio. The dies for slot die coating are typically designed for the specific ink. This allows very uniform distribution of the coated fluid, based on the fluid's rheology. The slot die support system is just as important as the slot die. The support system accurately and precisely positions the slot die relative to the backup roll or substrate. For elevated temperature applications, dies are designed with multiple heating zones for optimum heat control. Increased heat control promotes uniform fluid viscosity contributing to uniform cross-web distribution. Slot die process optimization occurs by precisely matching process variables, such as flow rate, line speed, and vacuum pressure, with the properties of coated fluid and the substrate properties, such as viscosity, surface tension, and adhesiveness. Slot die system is a closed system, which

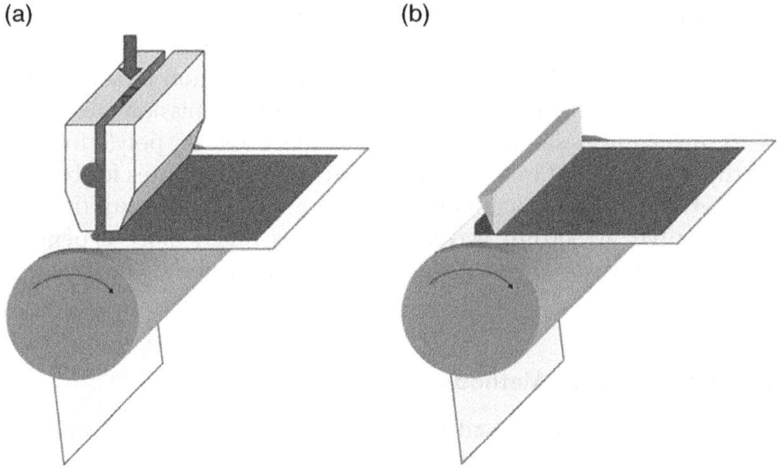


**Figure 11.1** Schematic of slot die coating (a) and knife over roll coating (b).

reduces coating fluid contamination and prevents evaporation of the solvent, which can lead to changes in concentration and, as a result, to a change in layer thickness. Additionally, slot die allows coating very low viscosity inks, which is very typical for OPV inks.

Knife over Roll Coating ("Gap Coating"): This technique [29, 30] is similar to doctor blading. The difference is that doctor blading is a sheet-to-sheet (S2S) coating technique, where the substrate is stationary and the blade (or knife) is moving along the substrate. In the "knife over roll" technique, the knife is stationary and the web is moving (Figure 11.1b). This process relies on a coating being applied to the substrate, which then passes through a gap between a knife and a support roller. The excess of fluid is scraped off. This process can be used for high viscosity fluids and for very high weight coatings.

Flexographic Printing: Flexography is one of the most common types of printing. It relies on a flexible relief plate to carry the image, similar to a rubber stamp (Figure 11.2a). The use of a flexible plate is the reason for the term flexography. While some flexographic plates are made of rubber, like a rubber stamp, most are made of photopolymeric materials. This flexible plate, carrying the image, comes in contact with a rigid cylinder, so printing can be performed in a rotary fashion, as opposed to stamping like with a rubber stamp. The plate's raised image is covered with a uniform layer of ink using an ink-metering device called an anilox roll, which has millions of small cells or wells on its surface to carry a precise volume of ink that is transferred to the flexographic plate. After the plate has been inked, it is then pressed against the substrate, which is supported by an impression roll. The pressure between the anilox roll

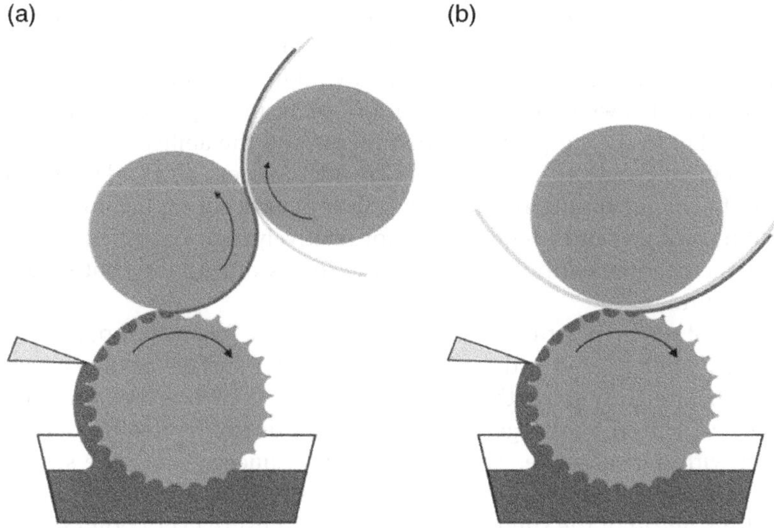


**Figure 11.2** Schematics of the flexographic printing process (a) and the gravure printing process (b).

and printing plate and between the printing plate and substrate is mechanically controlled using fine pitch adjustment screws. Flexographic printing has traditionally not been thought of as a high-quality printing method because of variability in the flexographic plates and anilox rolls and the fact that the image on the plate can be distorted easily with pressure.

Gravure Printing: The basic gravure process is comprised of an engraved metal image carrier, a sharp blade called a doctor blade, and a rubber-covered impression roll (Figure 11.2b). The engraved image carrier is wetted with ink, and the doctor blade wipes the excess ink from the cylinder surface, leaving only ink in the recessed image. The ink is then transferred to the substrate by pressing the substrate against the cylinder with the impression roll.

While flexography uses a relief image on a flexible carrier, gravure prints with a recessed image on a rigid cylinder. In other words, gravure is a printing process that uses an engraved surface to carry an image that will be printed. It is the only high volume printing process requiring a recessed image on the image carrier (plate or cylinder). Originally the images were engraved or carved into a flat metal plate, and the associated printing process was called intaglio printing. The modern image carrier is almost exclusively a cylinder instead of a flat plate and is often referred to as rotogravure or simply gravure. While the term gravure and intaglio are both used, intaglio is usually associated with security printing where the artwork is developed from hand-engraved lines and high viscosity inks are used.

Screen Printing: Screen printing has a unique place in the printing industry since it encompasses everything from the simplest printers printing one short runs to large industrial operation printing many millions of units under clean room conditions. This wide range of operation is a result of process flexibility that is largely due to the simplicity of the process and the ability to print ink films from sub micrometer to hundreds of micrometers thick. The main distinction of screen printing from other printing and coating techniques is large wet film thickness and a requirement for relatively high viscosity and low volatility of the coating solution. The material losses during the coating are very insignificant.

A schematic of screen printing is shown in Figure 11.3. A screen is patterned with an image, while the ink is deposited on the upper surface of the screen. The screen is held slightly above the substrate, which is known as the of contact gap. To make a print, the squeegee blade is brought down, forcing the screen into contact with the substrate. The passage of the squeegee over the screen forces the ink through the screen onto the substrate.

Sufficient ink ahead of the squeegee is required to ensure every point of the image is printed. The two most commonly cited printing mechanisms that influence printing quality are (i) hydrodynamic pressure in the nip between the squeegee and the screen and (ii) viscous forces created between the ink and the screen when combined with the attraction forces of the substrate. At the end of the print stroke, the squeegee blade is raised and returned to the starting point of the print. During this reciprocating action, a flow coat blade follows the squeegee across the screen to redistribute the ink that has gathered in the ink roll in preparation for the next print.

Screen printing can be also adapted for R2R manufacturing. In this case the screen is modified to a cylinder shape. The basic principle of operation for rotary screen printer is the same as for flatbed screen printer. As the screen

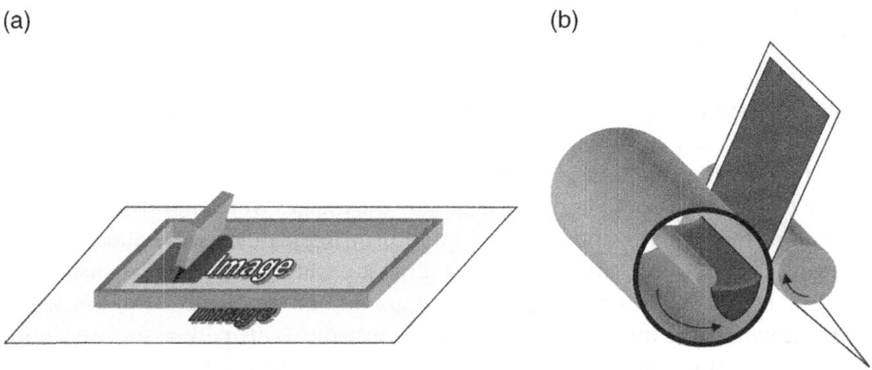


**Figure 11.3** Schematics of the flatbed screen printing (a) and the rotary screen printing (b) processes.

rotates, the squeegee device pushes print paste through the design areas of the screen onto the substrate (Figure 11.3b).

Inkjet Printing: The only digital process that has been successfully applied to the manufacture of electronic components is inkjet printing. An inkjet printer operates by electronically controlling the production of droplets of ink that, when applied to the substrate, form the image. Inkjet printers can be grouped into three main categories depending on the method by which the droplet of ink is formed and directed toward the target substrate: continuous inkjet, drop-on-demand (DoD) inkjet, and bubble jet.

Continuous inkjet systems operate by electrically deflecting a droplet from a stream of droplets when a dot is required to be printed. These printers are most commonly used in applications where the quality of the final image is not critical such as product numbering or automatic mailing. A continuous stream of droplets passes through a set of electronic signal-driven discharge electrodes, which selectively deflect the droplets to one side where it is directed toward the substrate. Droplets that are not deflected pass into a gutter that recirculates the ink back to a reservoir via an in-line pump.

In the DoD inkjet system, ink is supplied to the chamber with a nozzle at one end and a piezoelectric crystal (PZT) at the other (Figure 11.4a). When a dot is required on the substrate, the PZT is excited and an ink droplet is formed to be printed on the substrate. This applied voltage causes the PZT to expand and contract. The pressure wave that is generated during the PZT expansion/contraction creates ink droplet that is ejected through the nozzle at the end of the ink chamber. The consistency and control of the DoD inkjet system is generally higher than that achieved by continuous inkjet since the flight distance and complexity of the droplet allocation magnesium are reduced.

Bubble jet can also be considered as a DoD inkjet technology as the ink droplet is produced on demand. However, the droplet formation mechanism in

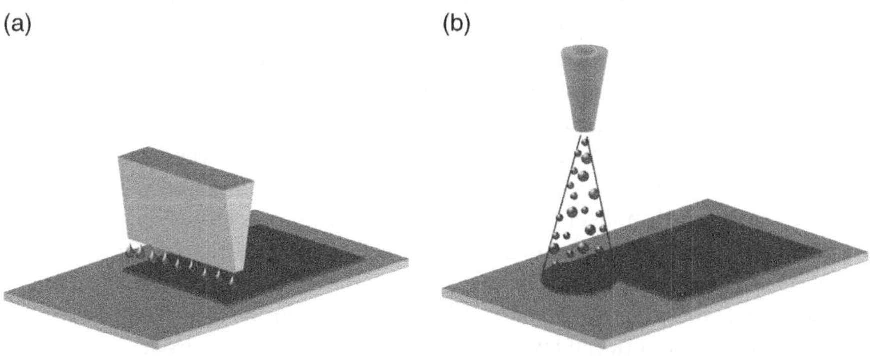


**Figure 11.4** Schematics of the drop-on-demand inkjet printing (a) and spray coating (b) processes.

bubble jet is different than in the DoD inkjet system driven by PZT. The ink droplet formed by bubble jet is forced through the nozzle by the expansion of a bubble at the rear of the chamber, hence the name bubble jet. This gas bubble is produced by locally boiling the ink using a high speed digitally controlled micro-heater.

Spray Coating: Spray coating denotes the formation of a thin film via millions of micron-sized droplets sprayed toward the substrate. In spray coating, the liquid is fed to a nozzle that atomizes it by either ultrasonic vibration or pressure. The micron-size droplets are then directed to the substrate and land on it with random displacement (Figure 11.4b). The thickness of the layer is determined by the amount of liquid dispensed per area. The deposition is often labeled as rough, since the droplets dry in extremely short times due to their large surface-to-volume ratio. This coating technique allows a lower material consumption as compared with spin coating, but not very efficient materials usage is achieved compared with other techniques such as slot die and gravure printing. The main advantage, however, is the possibility to coat arbitrary-shaped substrates where standard coating cannot be applied.

## 11.3 Organic Photovoltaics

OPV have undergone significant development over the past few years. This solar cell technology offers the possibility of inexpensive R2R fabrication on flexible substrates and a wide choice of materials for applications where flexibility and color are important. The main objectives in the field of OPV are achieving high efficiency, long-term stability, and low cost. Low cost can be achieved by combining inexpensive materials with fast solution-based R2R manufacturing techniques. Indeed, compared with conventional Si-based solar cells, organic solar cells should be less expensive and easier to manufacture due to lower energy consumption, low temperature and vacuum-free deposition, and possibility of direct patterning during the deposition. R2R coating and printing of organic semiconductors are the focus of many companies and research groups. Practical applications based on organic semiconductors, such as organic light-emitting diodes and OPV devices, place strict requirements on the layer properties, such as thickness, uniformity, and over-layer accuracy. Many different printing and coating techniques are applied and developed for the deposition of thin layers of organic semiconductors. However, besides the layer organic semiconductor, organic solar cells contain other functional layers such as electron- and hole-transport layers (HTL) and two electrodes. For industrial manufacturing of organic solar cells, an R2R technology for the deposition of all functional layers in the devices stack is required.

Figure 11.5 illustrates two typical device architecture used in the manufacturing of organic solar cells. Transparent electrode, typically indium tin oxide

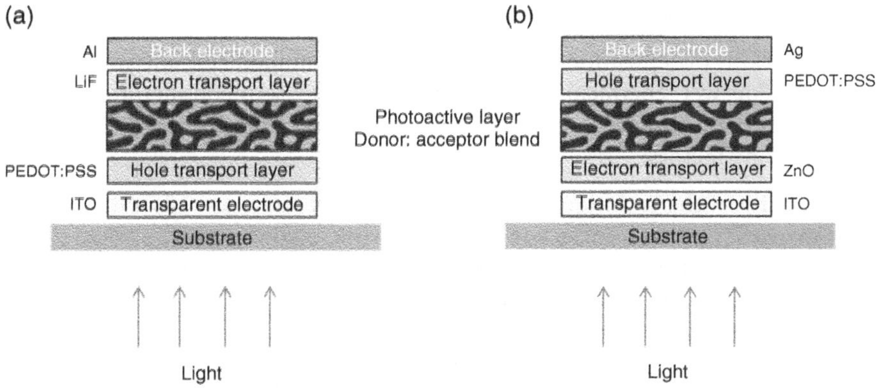


**Figure 11.5** Schematic illustration of conventional (a) and inverted (b) organic solar cells.

(ITO), allows light to penetrate through the device stack to reach PAL. In most of the cases, buffer layers are applied between the electrodes and PAL to ensure efficient charge-selective transport. If HTL is located between transparent ITO electrode and PAL, such device architecture is typically called conventional architecture (Figure 11.5a). The reverse layer order and positioning of electron-transport layer (ETL) between ITO and PAL result in the so-called inverted device architecture (Figure 11.5b). Typical HTL materials used in organic solar cells are poly(3,4-ethylenedioxythiophene):poly(styrenesulfon ate) (PEDOT:PSS) or p-type metal oxides. Common ETL materials in OPV devices are n-type metal oxides, for example, zinc oxide (ZnO). The PAL typically contains the mixture of donor (semiconducting polymer) and acceptor (fullerene derivative). Mixing donor and acceptor materials in a bulk allows creating large area of the interfaces responsible for effective charges separation. Back metal electrodes are typically used to complete organic solar cells.

In order to develop a viable manufacturing process for organic solar cells, a comprehensive technology assessment for each layer in the device stack is required. The technology development includes inks development, solution-based deposition of functional layers, and drying and patterning of the layers. First, inks with the required properties must be formulated, and the deposition method appropriate for this ink selected. The influence of the deposition method and parameters on the layer properties has to be determined. In particular, the influence of drying conditions on the layer properties, such as morphology and uniformity, has to be analyzed and understood. Depending on the deposition method, it is important to determine how the patterning of the layers will be applied. The design and the cell architecture should be optimized. It is important to determine the correlation between all these parameters and the device performance, find a balance, and transfer this knowledge to an R2R process.

### 11.3.1 Technology Assessment

Benchmarking of deposition techniques for R2R manufacturing is a complicated process. For wet chemical deposition of organic semiconductors, a large number of deposition techniques can be used. The choice is mainly based on the following factors:

- Suitability of the technique for the deposition of the materials with desired properties (viscosity, surface tension of the ink, etc.)
- The capacity of the technique to print/coat the desired feature sizes (lateral resolution, thickness, and uniformity)
- Throughput of the process

Each layer in OPV devices has its own requirements for the deposition technique. The choice of the technique should be done individually for each layer. Although most of the techniques mentioned previously in the Section 11.2.2 have been tested for the deposition of PAL in OPV devices, not all of them are suitable for R2R mass production. The factors that limit the usage of some deposition techniques are poor compatibility with R2R processing, high materials waste, low speed, high contamination during the processing, low stability and robustness of the process, poor uniformity of the printed layer, and minimum thickness of the wet layer. Often the viscosity of the ink limits the usage of a certain technique for the deposition of the functional layer. In this case the viscosity of the ink will be the limiting factor. Typically, viscosity of inks for PAL is very low, for example, standard P3HT/PCBM solution has a viscosity in the range of 1–5 mPa·s. Viscosity of the PEDOT:PSS solution is typically in the range of 10–30 mPa·s (however now high viscous pastes are also available with viscosities higher than 50 mPa·s). ZnO solution typically also has very low viscosity of 1–2 mPa·s. The usage of such deposition techniques as screen printing, offset printing, or curtain coating is probably not possible for the deposition of these functional layers without modification of the inks. Silver-based inks are available as both low viscosity inks and in paste form, which is typically used for screen printing. The ink viscosities required for the different deposition techniques are given in Table 11.1.

The viscosity of the ink is one of the critical factors for the choice of the deposition technology. Viscosity affects the inks flow and quality of the layer. Some techniques require relatively low viscosity, allowing individual printed pixel to merge together to form a uniform layer. However, sometimes, too low viscosity can be problematic because it does not allow for very sharp edges and the ability to print very small features.

Inkjet printing is a well-studied method for the deposition of PEDOT:PSS [31, 32] and polymer–fullerene blends [33–35]. The advantage of this method is the possibility to print patterned layers in one printing step. The challenge is

**Table 11.1** Comparison of different deposition techniques used for organic semiconductor layers.

| | Possibility of patterning | Materials waste | Layer thickness accuracy (nm scale) | Viscosity requirements (mPa·s) |
|---|---|---|---|---|
| Slot die coating | ±[a] | Low | Very good | 1–10,000 |
| Knife over roll coating | – | Moderate | Moderate | 50–1,000 |
| Flexographic printing | + | Low | Moderate | 15–200 |
| Gravure printing | + | Low | Good | 15–500 |
| Screen printing | + | Low | Moderate | 50–50,000 |
| Inkjet printing | + | Low | Good | 1–40 |
| Spray coating | ±[b] | High | Low | 1–40 |

[a] Slot die as a coating method typically produces non-patterned layers. However, using thin inserts in the die (shims) or using pre-structured die configuration (notch die), the coating can be structured in one dimension, for example, producing stripe pattern. Using intermittent system with accurate and controlled start/stop option, the coating can be patterned in two dimensions, producing, for example, rectangular patterns.
[b] Patterning can be done using masked deposition.

to find appropriate solvent systems for polymer–fullerene blends, which will provide appropriate spreading and wetting of the ink on the substrate and homogeneous drying with required morphology of the active layer. Gravure printing, which is widely used for the printing of organic transistors [36, 37], has also been applied for the deposition of OPV layers [38]. The main difficulty in gravure printing is viscosity of the ink, which for OPV blends is slightly lower than required for gravure printing. There are a number of publications on spray coating of PV inks [31, 39–42]. This deposition method is very efficient especially for low viscosity fluids. Screen printing has been applied for the deposition of PAL based on MDMO-PPV:C60-PCBM [43] and MEH-PPV:C60-PCBM [44, 45]. A complete process for the production of flexible large area polymer solar cells based entirely on using screen printing has been demonstrated [46]. There are many publications where slot die coating was chosen for the deposition of several layers, including active layers, in polymer-based solar cells [47–52]. Slot die deposition is one of the most common coating techniques, and it can provide very thin, uniform non-patterned layers. Figure 11.6 demonstrates R2R slot die-coated PAL on a plastic substrate. One-dimensional patterning is possible by coating stripes that can be used for making OPV modules. Using intermittent system with accurate and controlled start/stop option, the coating can also be patterned in two dimensions producing rectangular patterns.

**Figure 11.6** Slot die coating of photoactive layer on plastic substrate.

Each deposition technique has advantages and disadvantages. The main issue with many coating techniques is the inability to pattern the deposited layers, which is required for device and module manufacturing. In this respect printing methods are preferred. However, many coating techniques can provide very thin and uniform layers, while printing methods are inferior in achieving desired layer uniformity. The patterning of coated layers in most cases can be done in a separate process step that, however, increases the manufacturing cost. Typical post-process patterning methods are laser ablation [53], photolithography [54], plasma etching [55], or solvent etching. Patterning of coated layers can be applied in one step, together with coating, by self-assembly based on wetting and de-wetting processes. Finally, as mentioned previously, stripes and intermittent coating can be applied for the structuring of coated layers using standard coating methods.

The choice of a deposition technique depends on the main characteristics of the method and the features that made this method attractive. The main selection criteria are based on the following:

- Material factors, such as viscosity, surface tension of the ink, and surface energy of the substrate
- Product factors, such as layer uniformity, layer thickness, and patterning requirements
- Process factors, such as R2R compatibility, line speed, stability of the process, robustness and simplicity of the process, and manufacturing cost

However, selection of the particular deposition technique depends on process expertise and availability of the necessary equipment. Collaborations and partnerships very often make some of the deposition techniques more attractive. Economic factors such as market size and manufacturing cost are also important considerations in the selection of the deposition method.

### 11.3.2 Roll-to-Roll Printing and Coating of Electrode Materials

The growing interest in OPV technology was prompted by its promise for low-cost energy conversion. In order for OPV technology to have an impact on the power generation market in the long term, it must combine high PCE with low production costs and long-term stability. The low-cost potential is based on the use of low-cost materials and substrates and the very high production speeds that can be reached by R2R printing and coating [56–62].

The presence of a transparent conducting electrode such as ITO limits the reliability and significantly increases the cost of OPV devices as it is brittle and expensive [63–65]. In order to enable low-cost processing, much effort has been put into the development of an alternative to ITO as a front electrode [66–74]. In the past decade, many alternatives have been proposed. Researchers have explored many potentially conductive coatings of ultrathin metal films [75, 76], highly conducting carbon nanotubes (CNTs) [77], graphene [78], metal nanowires [79–81], and so on.

PEDOT:PSS has been intensively investigated and reported [82–85] as a promising electrode material. Replacing ITO by high conductivity PEDOT:PSS offers benefits in high transparency, long-term stability [86], solution processability, and mechanical stability [84, 87]. Although replacement of ITO by high conductivity PEDOT:PSS has been successfully demonstrated, it is only a good alternative for very small solar cells. Increasing the cell dimension leads to rapid decrease of the solar cell performance, mainly due to a decrease in the fill factor (FF) [85], which is caused by increasing power losses ($P_l \sim R_s I^2$) due to the series resistance ($R_s$) and current $I$, which increases proportionally to the cell area $A$. Highly conductive PEDOT:PSS has much lower conductivity than ITO, and therefore a decrease in efficiency is much higher with increasing cell dimension. Increasing the thickness of the PEDOT:PSS layer not only reduces its resistance but also leads to a significant decrease in optical transmission [82, 83] that has negative effect on the device performance. Various modifications of PEDOT:PSS with enhanced conductivity have been intensively investigated and reported [82, 84, 88]. Although most of these studies have successfully demonstrated improved conductivity of PEDOT:PSS, it is still too low in comparison with ITO. Improvements in conductivity of polymeric electrodes are possible by combining them with metal current collecting grids. Current collecting grids can be either thermally evaporated through microstructured shadow masks [73, 89], made by lithographic methods [69, 90, 91], or printed [68, 69, 92]. Larger amounts of metal in the grids typically provide better conductivity, and, as a result, the efficiency losses due to an increase of the cell dimension are smaller [68]. The efficiency of a metal grid current collector is determined by the conductivity of the metal grid and the distance between the grid lines [70]. Together, these two parameters determine the total conductivity of the electrode and therefore set the boundary for the achievable efficiency when combined with organic solar cells. A disadvantage of this approach is that the topology of the grids prevents uniform deposition of subsequent layers. However this issue can be overcome by embedding the current collecting grids into the substrates [59, 72, 92]. The image of the large area organic solar cell with printed current collecting grids combined with PEDOT:PSS as a transparent electrode is shown in Figure 11.7.

For low-cost manufacturing, back electrodes also have to be produced from solution of PEDOT:PSS as an HTL followed by the deposition of Ag. This option is a typical back electrode in organic solar cells with an inverted configuration. Although it is well accepted that this electrode can be solution processed, often the Ag is still deposited by thermal evaporation. Alternative solution processing deposition methods for Ag back electrodes are spray coating [93–96], inkjet printing [97–99], flexographic printing [97], or screen printing [52, 97, 100]. A significant limitation in the solution processing of the back electrode is the requirement of sintering at high temperature in air, which can lead to the degradation of the PEDOT:PSS and PAL [101].

Screen printing under ambient condition is the most explored deposition option so far in both lab scale devices and R2R production. While only limited

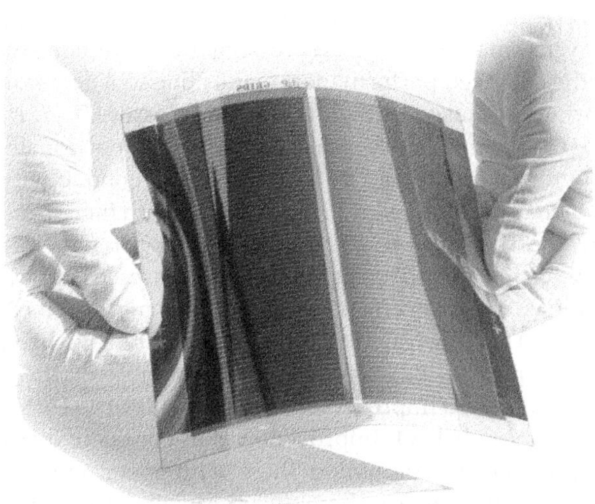

**Figure 11.7** Large area ITO-free OPV module. Transparent ITO electrode is substituted by a combination of printed current collecting grids and PEDOT:PSS semiconducting polymer.

sintering time is needed (3 min at 140°C) to yield high conductivity, the majority of defects are caused by an interaction of the solvents from the screen printed Ag ink with the underlying layers [98]. Moreover, screen printing is a contact method, which may also lead to an increasing number of defects in underlying layers. To protect underlying layers from the aggressive effects of solvents and to avoid the mechanical stress from the screen printing process, the thickness of the PEDOT:PSS layer is significantly increased to provide a kind of "buffer layer" protecting the PAL [52, 100].

Inkjet printing is a non-contact technique and might allow the use of thinner PEDOT:PSS layers, which lowers the cost of the OPV device. An inkjet-printed back electrode has been successfully demonstrated with an 800 nm thick PEDOT:PSS layer in organic solar cells with an inverted cell configuration [98]. Thick PEDOT:PSS layer was required to prevent penetration of the solvents from the Ag ink, which would lead to the degradation of the device and a drop in its performance. In recent studies inkjet printing of the back electrode was combined with a very thin crosslinked PEDOT:PSS layer [102]. Reducing the thickness of PEDOT:PSS layer from 800 to 40 nm is a very significant step toward reducing the manufacturing cost of the OPV device. In addition, combining thin PEDOT:PSS layer with an inkjet-printed Ag grid structure allows R2R manufacturing of semitransparent devices.

### 11.3.3 Patterning and Module Manufacturing

One of the issues toward the upscaling and mass production of OPVs is an increase in the active area of the modules. Currently R2R manufacturing of OPV modules is mostly done by coating stripes, which are then interconnected into the modules. The relatively high resistance of typical electrode materials (high sheet resistance) limits the width of single cell. Narrow cell width and rather low resolution of stripe coating technologies [103, 104] are responsible for the large interconnection area (non-active area). Typical geometrical fill factor (GFF, the ratio between the active area and total area of the OPV module reported for coated modules) is in the range of 50–75% [63, 73, 105]. However, even then, very precise printing or coating on large areas with a moving substrate is required to achieve this relatively low GFF, which still presents a significant challenge. It requires a very high degree of technical skill combined with scientific knowledge of possible physical and chemical processes occurring during the processing [58]. Therefore, the development of a stable process that can provide a high GFF is of great interest. The GFF can be increased either by increasing the cell width by utilizing electrodes with low sheet resistance [59, 69] or by minimizing the interconnection area using very precise patterning techniques. Figure 11.8 shows OPV modules made using the stripe coating technique and a module made via laser patterning.

**Figure 11.8** OPV modules made using stripe coating technique and a module made via laser patterning.

High resolution laser patterning can resolve this problem and significantly contribute to an increased GFF [53, 106–109]. Recent studies show OPV modules with GFF > 95% manufactured by a S2S process [107]. The standard method of manufacturing such modules involves the alternation of laser patterning and coating. Successful demonstration on the S2S scale does not guarantee the direct extension of this technology to the R2R scale, because the top electrodes are still thermally evaporated.

The replacement of an evaporated electrode by a printed one is theoretically possible, but laser patterning of a printed silver layer becomes challenging due to the high roughness of the last. A patterned printing of top (back) silver electrodes appears to be an ideal solution, but this is still a challenging task because the resolution and alignment must be very precise. Thus, scientists in the field of organic electronics, and OPV in particular, are still looking for an effective printing technique that allows the deposition of materials at high precision and high speed. Combination of a laser ablation of coated layers and very precise printing of the top electrodes was introduced [110]. This innovation shows feasibility for industrial manufacturing of modules with high GFF of 92.5%. Photographic image of such module is shown in Figure 11.9.

### 11.3.4 Current Progress in R2R Manufacturing of Organic Photovoltaics

Thin-film PVs based on solution processable semiconductor materials have attracted a lot of attention because of their potential for very low-cost manufacturing. Low manufacturing cost can be achieved by adapting printing and coating techniques owing mainly to their high throughput. Although the efficiency and lifetime of OPV modules require further optimization, many

**Figure 11.9** R2R-produced OPV module, which combines laser patterning of slot die-coated layers and very precise printing of the top electrodes. *Source:* Reproduced from Ref. [110] with permission © 2015 Wiley-VCH Verlag GmbH&Co. KGaA, Weinheim.

research groups have already demonstrated preindustrial scale R2R process for OPV modules manufacturing. Various printing and coating techniques were utilized for the R2R deposition of functional layers.

The most often reported OPV modules are manufactured using slot die coating for most of the functional layers [9, 52, 110–113], where the back electrodes are screen printed [110, 112], flexographic printed [112, 114], inkjet printed [110], or thermal evaporated [9, 115]. Krebs et al. have demonstrated numerous R2R processes for the OPV modules, with different device stacks employing different deposition methods. In the so-called "ProcessOne" [52], OPV modules were manufactured on ITO-coated PET substrate. The first three layers (zinc oxide (ZnO), PAL, and PEDOT:PSS) were processed using slot die coating, and a silver back electrode was flatbed screen printed to finalize the stack before it was laminated. Since then, numerous productions of solar cells and modules based on this method have been published by the Danish group [63, 112, 116–118]. This process was further adapted for other photoactive materials [116–118].

Slot die coating was utilized in studies by several groups worldwide for the manufacturing of the several layers in the OPV cells and modules [74, 115, 119]. The systematic investigation of thermal effects during the drying step as well as effect of thermal annealing on morphology and performance of inverted organic solar cells fabricated by R2R slot die coating has been reported by Huang et al. [115]. Improved drying in the R2R process was demonstrated to transform the wet active layer into solid-state film (an early polymer crystallization step).

A number of ITO-free approaches have been demonstrated by different groups [68–70, 73, 74, 89, 92, 120–123]; however, not many of these were scaled up to the R2R scale. The solar cell technology known as "IOne" [111] reported by DTU represents real progress in all-solution and vacuum-free R2R manufacturing of OPV. The process "IOne" is a lower-cost and significantly faster process that presents no disadvantages when compared with "ProcessOne" while both being vacuum- and ITO-free and presenting a much better operational stability and a significantly lower materials and overall cost [124]. It is noted that the "IOne" process has been demonstrated in a silver-free version based only on carbon-based electrode with the same performance as silver-based version, thus indicating that neither ITO nor silver is required to make efficient and scalable OPV [67]. The efficient replacement of silver was achieved with carbon paste, which yields superior printing and stability performance over printed silver conductors while requiring somewhat thicker printed layers to achieve the same conductivity.

A research group from the University of Newcastle has demonstrated a pathway for ITO-free fully R2R-prepared organic solar cells in a conventional geometry with an R2R-sputtered aluminum top electrode [8]. This shows that solution processable printing and coating methods can be combined [9] with R2R vacuum techniques [10] for electrode deposition in manufacturing complete OPV modules.

Most of the slot die-coated modules mentioned previously are produced using stripe coating to make interconnection of individual cells in the modules. Galagan et al. [110] have reported R2R-manufactured modules with slot die-coated three layers (ZnO/PAL/PEDOT:PSS) and either screen-printed or inkjet-printed back electrodes. Interconnections in such modules are made by laser patterning, which provides very high GFF. The photographs of such modules are shown in Figure 11.9.

The use of gravure and flexographic printing was also extensively reported as a promising technique for the preparation of organic solar cells [38, 125–133]. These two-dimensional printing techniques are differentiated from other coated OPVs by enabling the direct patterning of arbitrarily shaped and sized features and enhancing the freedom to connect the cells in modules. Up to four OPV functional layers have been successively gravure printed on ITO-coated PET flexible substrate, allowing the high production throughput in an R2R printing process [132]. Välimäki et al. [130] have described the fabrication of R2R-printed OPV modules using gravure printing and rotary screen printing processes.

One of the remaining issues toward the application and integration of OPV modules is the possibility to customize shape and dimensions of the modules. Slot die coating, which is widely used for the R2R manufacturing of OPV, allows stripe patterning in one direction, thus providing only rectangular-shaped modules. R2R production technology, which can change the module design per customer request, often involves a combination of different deposition techniques and post-process structuring using, for example, laser ablation [107, 134]. This requires significant capital investment in the appropriate equipment. Inkjet printing [135] is a deposition technique whose main advantage is that it is DoD: deposition patterns are simply changed through software, allowing in-line changes of the device design being processed. Using inkjet printing for all layers of an OPV provides the ultimate design flexibility. This is especially beneficial for the production of prototypes and up to medium-scale production volumes. Also, inkjet printing requires small start-up volumes of inks and is highly economical in materials use.

Inkjet printing is well known from the graphics, textile, and ceramics industries, where it is capable to run under demanding conditions at line speeds far exceeding 10 m/min. The value of inkjet printing in the field of organic electronics also has been recognized. The production of high-quality conductive silver structures using inkjet printing in high speed R2R production has been recently demonstrated for different electronic applications [136, 137] as well as part of the semitransparent electrode in ITO-free OPV devices [57]. These latest developments bring R2R inkjet printing production of OPVs a step closer by demonstrating the feasibility of creating these devices reliably using a high volume R2R process.

A strong interest in the inkjet printing of OPV has been demonstrated over the last few years. It has been utilized for the deposition of different layers in the OPV stack. A lot of studies were focused on the inkjet printing of PEDOT:PSS [31, 138] and on its use as replacement of the ITO electrode by combination with inkjet-printed Ag current collecting grids [57, 68, 99, 139, 140]. Inkjet printing of OPV back electrodes has also been successfully demonstrated by different groups [98, 102]. However, many studies over the last few years were focused on the inkjet printing of the PAL [33, 34, 141, 142]. Screening of different solvents and solvent mixtures for efficient inkjet printing of P3HT/PCBM was of particular interest. The best performance for an inkjet-printed P3HT/PCBM blend has been reported by Jung et al. [143] reaching 3.8% and was obtained by using chlorobenzene (CB) as a solvent. Recently, Jung et al. [144] have reported all-inkjet-printed solar cells with PCE of 2%. Although the results show the great potential of inkjet printing as a digital printing technology for manufacturing OPV devices, the reported approach is only suitable for few-millimeter-size solar cells, because of a very high resistance of the used PEDOT:PSS electrode. Scaling up of such a technology requires another printable electrode allowing large area devices. Furthermore, the most critical factor hindering the scale-up of the approach of Jung et al. is the use of chlorinated solvent mixtures for the deposition of functional layers. Thus, the most challenging issues toward scaling up all-inkjet-printed solar cells remain: increase of a PCE of large area devices, printing of large area and low resistance electrodes, printing of all layers from non-chlorinated solvents, utilization of multiple industrial nozzle print heads, and demonstration of flexibility to produce various shapes and designs.

There are several examples where OPV cells and modules were produced entirely from solution. In these cases ITO, which is a typical transparent electrode, is substituted by a printed/coated transparent electrode [72, 111, 145, 146]. In a few examples, the OPV modules were produced entirely by slot die coating [145] or by combination of slot die coating, screen printed, inkjet printed or flexo printed [72, 111].

Large area fully inkjet-printed organic solar cells were demonstrated by Solliance [146]. The demonstrated technology allowed direct patterning of all the layers, including the electrodes, offering full freedom of design without the use of masks or hardware structuring (Figure 11.10). The semitransparent front and back electrodes consist of PEDOT:PSS and conductive Ag current collecting grids, avoiding the use of ITO. The inkjet printing of six functional layers [146] demonstrated minimal losses in performance as compared with the lab scale standard, spin-coated devices. All-inkjet-printed large area (>1 $cm^2$) organic solar cells with PCE of 4.1% deposited from environmentally friendly solvents in an air atmosphere are demonstrated for the first time. Organic solar cells were fabricated using industrial scale (512 nozzles) print heads, compatible with R2R technology. To demonstrate the great advantage of inkjet printing as a digital technology allowing freedom of forms and designs, large area organic solar cells with different shapes were fabricated (Figure 11.10). The reported results confirm that inkjet printing has high potential for the processing of OPV, allowing quick changes in design as well as in materials.

R2R printing and coating of OPVs show great potential for high throughput production of cheap flexible solar cells [119, 129, 148–151]. Although many successful examples on integration of R2R-coated and R2R-printed modules

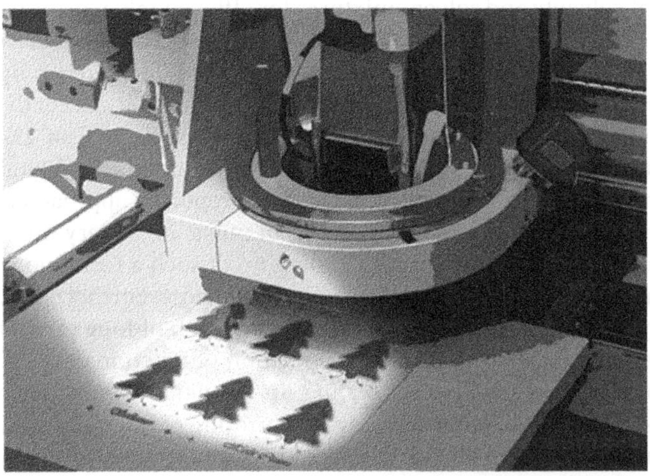

**Figure 11.10** Inkjet printing of free-form organic solar cells. *Source:* Reproduced from Ref. [147] with permission from The Royal Society of Chemistry.

were demonstrated [152, 153], there are still many issues that need to be resolved before full-scale commercialization of these technologies [154, 155]. A lot of efforts are directed toward quality control strategies for the printed layers [124, 152]. The analysis of layer quality during the coating and a better prediction of whether the module will work or fail can significantly increase the yield of the produced modules and therefore lower manufacturing cost. Another important issue concerns the environmental impact of the various process steps. Indeed, typical solvents used for photoactive materials are halogenated aromatic solvents, for example, CB or ortho-dichlorobenzene (o-DCB). OPV devices show highest performance when produced using these solvents. Driven by environmental concerns and the potential commercialization of OPV technologies, several groups have demonstrated highly efficient OPV modules produced from non-halogenated solvent using R2R compatible techniques [62, 119, 156]. High GFF demonstrated in R2R-coated modules [110] and freedom of forms demonstrated with inkjet-printed process [146] will bring the OPV technology closer to full-scale commercialization. The remaining issues before mass production of OPVs are still relatively low PCE [6] and insufficient stability [157, 158]. Efficiency is becoming an increasingly important driver for reducing the cost of large area PV systems [5]. Solving these issues will open a lot of new perspectives for organic solar cells.

## 11.4 Perovskite Photovoltaics

The interest in PSCs has significantly increased over the last few years [159, 160]. High PCE and low-cost potential make PSC a very promising candidate for future applications [161–163]. This type of solar cell is based on hybrid organic–inorganic lead halide-based material (perovskite), for example, methylammonium lead halides, as the light-harvesting active layer. Perovskite-based solar cells can be produced in different device architectures. The conventional and the most common is "mesoscopic" structure (Figure 11.11a), which utilizes a nanocrystal-oxide scaffold layer. This structure has a lot of similarities with solid-state dye-sensitized solar cells (DSSC). However, same as organic solar cells, PSCs can be produced also in "planar" structure with conventional (Figure 11.11b) and inverted (Figure 11.11c) architectures. Conventional or "n-i-p" structure contains n-type material (ETL) between transparent electrode and perovskite PAL and p-type material (HTL) between the perovskite and back metal electrode. For "n-i-p" structure, n-type of the semiconductors are typically titanium oxide ($TiO_2$) or tin oxide ($SnO_2$), and p-type of the semiconductors are typically organic materials spiro-OMeTAD (2,2′,7,7′-tetrakis[*N*,*N*-di(4-methoxyphenyl)amino]-9,9′-spirobifluorene), poly(triaryl amine) semiconductor) or P3HT (poly(3-hexylthiophene-2,5-diyl)). In inverted or "p-i-n" structure, the order of the layers is opposite. For this

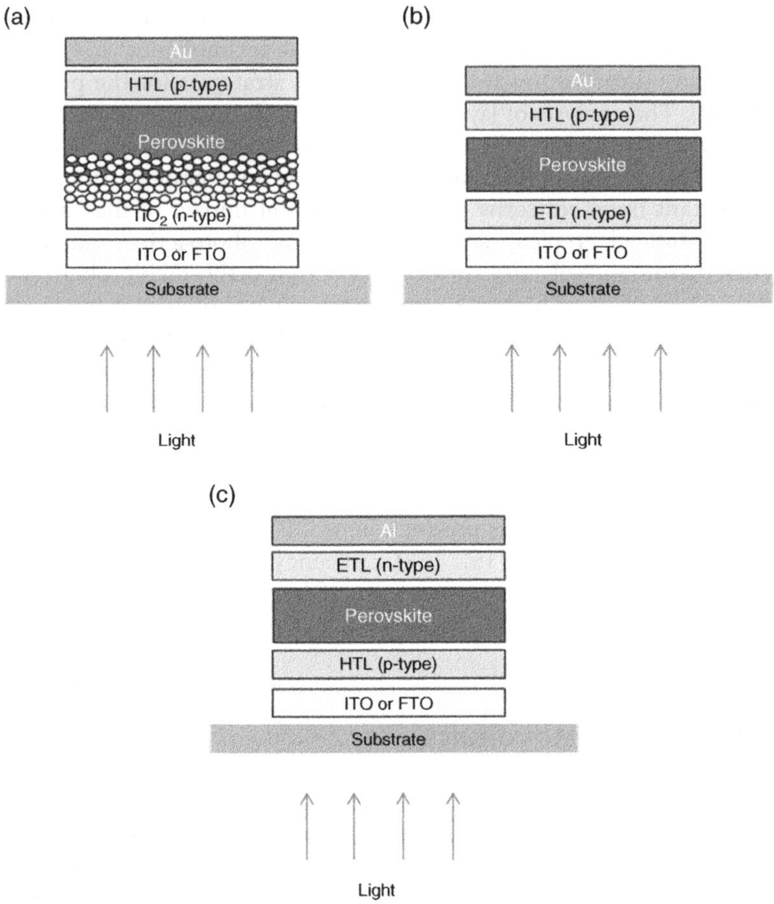


**Figure 11.11** Schematic illustration of mesoporous (a) and planar "n-i-p" (b) and "p-i-n" (c) architectures of perovskite solar cells.

device architecture NiO (nickel oxide) or PEDOT:PSS are used as p-type materials, and PCBM ([6,6]-phenyl C61 butyric acid methyl ester) and/or ZnO (zinc oxide) are used as n-type materials. Both device architectures utilize either ITO or FTO (fluorine doped tin oxide) as a transparent electrode and an evaporated metal back electrode. All device architectures currently reach high level of development on a lab scale.

A certified record efficiency of PSC has already reached 22.1% [6], making this solar cell technology competitive with traditional silicon technologies. However, the active area of PSC reported in the literature is mostly limited to few square millimeters. The scale-up of the active area of the devices is expected to lead to an efficiency drop due to series resistance of the electrodes. Indeed, Yang et al. [164] have demonstrated that scaling from 0.12 to 1.1 $cm^2$ led to an efficiency drop from 17.5 to 15.5% in the devices with FTO transparent electrode and evaporated Ag back contact. The PCE in these devices dropped due to a decrease in FF with increasing cell area. This can be critical for the manufacturing of perovskite PV modules [165–169], where the dimension of the single sub-cell should be selected in such a way that the series resistance of the electrodes does not affect the performance of the module. A perovskite module of $10 \times 10\,cm^2$, reported by J. Seo et al. [168], has a width of ITO patches of 0.8 cm (0.6 cm active area and 0.2 cm interconnection). Such modules demonstrate a PCE of 8.7%, with FF of 0.57. The modules reported by S. Razza et al. [167], with an area of 10 $cm^2$, demonstrate PCE of 10.3% with FF of 0.58, while scaling up to 100 $cm^2$ led to a drop in efficiency to 4.3% with FF of 0.54. Not only the dimensions and the resistance of the electrodes play a role, but also the uniformity of the layers and quality of the interconnection are also very critical. Nevertheless, the key question for scale-up is "what should the optimal sub-cell width be to avoid resistive losses in the module?" The theoretical modeling performed by Galagan et al. [170] reveals the critical influence of the electrode sheet resistance on the performance of PSCs and modules when scaling up of the active area of the sub-cells. Simulated results show that if the sheet resistance of one of the electrodes (e.g., transparent conducting oxide (TCO)) is $10\,\Omega^{-2}$, the optimum width of the sub-cells in the module is in the range of 0.3–0.7 cm. However, this value decreases to 0.3–0.4 cm, if the sheet resistance rises to $40\,\Omega^{-2}$, while for $60\,\Omega^{-2}$, the optimum width of the sub-cell is only 0.3 cm. The results identified several critical issues in the large area processing and scaling up of perovskite PV technology. First, because the width of the sub-cells in the modules is relatively small, the efforts should be focused on reducing the interconnection (dead) area. Laser technology seems to be most promising to achieve a high aspect ratio between the active area and interconnections zone [53].

Furthermore, the low-cost potential of PSCs is based on the low cost of materials and the possibility of solution R2R processing [171, 172]. However, R2R processing introduces some limitations in materials selection and limits the process temperature to approximately 140°C in case of plastic substrates. Plastic substrates set a temperature limit and reveal the necessity of alternative low temperature sintering for the functional layers. Alternatively, substrates that can tolerate high temperature processing, for example, metal substrates, can be explored for application in R2R perovskite processing. Recent development demonstrated possibilities to manufacture PSCs on plastic substrates with low process temperatures [173–175]. The possibility to use plastic substrates and low temperature solution processable methods for manufacturing PSCs makes this technology very promising for future commercialization.

### 11.4.1 Scalable Processing Techniques for Manufacturing Perovskite Solar Cells

Rapid progress in the development of PSC has been demonstrated in recent years. Although many issues such as stability, toxicity, current–voltage (I-V) hysteresis, and so on need to be resolved before the commercialization of this technology, scale-up and large area processing have become the focus of the latest research. Spin coating provides an excellent route for the thin-film deposition of the layers in PSC devices. Based on successful experience with organic solar cell manufacturing, many printing and coating techniques were adapted for the manufacturing of PSCs. Because different layers in the perovskite devices have different physical properties, the selection of deposition methods must be guided by the same considerations as in OPV in optimizing each layer in the device stack. For example, screen printing was often utilized for the deposition of mesoscopic $TiO_2$ and $Al_2O_3$ materials [176, 177], while inkjet printing was successfully applied for the deposition of the nanocarbon hole-extraction layer [178].

Manufacturing of PSCs is a complex process that includes the optimization of each layer. The first attempts to replace spin coating for the processing of the perovskite layer were done using blade coating technique [179–182]. Although blade coating is not strictly an R2R deposition method, it can be considered as an intermediate step for further scale-up. This technique was widely used for fabrication of organic solar cells and is proven to be simple, environmentally friendly, and low-cost method for solution-processed PVs. Blade coating was used extensively for the processing of perovskite PAL [180–182] as well as for fully printed PSCs [179].

A high throughput ultrasonic spray coating was used for the fabrication of perovskite layer [183], whereby perovskite films with high uniformity, crystallinity, and surface coverage were obtained in a single step. In order to achieve a high-quality perovskite film for high-performance devices, the optimization of the ultrasonic spray coating deposition was done by comparing different solvents, tuning the substrate temperature, and optimizing the perovskite layer thickness. In order to extend this process to flexible plastic substrates, high temperature annealing (often required for curing $TiO_2$ layer) must be avoided. Photonic curing technique [118], which is compatible with the use of plastic substrates, was used for the curing of the $TiO_2$ layer. This technique enables R2R high temperature processing of thin-film materials on low temperature substrates.

Significant results were demonstrated on scaling up PSCs using flexible substrates and R2R deposition techniques such as slot die coating under ambient conditions [172, 184]. Both normal and inverted device geometries have been explored. It was found that the formation of the correct morphology for the perovskite layer depends heavily on the surface upon which it is coated which has significant implications for manufacturing. The big challenge is to develop scalable technology without efficiency losses compare to lab scale devices. However, early experiments show a significant loss in PSC performance when spin coating is substituted by R2R compatible deposition. For example, the PCE of the devices processed using spin coating on glass/ITO substrate with evaporated back electrode was 9.4%. While identical devices, with the same active area realized using slot die coating on flexible ITO/PET substrate with a printed back electrode, give a PCE of 4.9% [172].

Printed PSCs were demonstrated also with slot die coating as a scalable printing method by Hwang et al. [171]. A sequential slot die coating process was developed to produce efficient PSCs and to be used in a large-scale R2R printing process. Perovskite PV devices, where all layers excluding the electrodes were printed, demonstrated a PCE of up to 11.96%. This confirms that PSCs can be effectively produced by an R2R process.

### 11.4.2 Other Challenges in the Scale-Up of Perovskite Solar Cells

Although the scale-up of R2R fabrication of perovskite PVs is already gaining rapid progress, there are still many remaining challenges that have to be resolved before the full commercialization of this technology [185–189]. The control of the crystallization process over a large area is one of the critical issues for large area solution processing of perovskite thin films. Another challenging area is understanding and eliminating hysteresis in PSCs. The possible mechanisms of hysteresis are ion migration [190], trap states [191], ferroelectricity [192], capacitance [193], and structure defects at heterojunction interfaces that depend on a selection of carrier transport materials [194]. Although a lot of different hypotheses for the origin of hysteresis were proposed, the one dominant mechanism has not been identified yet.

Another major issue with perovskite PVs is stability. Currently the poor stability of PSCs is the main barrier to commercialization. The major cause of instability, according to Wang et al. [195], is the effect of moisture on the perovskite layer. However, the stability problem should be explored in the relationship of all layer and the interfaces in the devices. In an overview of the latest developments in the field of PSCs development, Ye et al. [189] have concluded that there are several effective ways to improve the device's stability, including improvement in the deposition technique and modification of perovskite film. The second approach to improve perovskite cell stability is the optimization of PAL and charges transport materials. To achieve the required stability for PSCs, future research should be focused on the careful selection of the device stack providing high intrinsic stability and the development of durable encapsulation strategy for the devices.

The presence of lead in PSCs and its toxicity is currently a major issue under investigation [159, 188, 196]. The deleterious effect of lead on human health and the potential environmental risks have prompted the exploration of alternative materials, for example, tin, in the PSCs. However, efficiency and stability of tin-based devices are much inferior to lead-based devices. Recent investigations with a focus on life cycle analysis (LCA) and an environmental impact analysis (EIA) of lead-containing PSCs show that lead has very little environmental impact [159, 163]. Furthermore, the comparison between the environmental impacts of tin- and lead-based PSCs shows that tin-based solar cells have in fact large impact [159]. Despite the toxicity of lead, the main contributors to the environmental impact are other compounds that comprise the solar cell rather than the lead iodide. The optimization of the manufacturing process of PSCs is essential in order to compete effectively against other thin-film solar cells or printed PV technologies such as OPV.

## 11.5 Conclusions

This chapter has shown that thin-film PVs can be produced by solution-based R2R printing and coating methods. The main benefits of R2R printing and coating include relatively easy processing, the potential of low cost and high production yield. Recent developments in the field of printed PVs have led to continuous improvement of the PCE of the manufactured PV modules. Although OPV have relatively low PCE, a variety of unique attributes (namely, nontoxicity, low weight, potential for low cost, and possibility of forming flexible modules of many different shapes, colors, and transparencies) enable applications that may not be achievable with other thin-film flexible PVs having higher efficiency. Research on PSCs is still in its early stages, but tremendous progress has already been made so far in terms of boost in efficiency, and it is fast becoming one of the most promising technologies. As for OPV technology, perovskite PV can be also produced from solution by R2R printing and coating. Thus, development of an R2R manufacturing technology for PSCs is becoming one of the hottest areas in the field. The establishment of reliable scaled up processes for manufacturing printed PVs, both OPV and PSCs, is the first crucial step toward the commercialization of these technologies.

## References

1 Green, M. A. *Nat. Energy* **2016**, 1, 15015.
2 *New Energy Outlook* **2015**, http://go.nature.com/aPwDU9 (Accessed on September 24, 2017).
3 Green, M. A.; Emery, K.; Hishikawa, Y.; Warta, W.; Dunlop, E. D. *Prog. Photovolt. Res. Appl.* **2015**, 23, 805.

4 Masuko, K.; Shigematsu, M.; Hashiguchi, T.; Fujishima, D.; Kai, M.; Yoshimura, N.; Yamaguchi, T.; Ichihashi, Y.; Mishima, T.; Matsubara, N.; Yamanishi, T.; Takahama, T.; Taguchi, M.; Maruyama, E.; Okamoto, S. *IEEE J. Photovoltaics* **2014**, 4, 1433.

5 Polman, A.; Knight, M.; Garnett, E. C.; Ehrler, B.; Sinke, W. C. *Science* **2016**, 352, aad4424.

6 Green, M. A.; Emery, K.; Hishikawa, Y.; Warta, W.; Dunlop, E. D., *Prog. Photovolt Res. Appl.* **2016**, 24, 905.

7 Sai, H.; Matsui, T.; Matsubara, K.; Kondo, M.; Yoshida, I. *IEEE J. Photovoltaics* **2014**, 4, 1349.

8 Andersen, T. R.; Cooling, N. A.; Almyahi, F.; Hart, A. S.; Nicolaidis, N. C.; Feron, K.; Noori, M.; Vaughan, B.; Griffith, M. J.; Belcher, W. J.; Dastoor, P. C. *Sol. Energy Mater. Sol. Cells* **2016**, 149, 103.

9 Griffith, M. J.; Cooling, N. A.; Vaughan, B.; Elkington, D. C.; Hart, A. S.; Lyons, A. G.; Quereshi, S.; Belcher, W. J.; Dastoor, P. C. *IEEE J. Sel. Top. Quantum Electron.* **2016**, 22, 112.

10 Griffith, M. J.; Cooling, N. A.; Vaughan, B.; O'Donnell, K. M.; Al-Mudhaffer, M. F.; Al-Ahmad, A.; Noori, M.; Almyahi, F.; Belcher, W. J.; Dastoor, P. C. *Energ. Technol.* **2015**, 3, 428.

11 Kessler, F.; Herrmann, D.; Powalla, M. *Thin Solid Films* **2005**, 480–481, 491.

12 Chuang, T. K.; Troccoli, M.; Kuo, P. C.; Jamshidi-Roudbari, A.; Hatalis, M. K.; Biaggio, I.; Voutsas, A. T. *Appl. Phys. Lett.* **2007**, 90, 151114.

13 Hu, X.; Song, Z.; Liu, W.; Qin, F.; Zhang, Z.; Wang, H. *Appl. Surf. Sci.* **2012**, 258, 5798.

14 Jeong, J. K.; Jin, D. U.; Shin, H. S.; Lee, H. J.; Kim, M.; Ahn, T. K.; Lee, J.; Mo, Y. G.; Chung, H. K. *IEEE Electron Device Lett.* **2007**, 28, 389.

15 Xie, Z.; Hung, L. S.; Zhu, F. *Chem. Phys. Lett.* **2003**, 381, 691.

16 Galagan, Y.; Moet, D. J. D.; Hermes, D. C.; Blom, P. W. M.; Andriessen, R. *Org. Electron.* **2012**, 13, 3310.

17 MacDonald, W. A.; Looney, M. K.; MacKerron, D.; Eveson, R.; Rakos, K. *Plast. Rubber Compos. Process. Appl.* **2008**, 37, 41.

18 MacDonald, W. A. *Polym. Int.* **2002**, 51, 923.

19 MacDonald, W. A.; Looney, M. K.; Mackerron, D.; Eveson, R.; Adam, R.; Hashimoto, K.; Rakos, K. *J. Soc. Inf. Disp.* **2007**, 15, 1075.

20 Fonrodona, M.; Escarré, J.; Villar, F.; Soler, D.; Asensi, J. M.; Bertomeu, J.; Andreu, J. *Sol. Energy Mater. Sol. Cells* **2005**, 89, 37.

21 Zardetto, V.; Brown, T. M.; Reale, A.; Di Carlo, A. *J. Polym. Sci. B Polym. Phys.* **2011**, 49, 638.

22 Choi, M. C.; Kim, Y.; Ha, C. S. *Prog. Polym. Sci.* **2008**, 33, 581.

23 Vergnat, C.; Landais, V.; Combet, J.; Vorobiev, A.; Konovalov, O.; Legrand, J. F.; Brinkmann, M. *Org. Electron.* **2012**, 13, 1594.

24 Yan, M.; Kim, T. W.; Erlat, A. G.; Pellow, M.; Foust, D. F.; Liu, J.; Schaepkens, M.; Heller, C. M.; McConnelee, P. A.; Feist, T. P.; Duggal, A. R. *Proc. IEEE* **2005**, 93, 1468.

25 Woo, K.; Kim, D.; Kim, J. S.; Lim, S.; Moon, J. *Langmuir* **2009**, 25, 429.
26 Lim, J. W.; Cho, D. Y.; Eun, K.; Choa, S. H.; Na, S. I.; Kim, J.; Kim, H. K. *Sol. Energy Mater. Sol. Cells* **2012**, 105, 69.
27 Lu, Y. H.; Hu, Z. Z.; Bian, J. M.; Wang, Y. F. *Adv. Mater. Res.* **2011**, 239–242, 1211.
28 Fahlteich, J.; Fahland, M.; Schönberger, W.; Schiller, N. *Thin Solid Films* **2009**, 517, 3075.
29 Sullivan, T.; Middleman, S.; Keunings, R. *AIChE J* **1987**, 33, 2047.
30 Gaskell, P. H.; Innes, G. E.; Savage, M. D. *J. Fluid Mech.* **1998**, 355, 17.
31 Steirer, K. X.; Berry, J. J.; Reese, M. O.; van Hest, M. F. A. M.; Miedaner, A.; Liberatore, M. W.; Collins, R. T.; Ginley, D. S. *Thin Solid Films* **2009**, 517, 2781.
32 Eom, S. H.; Senthilarasu, S.; Uthirakumar, P.; Yoon, S. C.; Lim, J.; Lee, C.; Lim, H. S.; Lee, J.; Lee, S.-H. *Org. Electron.* **2009**, 10, 536.
33 Aernouts, T.; Aleksandrov, T.; Girotto, C.; Genoe, J.; Poortmans, J. *Appl. Phys. Lett.* **2008**, 92, 033306.
34 Hoth, C. N.; Choulis, S. A.; Schilinsky, P.; Brabec, C. *J. Adv. Mater.* **2007**, 19, 3973.
35 Hoth, C. N.; Schilinsky, P.; Choulis, S. A.; Brabec, C. J. *Nano Lett.* **2008**, 8, 2806.
36 Kaihovirta, N. J.; Tobjörk, D.; Mäkelä, T.; Österbacka, R. *Adv. Eng. Mater.* **2008**, 10, 640.
37 Voigt, M. M.; Guite, A.; Chung, D.-Y.; Khan, R. U. A.; Campbell, A. J.; Bradley, D. D. C.; Meng, F.; Steinke, J. H. G.; Tierney, S.; McCulloch, I.; Penxten, H.; Lutsen, L.; Douheret, O.; Manca, J.; Brokmann, U.; Sönnichsen, K.; Hülsenberg, D.; Bock, W.; Barron, C.; Blanckaert, N.; Springer, S.; Grupp, J.; Mosley, A. *Adv. Funct. Mater.* **2010**, 20, 239.
38 Ding, J. M.; de la Fuente Vornbrock, A.; Ting, C.; Subramanian, V. *Sol. Energy Mater. Sol. Cells* **2009**, 93, 459.
39 Green, R.; Morfa, A.; Ferguson, A. J.; Kopidakis, N.; Rumbles, G.; Shaheen, S. E. *Appl. Phys. Lett.* **2008**, 92, 033301.
40 Hoth, C. N.; Steim, R.; Schilinsky, P.; Choulis, S. A.; Tedde, S. F.; Hayden, O.; Brabec, C. J. *Org. Electron.* **2009**, 10, 587.
41 Ishikawa, T.; Nakamura, M.; Fujita, K.; Tsutsui, T. *Appl. Phys. Lett.* **2004**, 84, 2424.
42 Vak, D.; Kim, S.-S.; Jo, J.; Oh, S.-H.; Na, S.-I.; Kim, J.; Kim, D.-Y. *Appl. Phys. Lett.* **2007**, 91, 081102.
43 Shaheen, S. E.; Radspinner, R.; Peyghambarian, N.; Jabbour, G. E. *Appl. Phys. Lett.* **2001**, 79, 2996.
44 Krebs, F. C.; Alstrup, J.; Spanggaard, H.; Larsen, K.; Kold, E. *Sol. Energy Mater. Sol. Cells* **2004**, 83, 293.
45 Krebs, F. C.; Spanggard, H.; Kjær, T.; Biancardo, M.; Alstrup, J. *Mater. Sci. Eng. B* **2007**, 138, 106.
46 Krebs, F. C.; Jørgensen, M.; Norrman, K.; Hagemann, O.; Alstrup, J.; Nielsen, T. D.; Fyenbo, J.; Larsen, K.; Kristensen, J. *Sol. Energy Mater. Sol. Cells* **2009**, 93, 422.
47 Blankenburg, L.; Schultheis, K.; Schache, H.; Sensfuss, S.; Schrödner, M. *Sol. Energy Mater. Sol. Cells* **2009**, 93, 476.

48 Krebs, F. C.; Norrman, K. *ACS Appl. Mater. Interfaces* **2010**, 2, 877.
49 Krebs, F. C. *Sol. Energy Mater. Sol. Cells* **2009**, 93, 1636.
50 Krebs, F. C. *Org. Electron.* **2009**, 10, 761.
51 Krebs, F. C. *Sol. Energy Mater. Sol. Cells* **2009**, 93, 465.
52 Krebs, F. C.; Gevorgyan, S. A.; Alstrup, J. *J. Mater. Chem.* **2009**, 19, 5442.
53 Mincuzzi, G.; Palma, A. L.; Di Carlo, A.; Brown, T. M. *ChemElectroChem* **2016**, 3, 9.
54 Lim, Y.-F.; Lee, J.-K.; Zakhidov, A. A.; DeFranco, J. A.; Fong, H. H.; Taylor, P. G.; Ober, C. K.; Malliaras, G. G. *J. Mater. Chem.* **2009**, 19, 5394.
55 Colsmann, A.; Stenzel, F.; Balthasar, G.; Do, H.; Lemmer, U. *Thin Solid Films* **2009**, 517, 1750.
56 Angmo, D.; Gevorgyan, S. A.; Larsen-Olsen, T. T.; Søndergaard, R. R.; Hösel, M.; Jørgensen, M.; Gupta, R.; Kulkarni, G. U.; Krebs, F. C. *Org. Electron.* **2013**, 14, 984.
57 Angmo, D.; Larsen-Olsen, T. T.; Jørgensen, M.; Søndergaard, R. R.; Krebs, F. C. *Adv. Energy Mater.* **2013**, 3, 172.
58 Søndergaard, R. R.; Hösel, M.; Krebs, F. C. *J. Polym. Sci. B Polym. Phys.* **2013**, 51, 16.
59 Van De Wiel, H. J.; Galagan, Y.; Van Lammeren, T. J.; De Riet, J. F. J.; Gilot, J.; Nagelkerke, M. G. M.; Lelieveld, R. H. C. A. T.; Shanmugam, S.; Pagudala, A.; Hui, D.; Groen, W. A. *Nanotechnology* **2013**, 24, 484014.
60 Wengeler, L.; Schmidt-Hansberg, B.; Peters, K.; Scharfer, P.; Schabel, W. *Chem. Eng. Process.* **2011**, 50, 478.
61 Wengeler, L.; Schmitt, M.; Peters, K.; Scharfer, P.; Schabel, W. *Chem. Eng. Process.* **2013**, 68, 38.
62 Kaduwal, D.; Schleiermacher, H. F.; Schulz-Gericke, J.; Kroyer, T.; Zimmermann, B.; Würfel, U. *Sol. Energy Mater. Sol. Cells* **2014**, 124, 92.
63 Krebs, F. C.; Tromholt, T.; Jorgensen, M. *Nanoscale* **2010**, 2, 873.
64 Ajuria, J.; Etxebarria, I.; Cambarau, W.; Munecas, U.; Tena-Zaera, R.; Jimeno, J. C.; Pacios, R. *Energy Environ. Sci.* **2011**, 4, 453.
65 Espinosa, N.; García-Valverde, R.; Urbina, A.; Krebs, F. C. *Sol. Energy Mater. Sol. Cells* **2011**, 95, 1293.
66 Angmo, D.; Gonzalez-Valls, I.; Veenstra, S.; Verhees, W.; Sapkota, S.; Schiefer, S.; Zimmermann, B.; Galagan, Y.; Sweelssen, J.; Lira-Cantu, M.; Andriessen, R.; Kroon, J. M.; Krebs, F. C. *J. Appl. Polym. Sci.* **2013**, 130, 944.
67 Angmo, D.; Krebs, F. C. *J. Appl. Polym. Sci.* **2013** 129, 1.
68 Galagan, Y.; Coenen, E. W. C.; Sabik, S.; Gorter, H. H.; Barink, M.; Veenstra, S. C.; Kroon, J. M.; Andriessen, R.; Blom, P. W. M. *Sol. Energy Mater. Sol. Cells* **2012**, 104, 32.
69 Galagan, Y.; Coenen, E. W. C.; Zimmermann, B.; Slooff, L. H.; Verhees, W. J. H.; Veenstra, S. C.; Kroon, J. M.; Jørgensen, M.; Krebs, F. C.; Andriessen, R. *Adv. Energy Mater.* **2014**, 4, 1300498.

**70** Galagan, Y.; Zimmermann, B.; Coenen, E. W. C.; Jørgensen, M.; Tanenbaum, D. M.; Krebs, F. C.; Gorter, H.; Sabik, S.; Slooff, L. H.; Veenstra, S. C.; Kroon, J. M.; Andriessen, R. *Adv. Energy Mater.* **2012**, 2, 103.

**71** Sondergaard, R. R.; Hösel, M.; Jørgensen, M.; Krebs, F. C. *J. Polym. Sci. B Polym. Phys.* **2013**, 51, 132.

**72** Yu, J. S.; Kim, I.; Kim, J. S.; Jo, J.; Larsen-Olsen, T. T.; Søndergaard, R. R.; Hösel, M.; Angmo, D.; Jørgensen, M.; Krebs, F. C. *Nanoscale* **2012**, 4, 6032.

**73** Zimmermann, B.; Schleiermacher, H. F.; Niggemann, M.; Würfel, U. *Sol. Energy Mater. Sol. Cells* **2011**, 95, 1587.

**74** Kaduwal, D.; Zimmermann, B.; Würfel, U. *Sol. Energy Mater. Sol. Cells* **2014**, 120, 449.

**75** Yambem, S. D.; Haldar, A.; Liao, K.-S.; Dillon, E. P.; Barron, A. R.; Curran, S. A. *Sol. Energy Mater. Sol. Cells* **2011**, 95, 2424.

**76** Yambem, S. D.; Liao, K.-S.; Curran, S. A. *Sol. Energy Mater. Sol. Cells* **2011**, 95, 3060.

**77** Barnes, T. M.; Bergeson, J. D.; Tenent, R. C.; Larsen, B. A.; Teeter, G.; Jones, K. M.; Blackburn, J. L.; van de Lagemaat, J. *Appl. Phys. Lett.* **2010**, 96, 243309.

**78** Huang, X.; Zeng, Z.; Fan, Z.; Liu, J.; Zhang, H. *Adv. Mater.* **2012**, 24, 5979.

**79** Miller, M. S.; O'Kane, J. C.; Niec, A.; Carmichael, R. S.; Carmichael, T. B. *ACS Appl. Mater. Interfaces* **2013**, 5, 10165.

**80** Yang, L.; Zhang, T.; Zhou, H.; Price, S. C.; Wiley, B. J.; You, W. *ACS Appl. Mater. Interfaces* **2011**, 3, 4075.

**81** Angmo, D.; Andersen, T. R.; Bentzen, J. J.; Helgesen, M.; Søndergaard, R. R.; Jørgensen, M.; Carlé, J. E.; Bundgaard, E.; Krebs, F. C. *Adv. Funct. Mater.* **2015**, 25, 4539.

**82** Ahlswede, E.; Mühleisen, W.; Bin Moh Wahi, M. W.; Hanisch, J.; Powalla, M. *Appl. Phys. Lett.* **2008**, 92, 143307.

**83** Chang, Y.-M.; Wang, L.; Su, W.-F. *Org. Electron.* **2008**, 9, 968.

**84** Hau, S. K.; Yip, H.-L.; Zou, J.; Jen, A. K. Y. *Org. Electron.* **2009**, 10, 1401.

**85** Jun-Seok, Y.; Jin-Mun, Y.; Seok-Soon, K.; Dong-Yu, K.; Junkyung, K.; Seok-In, N. *Semicond. Sci. Technol.* **2011**, 26, 034010.

**86** Galagan, Y.; Andriessen, R.; Rubingh, E.; Grossiord, N.; Blom, P.; Veenstra, S.; Verhees, W.; Kroon J. In Lope-C, Messe Frankfurt, Germany, May 31 to June 2, **2010**, p 88.

**87** Paetzold, R.; Heuser, K.; Henseler, D.; Roeger, S.; Wittmann, G.; Winnacker, A. *Appl. Phys. Lett.* **2003**, 82, 3342.

**88** Ouyang, J.; Chu, C. W.; Chen, F. C.; Xu, Q.; Yang, Y. *Adv. Funct. Mater.***2005**, 15, 203.

**89** Zimmermann, B.; Glatthaar, M.; Niggemann, M.; Riede, M. K.; Hinsch, A.; Gombert, A. *Sol. Energy Mater. Sol. Cells* **2007**, 91, 374.

**90** Tvingstedt, K.; Inganäs, O. *Adv. Mater.* **2007**, 19, 2893.

**91** Zou, J.; Yip, H.-L.; Hau, S. K.; Jen, A. K.-Y. *Appl. Phys. Lett.* **2010**, 96, 203301.

**92** Galagan, Y.; Rubingh, J. E. J. M.; Andriessen, R.; Fan, C. C.; Blom, P. W. M.; Veenstra, S. C.; Kroon, J. M. *Sol. Energy Mater. Sol. Cells* **2011**, 95, 1339.

93 Girotto, C.; Rand, B. P.; Steudel, S.; Genoe, J.; Heremans, P. *Org. Electron.* **2009**, 10, 735.
94 Hau, S. K.; Yip, H. L.; Leong, K.; Jen, A. K. Y. *Org. Electron.* **2009**, 10, 719.
95 Kopola, P.; Zimmermann, B.; Filipovic, A.; Schleiermacher, H. F.; Greulich, J.; Rousu, S.; Hast, J.; Myllylä, R.; Würfel, U. *Sol. Energy Mater. Sol. Cells* **2012**, 107, 252.
96 Krantz, J.; Stubhan, T.; Richter, M.; Spallek, S.; Litzov, I.; Matt, G. J.; Spiecker, E.; Brabec, C. J. *Adv. Funct. Mater.* **2013**, 23, 1711.
97 Hösel, M.; Søndergaard, R. R.; Angmo, D.; Krebs, F. C. *Adv. Eng. Mater.* **2013**, 15, 995.
98 Angmo, D.; Sweelssen, J.; Andriessen, R.; Galagan, Y.; Krebs, F. C. *Adv. Energy Mater.* **2013**, 3, 1230.
99 Van Franeker, J. J.; Voorthuijzen, W. P.; Gorter, H.; Hendriks, K. H.; Janssen, R. A. J.; Hadipour, A.; Andriessen, R.; Galagan, Y. *Sol. Energy Mater. Sol. Cells* **2013**, 117, 267.
100 Krebs, F. C.; Søndergaard, R.; Jørgensen, M. *Sol. Energy Mater. Sol. Cells* **2011**, 95, 1348.
101 Polino, G.; Shanmugam, S.; Bex, G. J. P.; Abbel, R.; Brunetti, F.; Di Carlo, A.; Andriessen, R.; Galagan, Y. *ACS Appl. Mater. Interfaces* **2016**, 8, 2325.
102 Galagan, Y.; Shanmugam, S.; Teunissen, J. P.; Eggenhuisen, T. M.; Biezemans, A. F. K. V.; Van Gijseghem, T.; Groen, W. A.; Andriessen, R. *Sol. Energy Mater. Sol. Cells* **2014**, 130, 163.
103 Hwiyong, C.; Dae-Hun, K.; Jaewon, L.; Sangyoon, L. *Jpn. J. Appl. Phys.* **2014**, 53, 05HC09.
104 Schmitt, M.; Scharfer, P.; Schabel, W. *J. Coat. Technol. Res.* **2014**, 11, 57.
105 Andersen, T. R.; Dam, H. F.; Hosel, M.; Helgesen, M.; Carle, J. E.; Larsen-Olsen, T. T.; Gevorgyan, S. A.; Andreasen, J. W.; Adams, J.; Li, N.; Machui, F.; Spyropoulos, G. D.; Ameri, T.; Lemaitre, N.; Legros, M.; Scheel, A.; Gaiser, D.; Kreul, K.; Berny, S.; Lozman, O. R.; Nordman, S.; Valimaki, M.; Vilkman, M.; Sondergaard, R. R.; Jorgensen, M.; Brabec, C. J.; Krebs, F. C. *Energy Environ. Sci.* **2014**, 7, 2925.
106 Kubis, P.; Li, N.; Stubhan, T.; Machui, F.; Matt, G. J.; Voigt, M. M.; Brabec, C. J. *Prog. Photovolt. Res. Appl.* **2015**, 23, 238.
107 Kubis, P.; Lucera, L.; Machui, F.; Spyropoulos, G.; Cordero, J.; Frey, A.; Kaschta, J.; Voigt, M. M.; Matt, G. J.; Zeira, E.; Brabec, C. J. *Org. Electron.* **2014**, 15, 2256.
108 Spyropoulos, G. D.; Kubis, P.; Li, N.; Baran, D.; Lucera, L.; Salvador, M.; Ameri, T.; Voigt, M. M.; Krebs, F. C.; Brabec, C. J. *Energy Environ. Sci.* **2014**, 7, 3284.
109 Lucera, L.; Kubis, P.; Fecher, F. W.; Bronnbauer, C.; Turbiez, M.; Forberich, K.; Ameri, T.; Egelhaaf, H.-J.; Brabec, C. J. *Energy Technol.* **2015**, 3, 373.
110 Galagan, Y.; Fledderus, H.; Gorter, H.; 't Mannetje, H. H.; Shanmugam, S.; Mandamparambil, R.; Bosman, J.; Rubingh, J.-E. J. M.; Teunissen, J.-P.; Salem, A.; de Vries, I. G.; Andriessen, R.; Groen, W. A. *Energy Technol.* **2015**, 3, 834.

111 Larsen-Olsen, T. T.; Sondergaard, R. R.; Norrman, K.; Jorgensen, M.; Krebs, F. C. *Energy Environ. Sci.* **2012**, 5, 9467.

112 Krebs, F. C.; Fyenbo, J.; Jorgensen, M. *J. Mater. Chem.* **2010**, 20, 8994.

113 Andersson, L. M. *Energy Technol.* **2015**, 3, 437.

114 Dam, H. F.; Andersen, T. R.; Madsen, M. V.; Mortensen, T. K.; Pedersen, M. F.; Nielsen, U.; Krebs, F. C. *Sol. Energy Mater. Sol. Cells* **2015**, 140, 187.

115 Huang, Y.-C.; Cha, H.-C.; Chen, C.-Y.; Tsao, C.-S. *Sol. Energy Mater. Sol. Cells* **2016**, 150, 10.

116 Søndergaard, R.; Manceau, M.; Jørgensen, M.; Krebs, F. C. *Adv. Energy Mater.* **2012**, 2, 415.

117 Bundgaard, E.; Hagemann, O.; Manceau, M.; Jørgensen, M.; Krebs, F. C. *Macromolecules* **2010**, 43, 8115.

118 Helgesen, M.; Carle, J. E.; Andreasen, B.; Hosel, M.; Norrman, K.; Sondergaard, R.; Krebs, F. C. *Polym. Chem.* **2012**, 3, 2649.

119 Galagan, Y.; de Vries, I. G.; Langen, A. P.; Andriessen, R.; Verhees, W. J. H.; Veenstra, S. C.; Kroon, J. M. *Chem. Eng. Process.* **2011**, 50, 454.

120 Manceau, M.; Angmo, D.; Jørgensen, M.; Krebs, F. C. *Org. Electron.* **2011**, 12, 566.

121 Lima, F. A. S.; Beliatis, M. J.; Roth, B.; Andersen, T. R.; Bortoti, A.; Reyna, Y.; Castro, E.; Vasconcelos, I. F.; Gevorgyan, S. A.; Krebs, F. C.; Lira-Cantu, M. *APL Mater.* **2016**, 4, 026104.

122 Choi, S.; Zhou, Y.; Haske, W.; Shim, J. W.; Fuentes-Hernandez, C.; Kippelen, B. *Org. Electron.* **2015**, 17, 349.

123 Zuo, L.; Zhang, S.; Li, H.; Chen, H. *Adv. Mater.* **2015**, 27, 6983.

124 Espinosa, N.; Lenzmann, F. O.; Ryley, S.; Angmo, D.; Hosel, M.; Sondergaard, R. R.; Huss, D.; Dafinger, S.; Gritsch, S.; Kroon, J. M.; Jorgensen, M.; Krebs, F. C. *J. Mater. Chem. A* **2013**, 1, 7037.

125 Hübler, A.; Trnovec, B.; Zillger, T.; Ali, M.; Wetzold, N.; Mingebach, M.; Wagenpfahl, A.; Deibel, C.; Dyakonov, V. *Adv. Energy Mater.* **2011**, 1, 1018.

126 Kopola, P.; Aernouts, T.; Sliz, R.; Guillerez, S.; Ylikunnari, M.; Cheyns, D.; Välimäki, M.; Tuomikoski, M.; Hast, J.; Jabbour, G.; Myllylä, R.; Maaninen, A. *Sol. Energy Mater. Sol. Cells* **2011**, 95, 1344.

127 Kopola, P.; Aernouts, T.; Guillerez, S.; Jin, H.; Tuomikoski, M.; Maaninen, A.; Hast, J. *Sol. Energy Mater. Sol. Cells* **2010**, 94, 1673.

128 Voigt, M. M.; Mackenzie, R. C. I.; Yau, C. P.; Atienzar, P.; Dane, J.; Keivanidis, P. E.; Bradley, D. D. C.; Nelson, J. *Sol. Energy Mater. Sol. Cells* **2011**, 95, 731.

129 Apilo, P.; Hiltunen, J.; Välimäki, M.; Heinilehto, S.; Sliz, R.; Hast, J. *Prog. Photovolt. Res. Appl.* **2015**, 23, 918.

130 Valimaki, M.; Apilo, P.; Po, R.; Jansson, E.; Bernardi, A.; Ylikunnari, M.; Vilkman, M.; Corso, G.; Puustinen, J.; Tuominen, J.; Hast, J. *Nanoscale* **2015**, 7, 9570.

131 Vilkman, M.; Apilo, P.; Välimäki, M.; Ylikunnari, M.; Bernardi, A.; Po, R.; Corso, G.; Hast, J. *Energy Technol.* **2015**, 3, 407.

132 Kapnopoulos, C.; Mekeridis, E. D.; Tzounis, L.; Polyzoidis, C.; Zachariadis, A.; Tsimikli, S.; Gravalidis, C.; Laskarakis, A.; Vouroutzis, N.; Logothetidis, S. *Sol. Energy Mater. Sol. Cells* **2016**, 144, 724.

133 Vak, D.; Weerasinghe, H.; Ramamurthy, J.; Subbiah, J.; Brown, M.; Jones, D. J. *Sol. Energy Mater. Sol. Cells* **2016**, 149, 154.

134 Kubis, P.; Li, N.; Stubhan, T.; Machui, F.; Matt, G. J.; Voigt, M. M.; Brabec, C. J. *Prog. Photovolt. Res. Appl.* **2013**, 23, 238.

135 Singh, M.; Haverinen, H. M.; Dhagat, P.; Jabbour, G. E. *Adv. Mater.* **2010**, 22, 673.

136 HolstCentre. Roll-to-roll inkjet printing and sintering at Holst Centre. https://www.youtube.com/watch?v=YzbENf208rA (Accessed on September 24, 2017).

137 Robert, A.; Pit, T.; Eric, R.; Tim van, L.; Romain, C.; Marcel, E.; Joost, V.; Sjoerd van de, G.; Pim, G. *Transl. Mater. Res.* **2014**, 1, 015002.

138 Xiong, Z.; Liu, C. *Org. Electron.* **2012**, 13, 1532.

139 Galagan, Y.; Coenen, E. W. C.; Abbel, R.; Van Lammeren, T. J.; Sabik, S.; Barink, M.; Meinders, E. R.; Andriessen, R.; Blom, P. W. M. *Org. Electron.* **2013**, 14, 38.

140 Huang, Y. C.; Hsu, F. H.; Cha, H. C.; Chuang, C. M.; Tsao, C. S.; Chen, C. Y. *Org. Electron.* **2013**, 14, 2809.

141 Ren, M.; Sweelssen, J.; Grossiord, N.; Gorter, H.; Eggenhuisen, T. M.; Andriessen, R. *J. Imaging Sci. Technol.* **2012**, 56, 40504.

142 Eom, S. H.; Park, H.; Mujawar, S. H.; Yoon, S. C.; Kim, S. S.; Na, S. I.; Kang, S. J.; Khim, D.; Kim, D. Y.; Lee, S. H. *Org. Electron.* **2010**, 11, 1516.

143 Jung, J.; Kim, D.; Lim, J.; Lee, C.; Yoon, S. C. *Jpn. J. Appl. Phys.* **2010**, 49, 05EB031.

144 Jung, S.; Sou, A.; Banger, K.; Ko, D.-H.; Chow, P. C. Y.; McNeill, C. R.; Sirringhaus, H. *Adv. Energy Mater.* **2014**, 4, 1400432.

145 Larsen-Olsen, T. T.; Machui, F.; Lechene, B.; Berny, S.; Angmo, D.; Søndergaard, R.; Blouin, N.; Mitchell, W.; Tierney, S.; Cull, T.; Tiwana, P.; Meyer, F.; Carrasco-Orozco, M.; Scheel, A.; Lövenich, W.; de Bettignies, R.; Brabec, C. J.; Krebs, F. C. *Adv. Energy Mater.* **2012**, 2, 1091.

146 Eggenhuisen, T. M.; Galagan, Y.; Biezemans, A. F. K. V.; Slaats, T. M. W. L.; Voorthuijzen, W. P.; Kommeren, S.; Shanmugam, S.; Teunissen, J. P.; Hadipour, A.; Verhees, W. J. H.; Veenstra, S. C.; Coenen, M. J. J.; Gilot, J.; Andriessen, R.; Groen, W. A. *J. Mater. Chem. A* **2015**, 3, 7255.

147 Eggenhuisen, T. M.; Galagan, Y.; Biezemans, A. F. K. V.; Slaats, T. M. W. L.; Voorthuijzen, W. P.; Kommeren, S.; Shanmugam, S.; Teunissen, J. P.; Hadipour, A.; Verhees, W. J. H.; Veenstra, S. C.; Coenen, M. J. J.; Gilot, J.; Andriessen, R.; Groen, W. A. *J. Mater. Chem. A* **2015**, 3, 7668.

148 Mulligan, C. J.; Wilson, M.; Bryant, G.; Vaughan, B.; Zhou, X.; Belcher, W. J.; Dastoor, P. C. *Sol. Energy Mater. Sol. Cells* **2014**, 120(Part A), 9.

149 Søndergaard, R.; Hösel, M.; Angmo, D.; Larsen-Olsen, T. T.; Krebs, F. C. *Mater. Today* **2012**, 15, 36.

150 Espinosa, N.; García-Valverde, R.; Urbina, A.; Lenzmann, F.; Manceau, M.; Angmo, D.; Krebs, F. C. *Sol. Energy Mater. Sol. Cells* **2012**, 97, 3.

151 Hösel, M.; Dam, H. F.; Krebs, F. C. *Energy Technol.* **2015**, 3, 293.

152 Krebs, F. C.; Jørgensen, M. *Adv. Opt. Mater.* **2014**, 2, 465.

153 Krebs, F. C.; Fyenbo, J.; Tanenbaum, D. M.; Gevorgyan, S. A.; Andriessen, R.; Van Remoortere, B.; Galagan, Y.; Jørgensen, M. *Energy Environ. Sci.* **2011**, 4, 4116.

154 Burgués-Ceballos, I.; Stella, M.; Lacharmoise, P.; Martínez-Ferrero, E. *J. Mater. Chem. A* **2014**, 2, 17711.

155 Po, R.; Bernardi, A.; Calabrese, A.; Carbonera, C.; Corso, G.; Pellegrino, A. *Energy Environ. Sci.* **2014**, 7, 925.

156 Machui, F.; Lucera, L.; Spyropoulos, G. D.; Cordero, J.; Ali, A. S.; Kubis, P.; Ameri, T.; Voigt, M. M.; Brabec, C. J. *Sol. Energy Mater. Sol. Cells* **2014**, 128, 441.

157 Lee, J. U.; Jung, J. W.; Jo, J. W.; Jo, W. H. *J. Mater. Chem.* **2012**, 22, 24265.

158 Gevorgyan, S. A.; Madsen, M. V.; Dam, H. F.; Jørgensen, M.; Fell, C. J.; Anderson, K. F.; Duck, B. C.; Mescheloff, A.; Katz, E. A.; Elschner, A.; Roesch, R.; Hoppe, H.; Hermenau, M.; Riede, M.; Krebs, F. C. *Sol. Energy Mater. Sol. Cells* **2013**, 116, 187.

159 Serrano-Lujan, L.; Espinosa, N.; Larsen-Olsen, T. T.; Abad, J.; Urbina, A.; Krebs, F. C. *Adv. Energy Mater.* **2015**, 5, 1501119.

160 van Reenen, S.; Kemerink, M.; Snaith, H. J. *J. Phys. Chem. Lett.* **2015**, 6, 3808.

161 Boix, P. P.; Nonomura, K.; Mathews, N.; Mhaisalkar, S. G. *Mater. Today* **2014**, 17, 16.

162 Park, N.-G. *Mater. Today* **2015**, 18, 65.

163 Espinosa, N.; Serrano-Luján, L.; Urbina, A.; Krebs, F. C. *Sol. Energy Mater. Sol. Cells* **2015**, 137, 303.

164 Yang, M.; Zhou, Y.; Zeng, Y.; Jiang, C.-S.; Padture, N. P.; Zhu, K. *Adv. Mater.* **2015**, 27, 6363.

165 Matteocci, F.; Cinà, L.; Di Giacomo, F.; Razza, S.; Palma, A. L.; Guidobaldi, A.; D'Epifanio, A.; Licoccia, S.; Brown, T. M.; Reale, A.; Di Carlo, A. *Prog. Photovolt. Res. Appl.* **2014**, 16, 3918.

166 Matteocci, F.; Razza, S.; Di Giacomo, F.; Casaluci, S.; Mincuzzi, G.; Brown, T. M.; D'Epifanio, A.; Licoccia, S.; Di Carlo, A. *Phys. Chem. Chem. Phys.* **2014**, 16, 3918.

167 Razza, S.; Di Giacomo, F.; Matteocci, F.; Cinà, L.; Palma, A. L.; Casaluci, S.; Cameron, P.; D'Epifanio, A.; Licoccia, S.; Reale, A.; Brown, T. M.; Di Carlo, A. *J. Power Sources* **2015**, 277, 286.

168 Seo, J.; Park, S.; Chan Kim, Y.; Jeon, N. J.; Noh, J. H.; Yoon, S. C.; Seok, S. I. *Energy Environ. Sci.* **2014**, 7, 2642.

169 Sfyri, G.; Kumar, C. V.; Raptis, D.; Dracopoulos, V.; Lianos, P. *Sol. Energy Mater. Sol. Cells* **2015**, 134, 60.

170 Galagan, Y.; Coenen, E.; Verhees, W.; Andriessen, R. *J. Mater. Chem. A* **2016**, 4, 5700.

171 Hwang, K.; Jung, Y.-S.; Heo, Y.-J.; Scholes, F. H.; Watkins, S. E.; Subbiah, J.; Jones, D. J.; Kim, D.-Y.; Vak, D. *Adv. Mater.* **2015**, 27, 1241.

172 Schmidt, T. M.; Larsen-Olsen, T. T.; Carlé, J. E.; Angmo, D.; Krebs, F. C. *Adv. Energy Mater.* **2015**, 5, 1500569.

173 Di Giacomo, F.; Zardetto, V.; D'Epifanio, A.; Pescetelli, S.; Matteocci, F.; Razza, S.; Di Carlo, A.; Licoccia, S.; Kessels, W. M. M.; Creatore, M.; Brown, T. M. *Adv. Energy Mater.* **2015**, 5, 1401808.

174 Dianetti, M.; Di Giacomo, F.; Polino, G.; Ciceroni, C.; Liscio, A.; D'Epifanio, A.; Licoccia, S.; Brown, T. M.; Di Carlo, A.; Brunetti, F. *Sol. Energy Mater. Sol. Cells* **2015**, 140, 150.

175 Sun, S.; Salim, T.; Mathews, N.; Duchamp, M.; Boothroyd, C.; Xing, G.; Sum, T. C.; Lam, Y. M. *Energy Environ. Sci.* **2014**, 7, 399.

176 Cao, K.; Zuo, Z.; Cui, J.; Shen, Y.; Moehl, T.; Zakeeruddin, S. M.; Grätzel, M.; Wang, M. *Nano Energy* **2015**, 17, 171.

177 Matteocci, F.; Cinà, L.; Di Giacomo, F.; Razza, S.; Palma, A. L.; Guidobaldi, A.; D'Epifanio, A.; Licoccia, S.; Brown, T. M.; Reale, A.; Di Carlo, A. *Prog. Photovolt. Res. Appl.* **2016**, 24, 436.

178 Wei, Z.; Chen, H.; Yan, K.; Yang, S. *Angew. Chem.* **2014**, 126, 13455.

179 Yang, Z.; Chueh, C.-C.; Zuo, F.; Kim, J. H.; Liang, P.-W.; Jen, A. K. Y. *Adv. Energy Mater.* **2015**, 5, 1500328.

180 Deng, Y.; Wang, Q.; Yuan, Y.; Huang, J. *Mater. Horiz.* **2015**, 2, 578.

181 Deng, Y.; Peng, E.; Shao, Y.; Xiao, Z.; Dong, Q.; Huang, J. *Energy Environ. Sci.* **2015**, 8, 1544.

182 Hu, Q.; Wu, H.; Sun, J.; Yan, D.; Gao, Y.; Yang, J. *Nanoscale* **2016**, 8, 5350.

183 Das, S.; Yang, B.; Gu, G.; Joshi, P. C.; Ivanov, I. N.; Rouleau, C. M.; Aytug, T.; Geohegan, D. B.; Xiao, K. *ACS Photonics* **2015**, 2, 680.

184 Gu, Z.; Zuo, L.; Larsen-Olsen, T. T.; Ye, T.; Wu, G.; Krebs, F. C.; Chen, H. *J. Mater. Chem. A* **2015**, 3, 24254.

185 Petrović, M.; Chellappan, V.; Ramakrishna, S. *Sol. Energy* **2015**, 122, 678.

186 Chen, Q.; De Marco, N.; Yang, Y.; Song, T.-B.; Chen, C.-C.; Zhao, H.; Hong, Z.; Zhou, H.; Yang, Y. *Nano Today* **2015**, 10, 355.

187 Habibi, M.; Zabihi, F.; Ahmadian-Yazdi, M. R.; Eslamian, M. *Renew. Sust. Energ. Rev.* **2016**, 62, 1012.

188 Williams, S. T.; Rajagopal, A.; Chueh, C.-C.; Jen, A. K. Y. *J. Phys. Chem. Lett.* **2016**, 7, 811.

189 Ye, M.; Hong, X.; Zhang, F.; Liu, X. *J. Mater. Chem. A* **2016**, 4, 6755.

190 Li, C.; Tscheuschner, S.; Paulus, F.; Hopkinson, P. E.; Kießling, J.; Köhler, A.; Vaynzof, Y.; Huettner, S. *Adv. Mater.* **2016**, 28, 2446.

191 Wu, X.; Trinh, M. T.; Niesner, D.; Zhu, H.; Norman, Z.; Owen, J. S.; Yaffe, O.; Kudisch, B. J.; Zhu, X. Y. *J. Am. Chem. Soc.* **2015**, 137, 2089.

192 Frost, J. M.; Butler, K. T.; Brivio, F.; Hendon, C. H.; van Schilfgaarde, M.; Walsh, A. *Nano Lett.* **2014**, 14, 2584.

**193** Kim, H.-S.; Park, N.-G. *J. Phys. Chem. Lett.* **2014**, 5, 2927.

**194** Jena, A. K.; Kulkarni, A.; Ikegami, M.; Miyasaka, T. *J. Power Sources* **2016**, 309, 1.

**195** Wang, D.; Wright, M.; Elumalai, N. K.; Uddin, A. *Sol. Energy Mater. Sol. Cells* **2016**, 147, 255.

**196** Benmessaoud, I. R.; Mahul-Mellier, A.-L.; Horvath, E.; Maco, B.; Spina, M.; Lashuel, H. A.; Forro, L. *Toxicol. Res.* **2016**, 5, 407.

# 12

# Field-Assisted Self-Assembly of Nanocomposite Films: A Roll-to-Roll Approach

*Saurabh Batra*[1] *and Miko Cakmak*[2]

[1] *3M Company, St. Paul, MN, USA*
[2] *Department of Materials & Mechanical Engineering, Purdue University, West Lafayette, IN, USA*

## 12.1 Introduction

Nanocomposites consist of two or more phases with at least one of the phases in the nanometer-scale range. Interest in these materials including polymer–metal [1–3], polymer–glass [4, 5], polymer–ceramic [6, 7], and so on [8], have been directed primarily toward the enhancement of properties of interest at low concentrations and/or low cost. Nanocomposites can also be differentiated with respect to their morphology or connectivity as has been described by Newnham in the late 1970s [9]. Newnham showed that multiphase composite connectivity plays an important role in the enhancement of piezoelectric and pyroelectric properties of materials by as much as several orders of magnitude, depending on the type of connectivity present in the materials [10]. In case of a two-phase composite taking into account that each phase can be self-connected in zero, two, and three dimensions, 10 different connectivities can be described. Thus, nanocomposites depend not only on the intrinsic properties of the matrix and particles but also on their morphology and interfacial characteristics. Traditional polymer processing techniques including extrusion, injection molding, blow molding, and so on cannot efficiently create anisotropic networks of particles organized in oriented morphologies in the thickness "Z" direction as they are primarily dominated by shear, extensional, biaxial, or a combination of these flows that tend to form such in-plane structures. To achieve directed assembly of such structures, out-of-plane external fields such as electric and magnetic fields may be used to enhance thickness direction properties useful for flexible electronics [11, 12], membranes [13], supercapacitors [14, 15], fuel cells [16], photovoltaics [17], and so on. The use of

*Roll-to-Roll Manufacturing: Process Elements and Recent Advances*, First Edition.
Edited by Jehuda Greener, Glen Pearson, and Miko Cakmak.

external fields to obtain their organization and preferred orientation can be more effective [18–20], creating beneficial directional anisotropy in properties needed in the thickness direction. However, most of the research involving field-assisted assemblies has been limited to small-scale bench-type studies. To tackle the issues of scalability and to better understand the impact of field-assisted assembly, a multipurpose novel film formation process to orient and align functional nanoparticles and polymer phases using external electric, magnetic, and thermal gradient fields was designed at the University of Akron by Cakmak et al. [21].

## 12.2 Process Overview

The 70 ft roll-to-roll (R2R) processing line described in this chapter consists of a casting system that deposits liquid film of a monomer and/or polymer solution in 10 nm to 1 mm thickness range and up to 6″ wide on to a flexible substrate. The substrate was carried by pretensioned steel belt through an *electric field* application zone. In this zone, a unique design of R2R electrical plate capable of potential difference of −15 to +15 KV was used to apply DC, AC, or a biased AC to orient the film in the vertical direction while subjecting it to UV through its built-in transparent conductive carrier. If *magnetic field orientation* was desired, an electromagnet capable of applying magnetic fields up to 2.2 T was used. This line also contained a movable *UV lamp* that could be used to freeze the structure of desired morphology when using photocurable resin and could also be used to etch one phase of a polymer blend or block copolymer (BCP) after electric field or magnetic field application zones. The final tool that is built on this machine was the *thermal alignment zone* that was designed to apply a "line of heat"-oriented transverse to the machine direction at nine different zones. Using this processing line the cost of manufacturing can be reduced by limiting the amount of functional fillers through directional alignment while enhancing the through-thickness properties [21].

An outline of electromagnetic processing line is shown in Figure 12.1, and a photograph of the actual equipment is shown in Figure 12.2. As shown in the figures, the processing line was designed to produce up to 6" wide functional films on the 70 ft long R2R machine. The R2R processing line consists of seven different zones: (i) unwinding/casting zone, (ii) electric field zone, (iii) magnetic field zone, (iv) UVcuring movable station, (v) thermal gradient zone, (vi) high temperature oven, and (vii) rewind station. The details on the individual zones and the equipment can be obtained from Cakmak et al. [21] As shown in Figure 12.1, the equipment consists of block-type structures where each block or zone can be used as a stand-alone or simultaneously depending on the task requirements.

All the zones discussed earlier lie on the frame created for the endless 10″ wide pretensioned stainless steel belt that rides on two drum rollers. This nonmagnetic stainless steel belt carries the polymer substrate and the cast

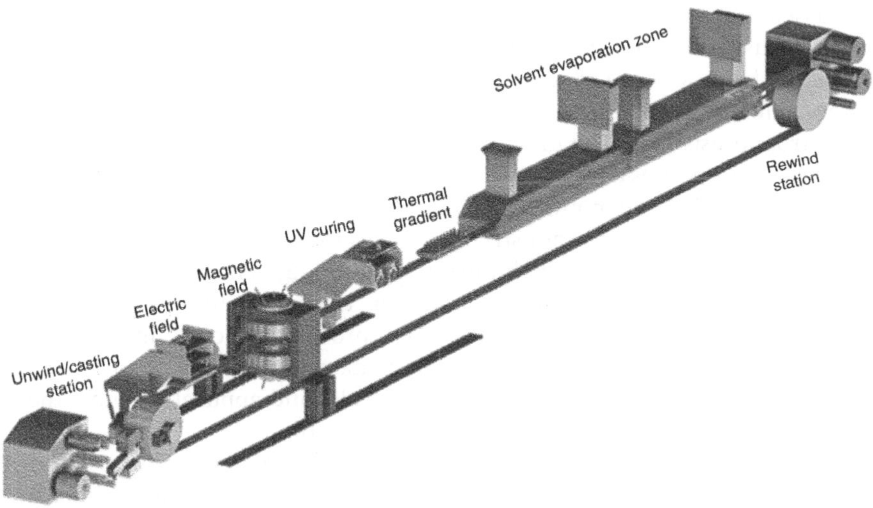


**Figure 12.1** Schematic of R2R electromagnetic processing line [21]. © John Wiley & Sons.

**Figure 12.2** Actual image of the electromagnetic processing line [21]. © John Wiley & Sons.

film through different zones for field-assisted assembly. A polymer solution is applied to a carrier via a casting blade. Solution casting was chosen over other techniques to make films in this processing line because the polymer molecules and particles in the solution cast films are mobile until the solvent evaporates. This provides a low viscosity system in which it is easier to orient particles in the case of electric and magnetic fields.

The following subsections detail how each of these fields is utilized for the right application to generate functional films.

## 12.3 Electric Field Alignment

First we will describe how the electric field alignment takes places in a solution/oligomer along with the governing equations necessary to select the right particle and the polymer matrix. Two different examples will then be discussed to create functional films creating ionic membranes as well as dielectric films using the electric field alignment technique in an R2R process.

Electric field can be used to manipulate particles in a dielectric matrix using three different techniques: electrophoresis [22], electroosmosis [23], and dielectrophoresis [24]. The direction and magnitude of the force depends on the electric properties of the particle and the matrix along with their geometry.

*Electrophoresis* is the motion of a charged particle in an electric field and has been used considerably in industrial processes for electrophoretic deposition of uniform, high speed, and high purity coatings [25]. This process is used to prepare coatings ranging from monolayer thickness, to millimeter-thick coatings, to self-assemble particles. The process typically includes two electrodes dipped in an aqueous solution containing particles. Depending on the charge on the particles, applied electric field, and application time, these particles deposit on top of one of the electrodes [26].

An important parameter that is involved in all electrokinetic flows is known as zeta potential ($\zeta$). Zeta potential is the charge that develops at the interface between a solid surface and the liquid medium [27]. This develops charges on particles in a liquid medium with applied DC field that leads to electrophoretic mobility, $\mu$, and is defined as the coefficient of proportionality between the velocity and the applied field given by

$$\mu = \frac{u}{E} \tag{12.1}$$

where $u$ is the velocity of particle under the applied electric field $E$. The electrophoretic mobility increases with increase in zeta potential and decreases with an increase in viscosity ($\eta$) of the liquid medium. The electrophoretic mobility of the particles is thus defined by Helmholtz–Smoluchowski equation:

$$\mu = \frac{\varepsilon\varepsilon_0\zeta}{\eta} \tag{12.2}$$

*Dielectrophoresis* is another alternative to manipulate particles in electric field and is caused by AC fields. The particles in a suspension with applied AC field are neither attracted nor repelled by the electrodes as the sign of the electrode polarization changes constantly. However, AC field leads to formation of induced dipoles on the particles, leading to particle–particle interaction [28]. These induced dipoles also interact with the gradient of the field, leading to dielectrophoretic force [29]:

$$F_{\mathrm{DEP}} = 2\pi\varepsilon_{\mathrm{m}} K r^3 \nabla E^2 \tag{12.3}$$

where $F_{\mathrm{DEP}}$ is the dielectrophoretic force, $\varepsilon_{\mathrm{m}}$ is dielectric constant of matrix, $K$ is the Clausius–Mossotti function, $r$ is the radius of the particle, and $E$ is the applied field.

The sign and magnitude of the dielectrophoretic force depends on the effective polarizability of the particles that is given by real part of Clausius–Mossotti function ($K$):

$$K = \frac{\varepsilon_p - \varepsilon_m}{\varepsilon_p + 2\varepsilon_m} \tag{12.4}$$

where $\varepsilon_p$, $\varepsilon_m$ are the dielectric constant of the particle and matrix, respectively [28].

This was one of the first R2R manufacturing processes for electric field-assisted alignment to make *Z*-oriented functional films. The most common approaches for electric field alignment through the thickness was a batch processes where two electrodes separated by spacers, with the solution poured in between the electrodes and cured using thermal or UV curing after the application of electric field. As has shown by other authors, for electric field alignment to be effective, it was necessary to have the two electrodes touching the solution. If the electrodes were separated by an air gap, the undulation occurs on the top of the polymer film, since the solution as a whole is polarized and attracted toward the electrode [30, 31].

### 12.3.1 Orienting Clay Particles in Electric Field

The first example talks about electric field alignment of clay particles (MMT) dispersed in photocurable resin through the thickness of the cured film for enhanced ionic conductivity [32].

The electric field setup consisted (Figure 12.3a) of two PVC rollers that were connected to a back plate that in turn sat on two rails connected to a pneumatic motor used to jog the whole assembly upward and downward (*Z*-direction), thereby controlling the distance between the rollers and the steel belt or any conducting substrate acting as ground. A continuous loop of transparent ITO-coated PET film is wrapped between these rollers; the PET film was pre-tensioned using a spring loaded assembly on one of the rollers. This ensemble acts as the top electrode. The rollers were also connected to a drive motor allowing a forward and backward motion of the ITO-coated PET at different speeds. In most cases the speed of the ITO-coated PET film on top of the rollers was matched with the speed of the stainless steel belt below. This was done to prevent any shearing effect on the cast film. After adjusting the speed, the whole assembly was lowered to match the distance between the PET film and the substrate to that of the thickness of the cast films. A copper roller was placed on top of the PET film, and this copper roller was connected to a high voltage power supply. The HV amplifier (Matsuda, model 20B20) can be used to generate a DC voltage; this amplifier was combined with an HP function generator and can generate three types of fields: AC, DC, and biased AC fields. The copper roller was spring loaded and hence was always in contact with the

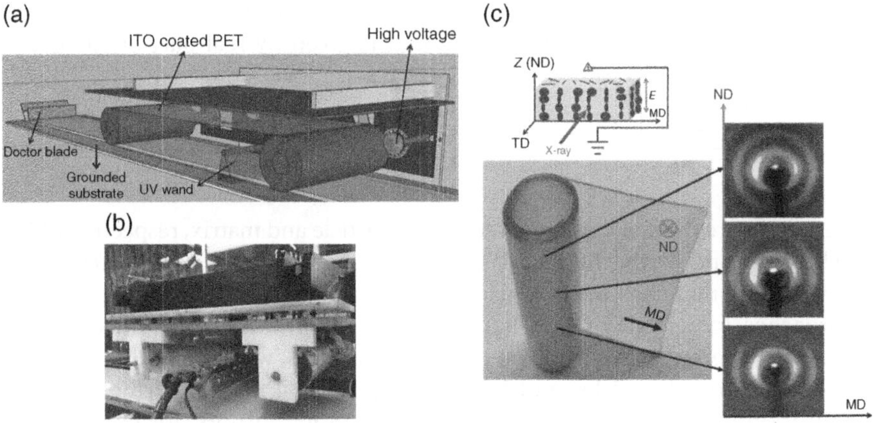


**Figure 12.3** (a) Schematic and (b) picture of roll-to-roll processing equipment developed to create indefinitely long oriented films using electric field alignment. (c) WAXS patterns at different locations for a roll of film with 6 wt % clay loading aligned at 724 V/mm [32]. © John Wiley & Sons.

conducting film while rotating with the film as it moves, creating a potential difference between the ITO-coated PET film on top and the substrate. A UV wand that radiates at the wavelength suitable for curing the resin to freeze the aligned morphology was placed in the center of the setup.

During a typical process, particles were first dispersed in the photocurable resin using sonication, magnetic stirring, high shear mixing, or other dispersion techniques. The monomer mixture was then cast onto the carrier using a slot die or a doctor blade assembly. The blade gap was set using the micrometers or feeler gauges to the desired thickness of the films. The gap between the top electrode film (ITO-coated PET film) and the conducting substrate was kept the same as that of the thickness of the cast film. The speed of the substrate and the top ITO-coated film was then matched. As the cast film approaches the electric field setup, it touched the transparent conducting film on top, which was maintained at high voltage using a copper contact roller. After the sample orients due to the potential difference between the two electrodes, it was necessary to freeze the structure before the film reached the end of the setup. This was carried out by irradiating UV light uniformly through the width of the sample using a UV wand, curing the cast film. Surface energy of the substrate was always kept higher than that of the transparent top electrode. This helped the film to always peel off from the top electrode, making the film stick to the bottom substrate. During the rewinding of the films, these oriented films can first be peeled off and rewound separately or can be wound with the substrate [32].

It was shown that the electric field-induced alignment can be accomplished using the R2R setup with realistic manufacturing conditions by casting large films of oriented clay (3″ wide and 10″ long) nanocomposites at 2, 4, and 6 wt % clay concentration with applied voltage of 724 V/mm at 10 cm/min speed. To verify the uniformity of orientation of the clay particles in the samples, WAXS patterns were obtained at different positions in the sample. Unlike shear orientation where the clay particles are oriented with their planes in the film plane, the electric field alignment orients these particles with their planes oriented normal to the film surface. Figure 12.3c shows the flexible-oriented clay nanocomposite film and WAXS patterns at three locations of the film.

The alignment was caused due to dielectrophoresis that leads to both rotational and translational motion due in a nonuniform field. An important manifestation of dielectrophoresis is that the dipole–dipole interaction that occurs even in a uniform electric field, a polarization field on one inclusion disturbs the electric field on the neighboring inclusion, hence producing attractive and repulsive forces between these inclusions. These attractive and repulsive forces lead to formation of chains of these inclusions in the matrix (Figure 12.4b). To observe the formation of chains and aligned orientation, the morphology of the clay nanocomposites was studied using electron microscopy as depicted in Figure 12.4.

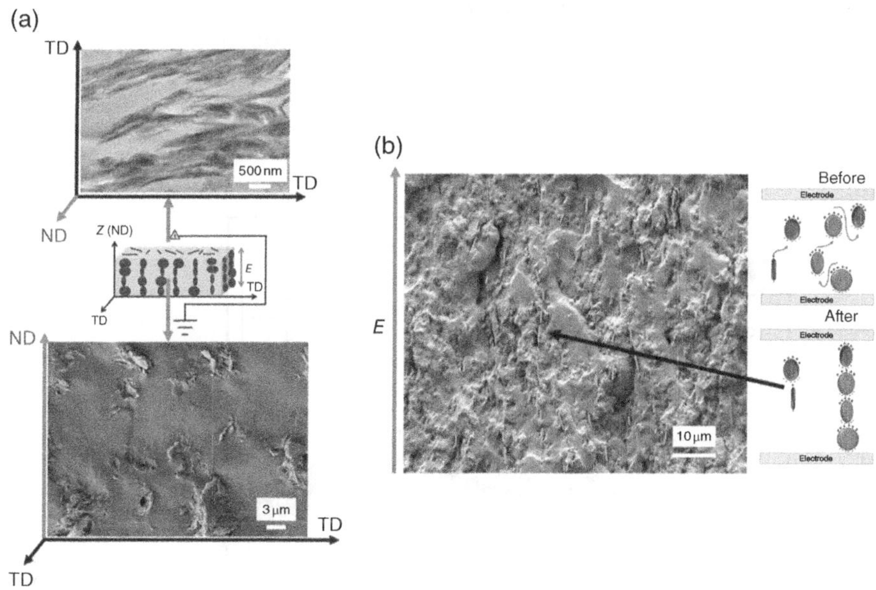


**Figure 12.4** (a) Orientation of clay was observed by TEM of microtomed samples microtomed through surface showing in-plane and three fractured cross-sectional SEM images showing out-of-plane orientation. (b) Chaining of clay platelets observed through cross-sectional SEM with visible clay edges marked with dotted lines and schematic of the dielectrophoretic chaining [32]. © John Wiley & Sons.

The enhancement in $Z$-direction properties of these clay nanocomposites were then demonstrated using dielectric measurements on the R2R-processed oriented films. The dielectric properties of both oriented and unoriented films with different clay concentrations were measured through the thickness of the film. Figure 12.5 compares the dielectric permittivity and dielectric loss of unoriented and oriented samples at room temperature. The dielectric constant above 100 Hz does not show any variation in between the oriented and

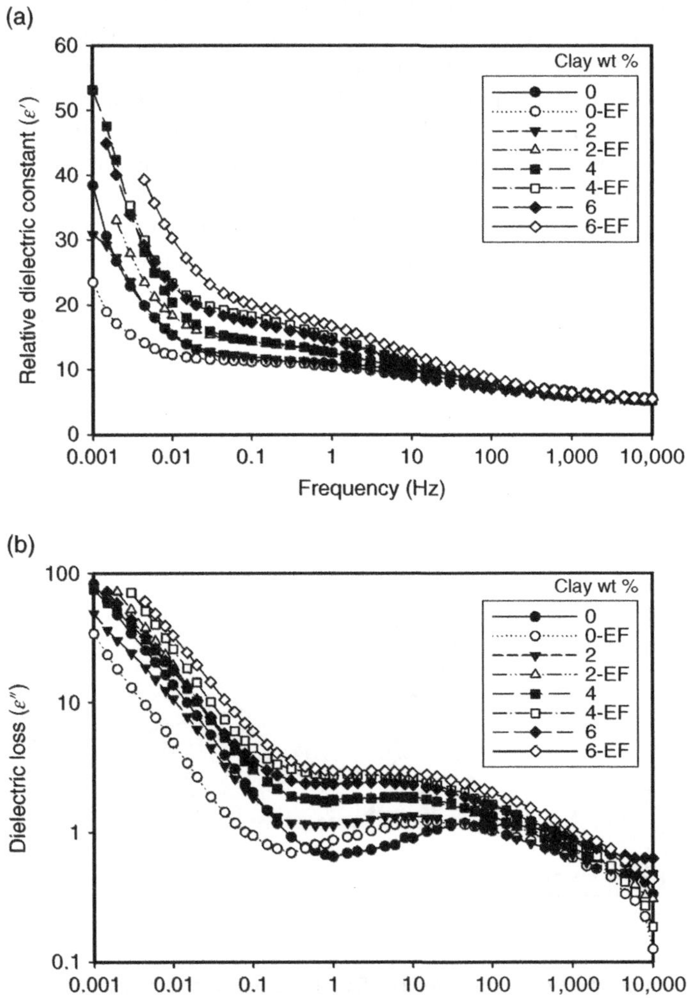


**Figure 12.5** Enhancement of dielectric properties electric field-oriented clay-filled nanocomposites (a) relative dielectric constant and (b) dielectric loses (EF, aligned nanocomposite film) [32]. © John Wiley & Sons.

unoriented clay nanocomposites due to the lack of dipole orientation. The dielectric constant starts increasing below 100 Hz till 0.03 Hz for filled systems as compared with unfilled systems as the clay-filled nanocomposites incorporate more ionic dipoles, leading to ionic polarization. The dielectric properties of oriented clay nanocomposites in this region were further enhanced due to the formation of oriented ionic dipoles. Below the frequency of 0.03 Hz, there was a drastic increase in dielectric constant with increasing clay concentration due to charge migration at the electrode–polymer interface, leading to electrode polarization [33]. The increase in dielectric properties below 0.03 Hz was attributed to ionic conductivity of clay platelets. Electrode polarization also causes an increase in the dielectric loss, and this loss dominates the true dielectric loss from the clay nanocomposites till 1 Hz.

Clay nanocomposites had anisotropic ionic conductivity in parallel and perpendicular oriented films [34]. It has been suggested that the difference in ionic conductivity of MMT nanocomposites is due to tortuosity of ionic migration that is caused by the orientation direction of MMT layers perpendicular to the measurement direction [35]. Therefore, MMT nanocomposites show higher ionic conductivity if the MMT layers are oriented in the direction of measurement compared with perpendicular orientation [36]. Below 0.03 Hz, oriented samples show much higher dielectric constant as compared with the unoriented samples. This was attributed to the fact that the oriented nanocomposites had higher ionic conductivity since the ionic migration was much easier when clays were oriented parallel to the direction of measurement. Also, the oriented nanocomposites had higher interparticle interaction due to chaining of clay platelets caused by dielectrophoretic alignment, making the ionic migration easier, leading to the higher dielectric constant for the oriented samples. Hence, *Z*-oriented clay nanocomposites can be used to increase the ionic conductivity at lower particle concentrations [32].

### 12.3.2 Orienting $BaTiO_3$ Particles in Electric Field

To enhance dielectric properties of polymer films, barium titanate particles are dispersed and organized with electric field undergoing both electrophoresis for skin layer formation and dielectrophoresis for chaining of these particles through the thickness of the film. In this case the particles were dispersed in the thermally curable resin PDMS. The setup remains similar to the one described in Figure 12.3a, except that for a few changes, the ITO-coated PET belt at the top was replaced with a metal belt and UV lamp was replaced by an IR lamp. A third iteration of the process included the use of a perforated belt; when the particles were dispersed in a solvent-based system, this would help in the solvent evaporation during the alignment process. The following example shows the alignment of $BaTiO_3$ particles; however, most these tests were conducted offline for simplicity.

Electrophoretic deposition of $BaTiO_3$ nanoparticles in PDMS was studied at different voltages. The electrophoretic mobility of particles was highly dependent on the zeta potential of the particles. The zeta potential of the $BaTiO_3$ nanoparticles was measured using a dilute dispersion of the particles in deionized water and was found to be negative ($\zeta = -26.3\,\text{mV}$) with the electrophoretic mobility of −20.63 μm/s. The electrophoretic deposition was studied in 10 wt % dispersion of $BaTiO_3$ nanoparticles in PDMS by applying varying DC voltages for 30 min. The dispersions were then cured with the applied voltage, and the resulting film was used for further characterization.

As can be observed from Figure 12.6, the electrophoretic deposition of the particles in the polymer matrix led to the formation of skin layer in the film. The skin layer observed had a clear demarcation at the interface as can be seen in the higher magnification SEM images. The $BaTiO_3$ nanoparticles were deposited on the positive electrode indicating they were negatively charged in the dispersion, which was in agreement with the negative zeta potential of these particles [37].

The skin layer observed in the SEM image was densely packed with the $BaTiO_3$ particles. With increasing voltage the skin layer thickness increased from 15 μm at 300 V/mm to 45 μm at 1400 V/mm as seen from Figure 12.6c. Thus, the thickness of the skin layer formed due to electrophoretic deposition was highly dependent on applied voltage and time of deposition, as has been shown by other researchers who have deposited nanoparticles from a colloidal solution onto an electrode [38, 39].

The chaining of particles was studied in the $BaTiO_3$ nanocomposites at the frequency of 100 Hz and an applied voltage of 1400 V/mm at different concentrations of the $BaTiO_3$. As can be seen from Figure 12.7, with the application of the AC field, $BaTiO_3$ chains were formed in the PDMS matrix. These chains were particularly long in the range of 3–5 μm at particle concentrations of 10 and 20 wt %; however, with the increase in the particle concentration, the average chain length decreased considerably. At the extremely high concentration of 50 wt %, the average chain length was down to 400 nm. This was attributed to the decrease in particle mobility with increase in particle concentration, as the viscosity increased considerably when the particle concentration was increased above 20 wt %. The dielectrophoretic force during the particle chaining competes with the inertial and drag forces that increase with increase in viscosity. The mobility of the particles undergoing dielectrophoretic motion is inversely proportional to the viscosity and is given by the following equation [40]:

$$\nu = \frac{\varepsilon_m \varepsilon_0 R^2 K}{12\pi\eta} \tag{12.5}$$

where $\varepsilon_m$ is the dielectric constant of the matrix, $\varepsilon_0$ is the permittivity of air, $R$ is the radius of the particle, $K$ is the Clausius–Mossotti function, and $\eta$ is the viscosity of dispersion.

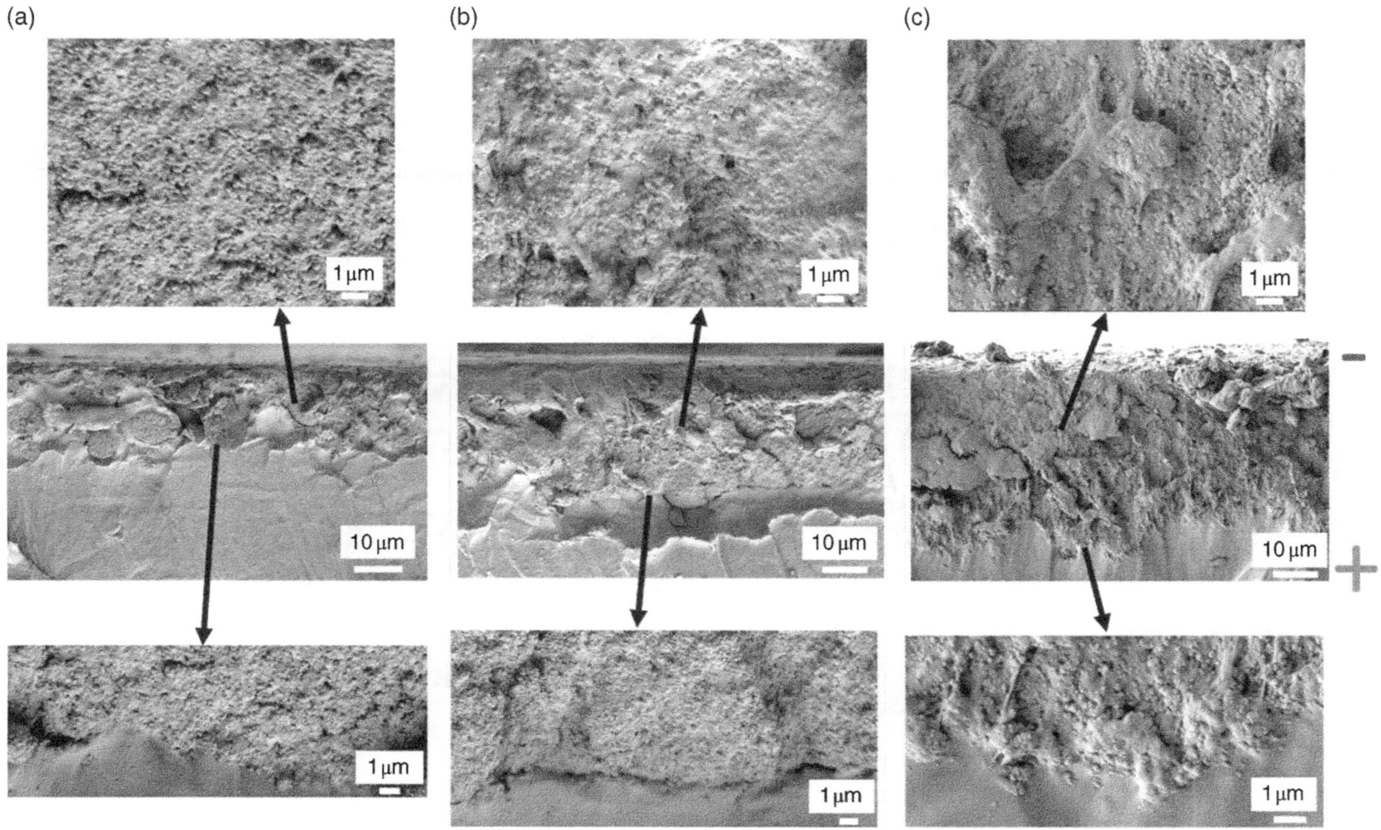


**Figure 12.6** Formation of skin layer or the BaTiO3-rich layer at different DC voltages: (a) 300 V/mm, (b) 800 V/mm, and (c) 1400 V/mm, the concentration of particles was kept constant at 10 wt % [37]. © Royal Society of Chemistry.

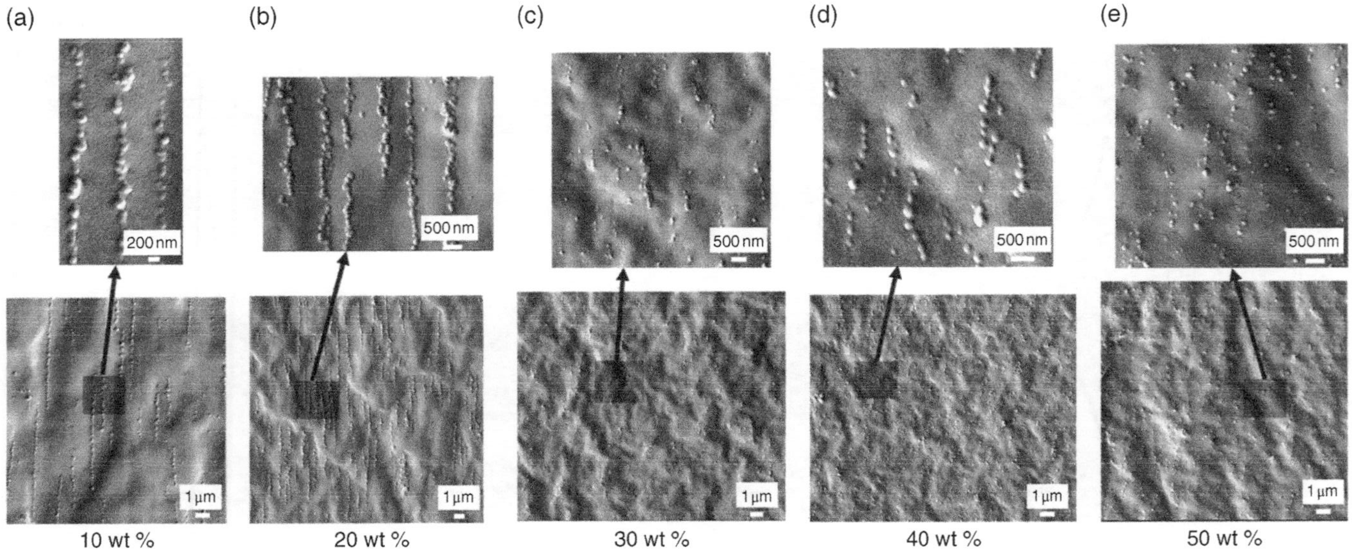


**Figure 12.7** Concentration effect on alignment of $BaTiO_3$ nanoparticles at 1400 V/mm and 100 Hz: (a) 10 wt %, (b) 20 wt %, (c) 30 wt %, (d) 40 wt %, (e) 50 wt % [37]. © Royal Society of Chemistry.

Thus, as the viscosity of the matrix increased due to higher particle concentration, the mobility of the particles decreased, leading to formation of shorter chains. Also with the increase in particle loading in the dispersion, there was an increased frustration between the particles because of restricted motion to form long chains. As a result, reduction in the distance between the chains and also the order of the chains was observed. As can be seen from Figure 12.7, at lower particle concentration the particle chains were much further apart and almost parallel to the direction of applied field. However, increasing concentration led to skewed chains and reduction in the distance between the two chains.

The orientation factor was determined using the order parameter *S*. If $\theta$ is the angle of a chain with respect to the electric field, the order parameter was determined by averaging the orientation angles of all chains according to

$$S = \frac{\left(3\cos^2\theta - 1\right)}{2} \tag{12.6}$$

The two limiting values from the previously mentioned equation are (i) $S = 1$ (in the case of perfect alignment where all chains are parallel to the direction of electric field) and (ii) $S = -0.5$ (if the chains are perpendicular to the field).

Thus, the average orientation of the chains was quantified from the SEM images by averaging more than 100 samples. For the control experiments, that is, without the application of any electric field, there was no chain formation, and all the particles were uniformly dispersed; hence the orientation parameter cannot be calculated.

The order parameter as a function of concentration is plotted in Figure 12.8. The orientation factors for lower particle concentration of 10 and 20 wt %

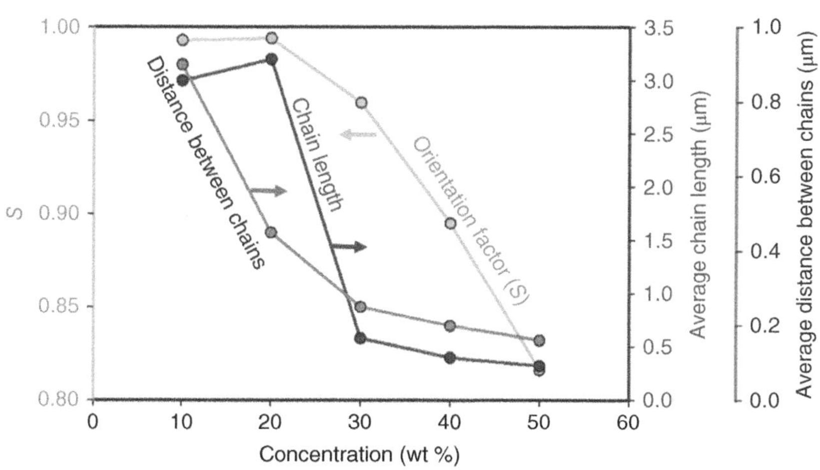


**Figure 12.8** Change in orientation factor (*S*), chain length and distance between chains as a function of concentration [37]. © Royal Society of Chemistry.

remained almost constant and were very close to perfect alignment with $S = 0.99$. As the concentration was increased, orientation factor decreased to $S = 0.86$ for 50 wt % particle concentration. This was due to the restriction of motion at high particle concentration and viscosity as has been discussed earlier.

Enhancement in the $Z$-direction properties of $BaTiO_3$ nanocomposite films was demonstrated using dielectric measurements. The dielectric properties of both oriented and unoriented films were measured through the thickness of the film. All the oriented film samples were subjected to 1400 V/mm. Figure 12.9 compares the dielectric permittivity and dielectric loss of unoriented and

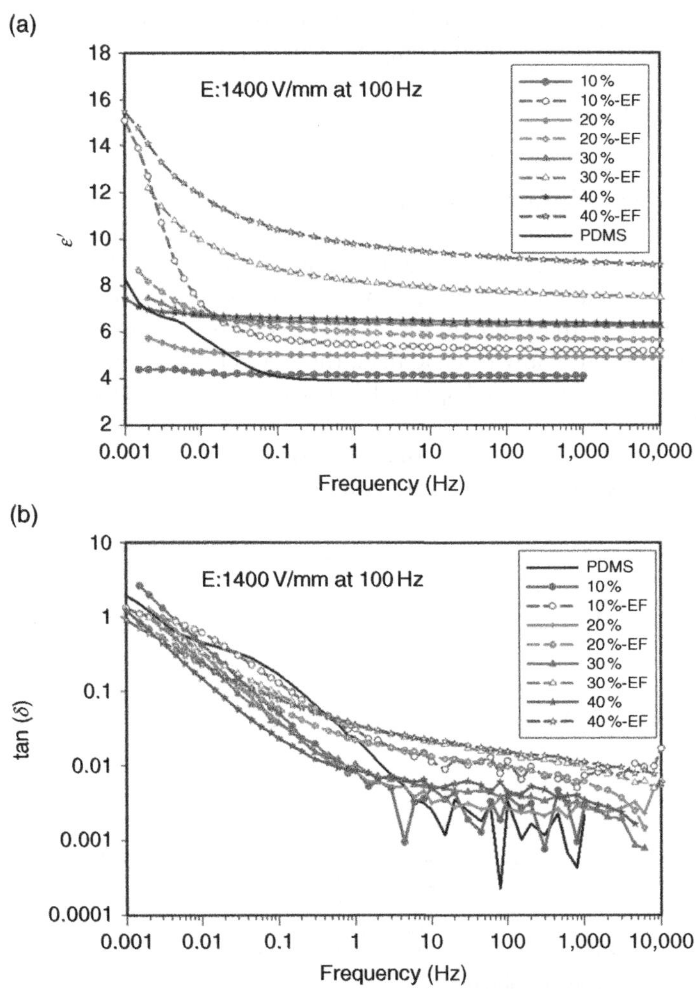


**Figure 12.9** Change in (a) dielectric constant and (b) dielectric loss for oriented and unoriented nanocomposite films of $BaTiO_3$ (EF, electric field oriented films) [37]. © Royal Society of Chemistry.

oriented samples at room temperature. The dielectric constant above 0.1 Hz remains constant; however it starts increasing below 0.1 Hz for oriented systems as compared with unoriented systems as the nanocomposites incorporate more ionic dipoles in the direction of measurement, leading to ionic polarization. The dielectric properties were also enhanced with addition of $BaTiO_3$ nanoparticles and also with their alignment.

Figure 12.10 compares three films at same concentration of particles (10 wt %), including the unoriented, oriented to form chains using dielectrophoresis, and

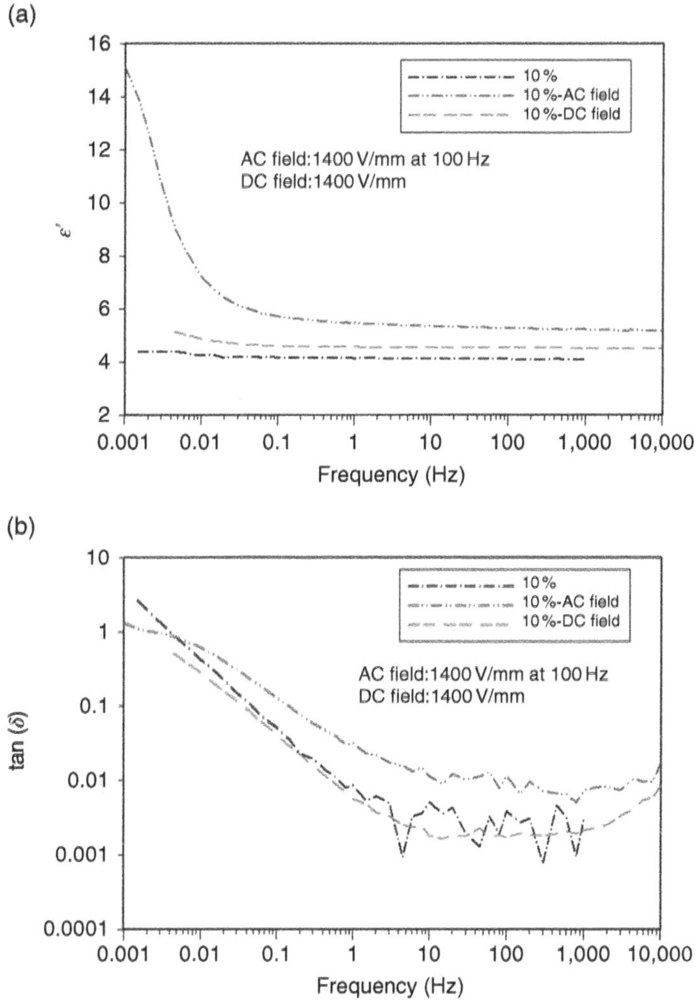


**Figure 12.10** Difference between (a) dielectric constant and (b) dielectric loss for oriented samples using dielectrophoresis and skin layer formation in electrophoresis [37].

phase separated films with skin layer consisting of densely packed particles. The dielectric properties of aligned composite were higher when compared with the skin layer and the unoriented sample.

As can be seen from Figure 12.11 where the dielectric properties between different films are compared at 10 Hz, the dielectric constant of pure matrix was 3.9 and increases with addition of $BaTiO_3$ nanoparticles to 6.3 till 30 wt %; however with further increase of nanoparticle concentration, only slight increase in nanoparticle concentration was observed. However, for the

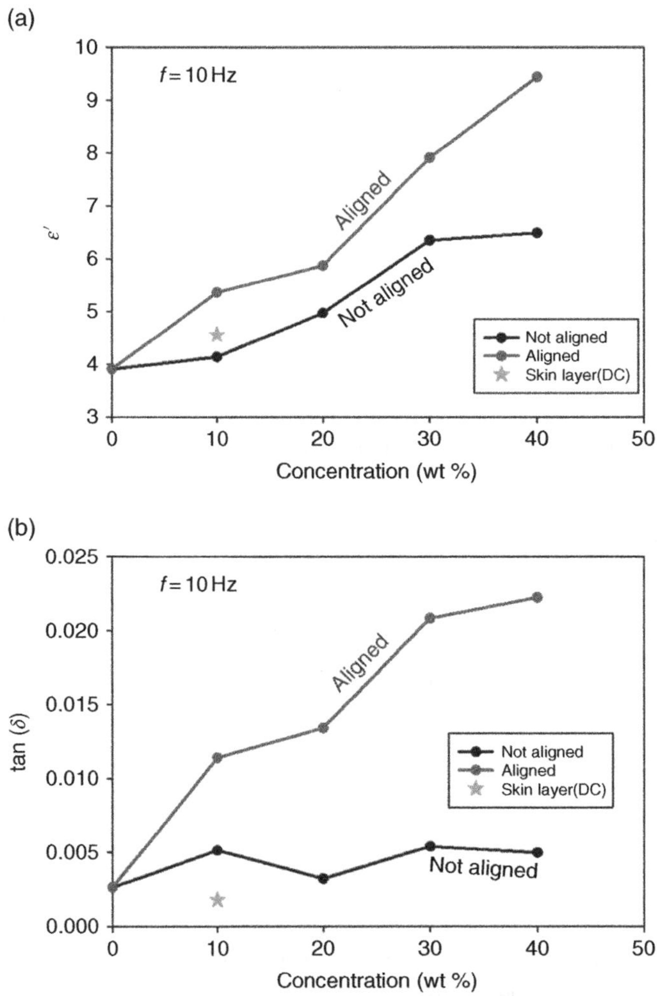


**Figure 12.11** Increase in (a) dielectric constant and (b) dielectric loss for oriented nanocomposite films of $BaTiO_3$ [37]. © Royal Society of Chemistry.

oriented films that were directionally percolated, there was an increased interaction between the particles, thus showing much higher dielectric constant of 9.5 with only 40 wt % particle concentration. Hence, alignment of particles increases the dielectric properties considerably; therefore, less amount of nanoparticles were required to achieve the similar dielectric properties, decreasing the cost of the nanocomposite films [37]. It should also be noted that dielectric loss tangent increases with increase in concentration for oriented samples, which can be attributed to the electrode polarization caused by alignment of particles. However, the dielectric loss remains constant for unoriented samples.

## 12.4 Magnetic Field Alignment

To illustrate the magnetic field effect on particles, we will use nickel nanoparticles dispersed in polystyrene solutions.

The electromagnet on the processing line is oriented in such a way that the stainless steel belt carrying the substrate along with the solution cast film lies in between the two poles of the magnet as shown in Figure 12.12. The two poles were kept very close to each other to get the magnetic field lines preferentially parallel to each other in the center. The solution cast film that contains magnetically susceptible particles or crystalline phases that are required to be aligned in the matrix are cast using either the doctor blade system for thick films or flow coating system for thin films. While using a doctor blade system, we obtain a wet film; hence rearrangement of particles and crystal phases is possible, and the alignment can be done by directly passing the film under the

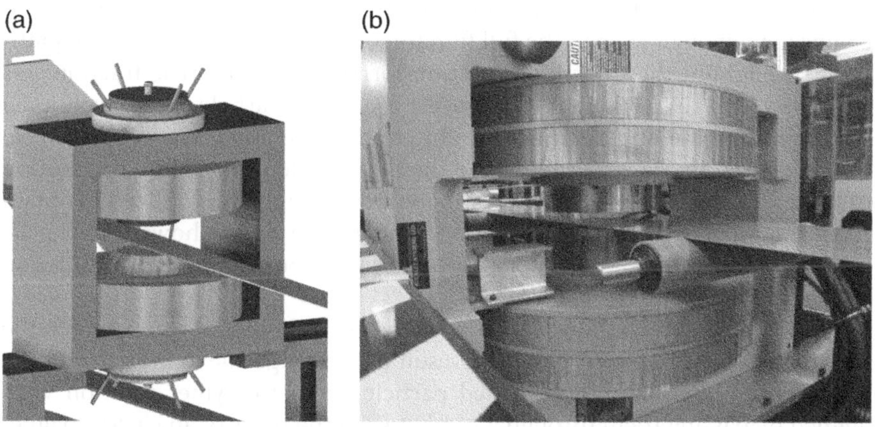


**Figure 12.12** Depiction of the magnetic field zone on the processing line [21]. © John Wiley & Sons.

magnetic field for $Z$-orientation alignment. The speed of the belt can be adjusted to employ freezing of the solution cast wet films; however, if a photocurable resin is used as a matrix, the structure can be frozen using a UV zone, which is movable and can be placed just after the magnetic field [21].

Magnetic field application in orienting the particles or phases is analogous to that of electric field as discussed earlier. Both magnetic and electric fields can be used for alignment of particles forming chains parallel to the applied field; however, in case of electric field, relative dielectric constant between matrix and dispersant determines the degree of alignment, whereas in case of magnetic field, relative magnetic susceptibilities determine the preferential orientation. An important distinction is electric field is limited in applicability by the amount of potential difference that can be applied between two parallel plates, the limit being electric breakdown of air, whereas magnetic field has no such limit. But there is a technological limitation on the strength of the magnetic field that can be generated. It is very difficult to achieve magnetic fields of more than a few teslas with an electromagnet. Hence, this limits to the use of highly susceptible particles including magnetic nanoparticles or phases such as liquid crystals for magnetic field alignment. However, super-magnets have been employed by researchers to orient weakly susceptible magnetic particles or polymer phases with different magnetic susceptibilities at lab scale, as the kinetics of these processes are very slow for the traditional R2R processing.

Figure 12.13 shows scanning electron micrographs of nonaligned and aligned films for selected particle fractions. Circles are placed on the Ni particles and aggregates to enhance the micrographs. For Ni/PS composites processed without a magnetic field (Figure 12.13a–e), no chain structure was observed. For low particle concentration samples (0.5 wt %), they form particle chains whose axes oriented in the thickness direction, but they do not span the full thickness (Figure 12.13f and g). Above about 2 wt %, the particles form aligned chains long enough to span the full film thickness (Figure 12.13h). More pathways throughout the thickness direction formed as the particle fraction was increased (Figure 12.13i, j). It was observe that these particle chains (columns) break through the upper surface, leading to the formation of protrusions like nano-needles. In order to understand the three-dimensional (3D) distribution of these chains (columns), micro computed tomography was utilized to obtain a 3D image of 10 wt % Ni powder/PS, and Figure 12.14 shows that the Ni powder stands up like rods whose axes oriented in the thickness direction of the PS matrix [41].

Figure 12.15 shows the schematic of the particle organization under the influence of a magnetic field and as a result of shrinkage due to the evaporation of solvent. The randomly dispersed particles in the polymer solution were aligned by passing them through a gap between the poles of an electromagnet that organizes these particles along nanocolumns. The shrinkage induced by solvent evaporation leads to the decrease of particle-to-particle distance and enhancement of electrical conductivity [41].

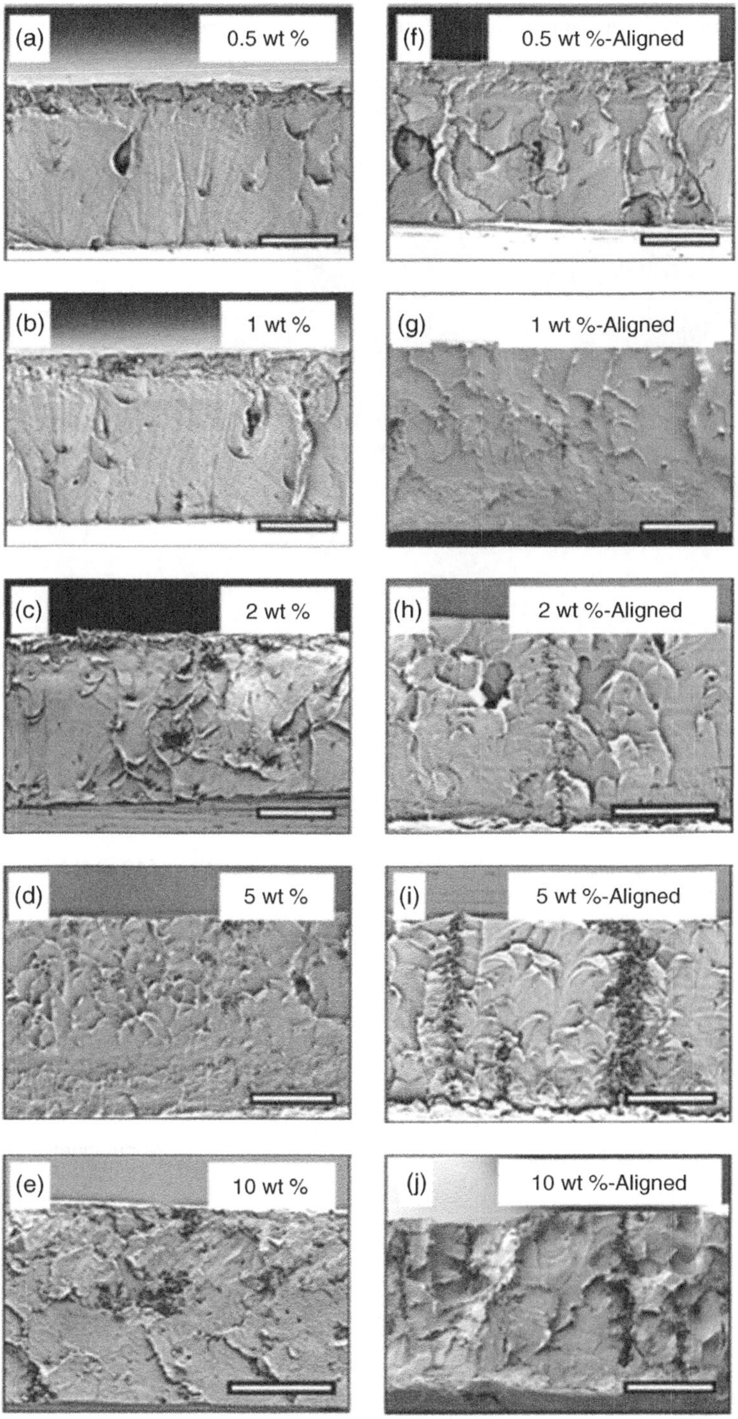


**Figure 12.13** Preferred orientation in Ni powder/PS as a function of Ni loading (scale bar: 50 μm) [41]. © Royal Society of Chemistry.

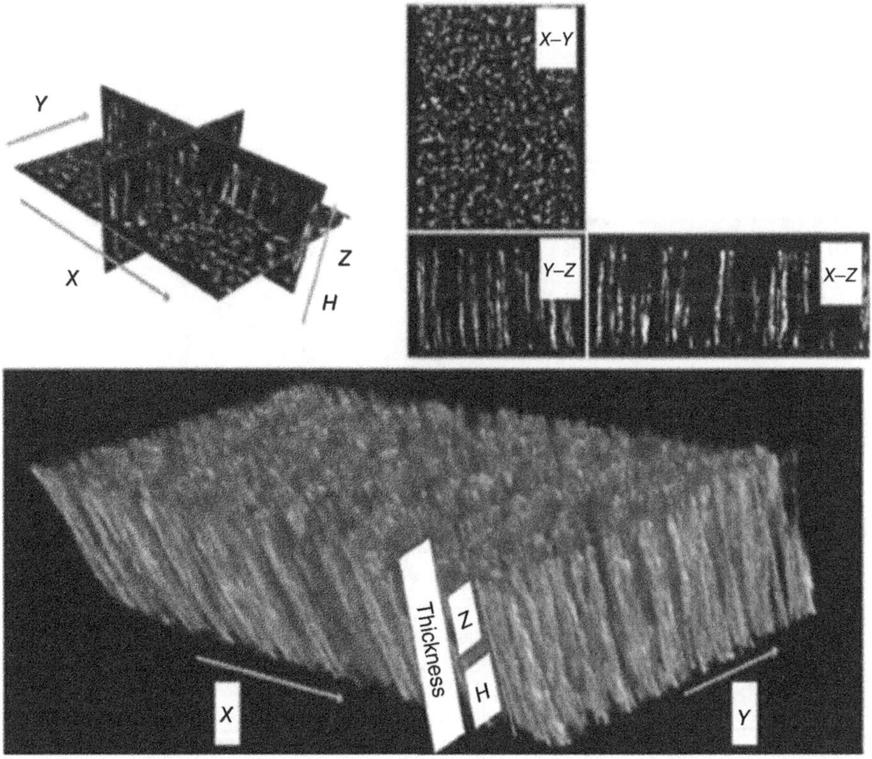


**Figure 12.14** Microcomputed tomography of the preferential orientation of 10 wt % Ni powder/PS [41]. © Royal Society of Chemistry.

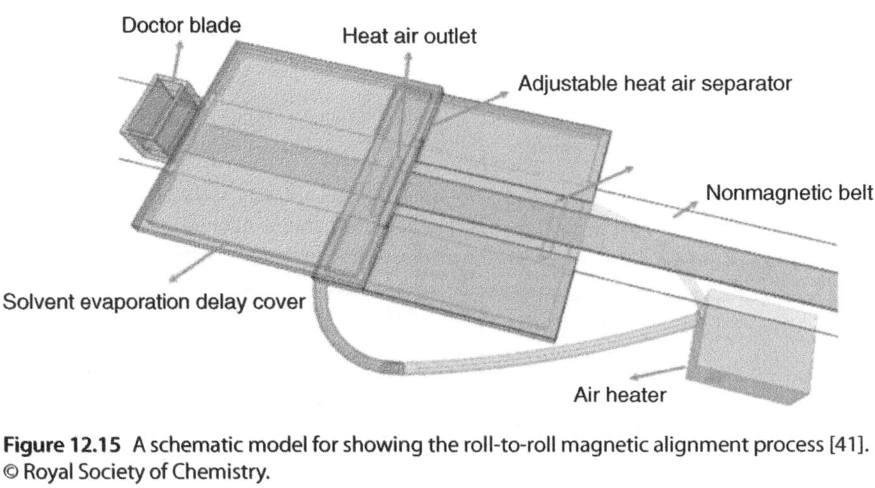


**Figure 12.15** A schematic model for showing the roll-to-roll magnetic alignment process [41]. © Royal Society of Chemistry.

In the presence of the magnetic field, each particle placed in solution is magnetized along the field direction. Since all dipoles are aligned in the magnetic field direction, they repel each other in the transverse direction and attract each other along the field direction. Based on magnetic hysteresis, the magnetization of Ni particle is increased with the increase of external magnetic field until the saturation field is reached. Since the saturation field of Ni particle is roughly 3T, magnetic dipole moments and the strength of the magnetic interaction between particles increase with the increase of magnetic field in the range of 0–3 T. The length of Ni chains increases as applied magnetic field strength increases. When the external magnetic field intensity is high enough, the chains are formed, and when they grow long enough, they span between the two surfaces. Particle organization and alignment along nanocolumns leads to a significant conductivity enhancement along the thickness direction as shown in Figure 12.16. The resistivity along the Ni chain direction (out of plane) is as low as $10^1\,\Omega m$ at 10 wt % Ni powder loading, while the in-plane resistivity is as high as $10^8\,\Omega m$. The resistivity anisotropy ratio is calculated to be greater than $10^8:1$. The percolation threshold of Ni powder/PS is about 2% compared with 80 wt % of randomly dispersed system. Ni powder/PS where a conductive network of Ni Powders is formed inside the polymer at that high particle loading, as shown in Figure 12.16. In Ni/epoxy resin composites and copper/epoxy resin composites, the critical percolation concentration are 8.5 and 5 vol %, respectively. In this work, high electrical conductivity can be achieved in the thickness direction with the use of as little as 2 wt % (0.2 vol %). The same conductivity was reached by adding 80 wt % of randomly distributed particles. The electrical resistance for pure Ni is $6.99 \times 10^{-8}\,\Omega m$, and $10\,\Omega m$ can be achieved for 10 wt % Ni/PS composites with aligned Ni particle chains.

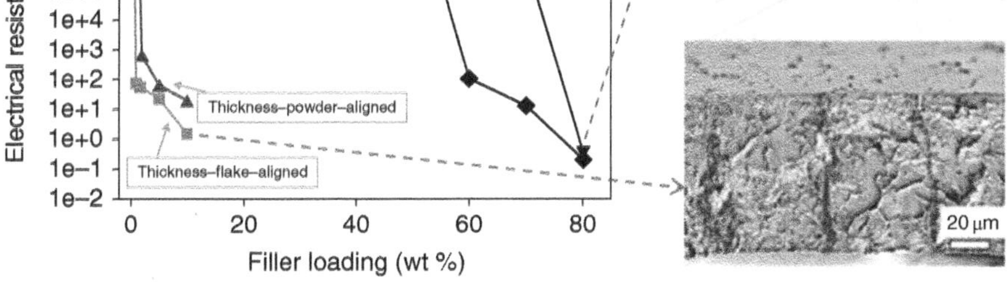


Figure 12.16 Resistivity of Ni/PS composite film [41]. © Royal Society of Chemistry.

The percolation threshold of Ni flake/PS is much lower than Ni powder/PS. It is due to flakes having a larger surface area that enable them to form an enhanced long-range network connectivity than powders. Ni flakes have broader surfaces, and this leads to higher contact and/or tunneling conduction efficiency as compared with irregularly shaped nickel nanoparticles. Thus Ni flakes can achieve a percolation threshold at a much lower particle concentration [41].

Formation of a thickness-oriented Ni chain nanocolumnar structure provides Ni/PS composites with not only anisotropic electrical resistivity but also anisotropic optical transmittance. The transmittance of Ni/PS composites prepared from both powders and flakes are measured over a wide range of filler loadings. Experimental data are shown in Figure 12.17 as a function of loading. For both

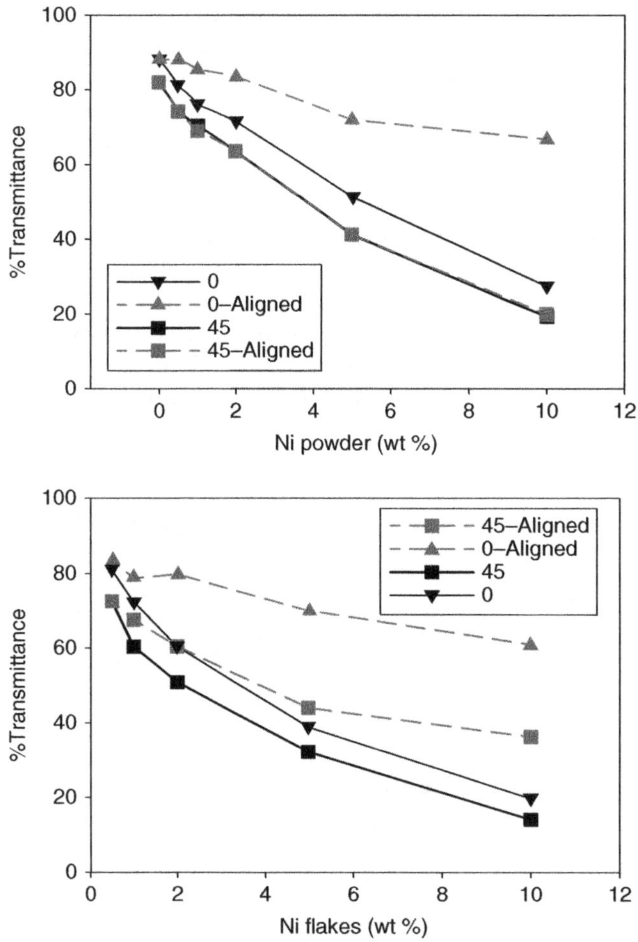


**Figure 12.17** Anisotropic optical transmittance of Ni/PS film (wavelength: 547 nm) [41]. © Royal Society of Chemistry.

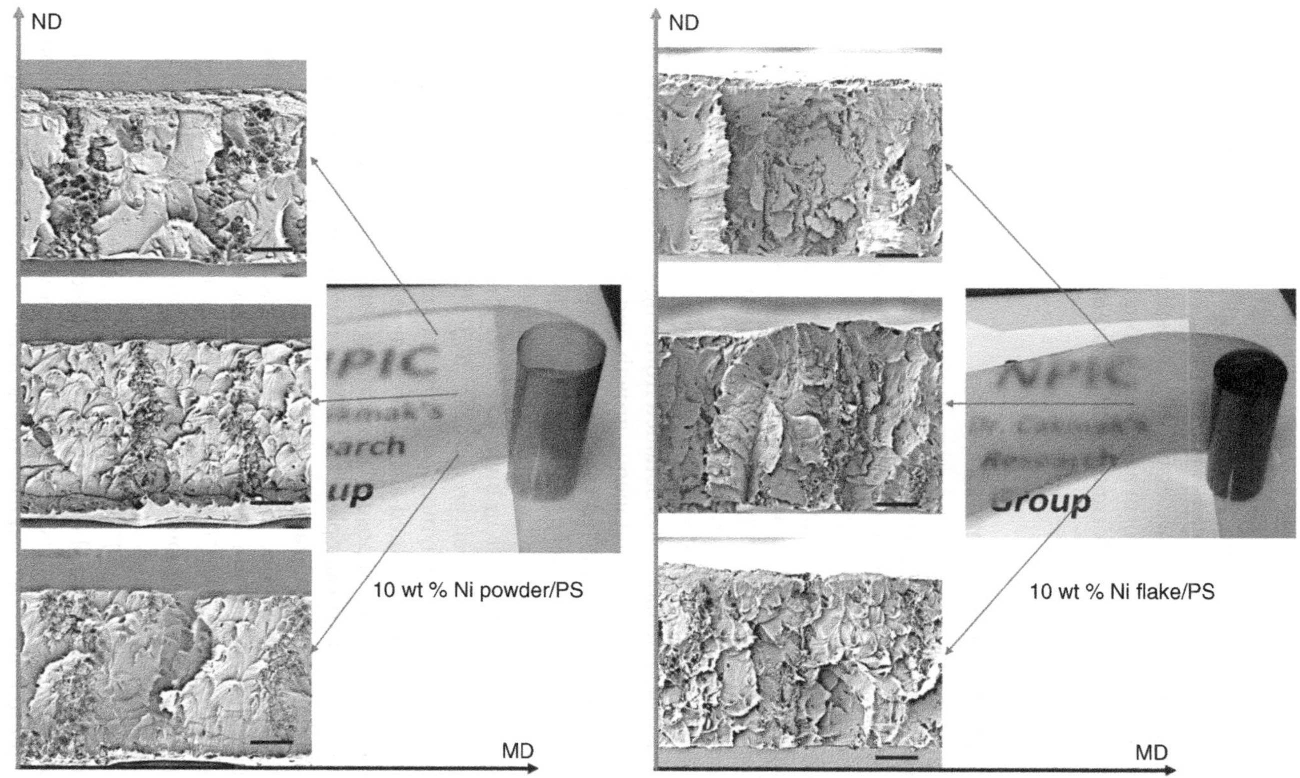


**Figure 12.18** Anisotropic structure of Ni/PS processed by roll-to-roll manufacturing line (scale bar: 20 μm) [41]. © Royal Society of Chemistry.

Ni powder/PS and Ni flake/PS, the normal transmittance (0° transmittance) of an aligned Ni/PS composite film was increased more than 60% as compared with a nonaligned composite containing the same 10 wt % Ni loading.

Figure 12.18 shows that the magnetic field aligned Ni/PS composites have a uniform alignment structure along the film thickness direction. The roll of films produced are also shown in Figure 12.18, which clearly show that at 10 wt % concentration, the films are essentially transparent and the nanocolumnar structures obtained are uniform across and along the film casting direction as also evidenced by the uniformity of transparency throughout the unrolled portions of the films.

## 12.5 Thermal Gradient

Thermal gradient, also known as zone solidification, zone melting was first developed by William Gardner Pfann at Bell Labs in the early 19th century to prepare high purity materials for manufacturing transistors. In this process, the material travels under a narrow heater slowly. When it encounters a thermal gradient, the crystal melts and subsequently recrystallizes when it encounters the other cooling side of this gradient, leaving impurities behind in the molten zone. Lovinger [42, 43] later applied this to semicrystalline polymers and obtained directional crystallization. Zone melting has also been used to recrystallize small organic molecules obtained from single crystals of millimeter-size thin films; due to the simultaneous growth and purification of organic [43] single-crystal thin films, it led to a large improvement of the optoelectronic properties. Thermal gradient has also been applied on BCPs. Hashimoto and coworkers [20, 44, 45] for the first time applied this method to a lamella-forming diblock copolymer in bulk, to reduce the number of defects and to form a single-crystal structure in the BCP films. A polystyrene/block/polyisoprene diblock copolymer was exposed to moving temperature gradient ($\Delta T = 70°\text{C/mm}$) and obtained a single crystal of lamella microdomain structure in the bulk film by imposing the moving gradient at a rate of 25 nm/s [46].

In this section two examples that discuss the use of thermal gradients to obtain directional crystallization and orientation of BCP phases are discussed.

The thermal gradient zone consisted of two setups: one was a conduction-based setup and other was radiation-based. The intent was to create instantaneous temperature changes of up to 350°C in the hot zone and −15°C in the cold zone in a space of about 0.5″. These zones were repeated 10 times to create sinusoidal gradients. The conduction setup consists of rod heaters encapsulated by copper bars of 0.5″ wide and 10″ long, and 1″ wide and 10″ long copper bars are used as cold zone by circulating water–ethanol mixture that was cooled using a chiller. The hot and cold zone copper bars are placed one after the other with ceramic spacer between them to prevent any direct conduction of heat and to create a larger gradients [46].

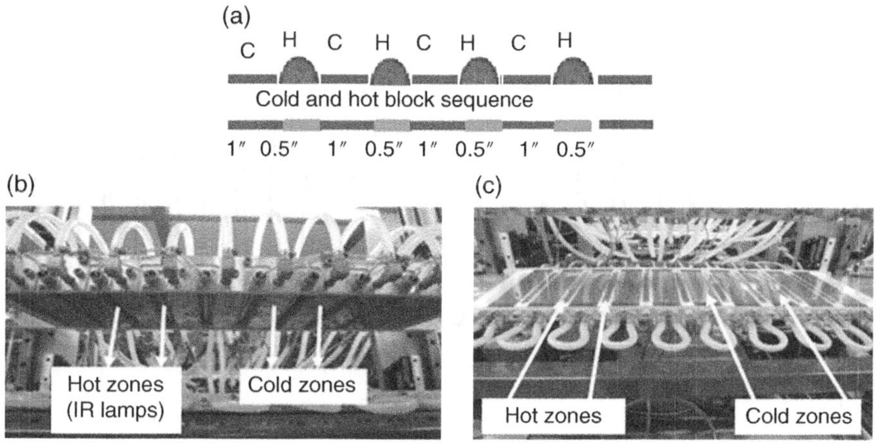


**Figure 12.19** (a) Schematic and pictures of (b) radiation- and (c) conduction-based thermal alignment zones [46]. © American Chemical Society.

The second assembly of the thermal gradient setup that works on radiation of heat lies just above the conduction setup separated by a distance of 0.5 cm. The top assembly that works on radiation heating consists of 10 IR lamps that focus the light on hot copper bars in the bottom zone. The IR lamps were placed one after the other so as to face the hot copper bars at the bottom, but they were also separated by cooled metal plates along the machine direction, hence creating the required gradient. The substrate containing the solution-cast dried film goes on top of the conduction setup and was separated by the radiation setup by distance of 0.5 cm. This zone consist of pyrometers placed strategically, so as to monitor the temperatures on the surface of the cast film, each hot and cold zone consists of three pyrometers measuring temperature uniformity across the width of the film. Figure 12.19 shows the image of both radiation and thermal gradient zone employed [21].

The two setups in the thermal gradient zones can either be used simultaneously or separately. The temperature in each of the hot zone at both bottom and top can be controlled separately. The ability to control the temperature in each zone helps to create a programmed gradient or a constant gradient. Figure 12.20 shows the ability to program required temperature gradient, in this study a thermocouple was placed on top of the kapton substrate to measure the change in temperature as it passes through various zones.

### 12.5.1 Directional Crystal Growth Using Thermal Gradient

Directional crystal growth of poly(ethylene oxide) was obtained using thermal gradient or zone solidification. 1 wt % PEO with Mw of 400,000 was first

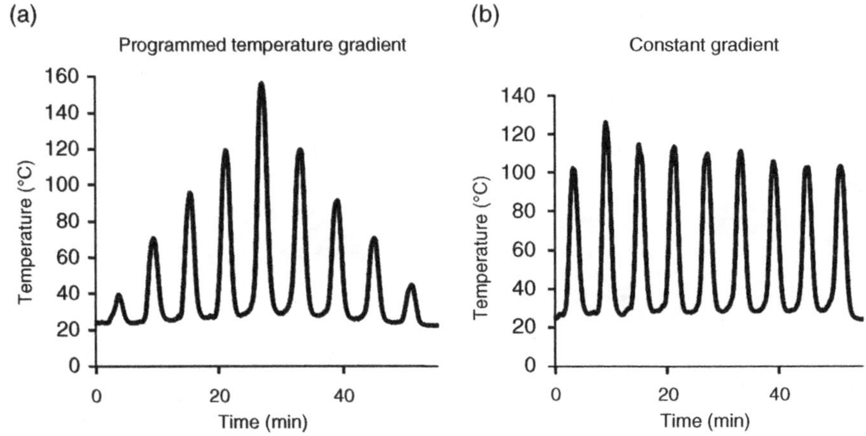


**Figure 12.20** Examples showing (a) programmed temperature gradient and (b) constant temperature gradient on the thermal alignment zone [21]. © John Wiley & Sons.

dissolved in dichloromethane. The solution was then cast on glass substrate using flow coater to obtain a thin film with dry thickness of 1 μm; the film was 7 cm wide and 25 cm long. The sample was then subjected to thermal gradient zone, in this example only the conduction heating zones were used. Also only one zone was utilized instead of all the nine available zones to create gradient. The hot zone was kept at 130°C to obtain 110°C on the film surface, and the cold zones were kept at 9°C; the profile of temperature gradient on top of the glass substrate is shown in Figure 12.21. The hot zone temperature was selected such that it was much higher than the melting point of PEO ($T_m$ = 63°C).

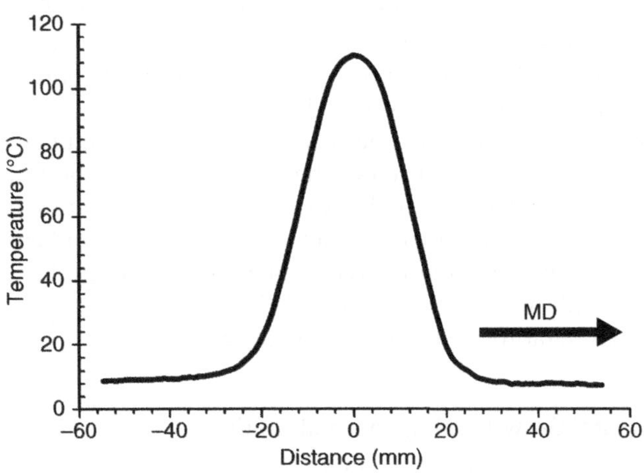


**Figure 12.21** Temperature gradient observed on top of the substrate [21]. © John Wiley & Sons.

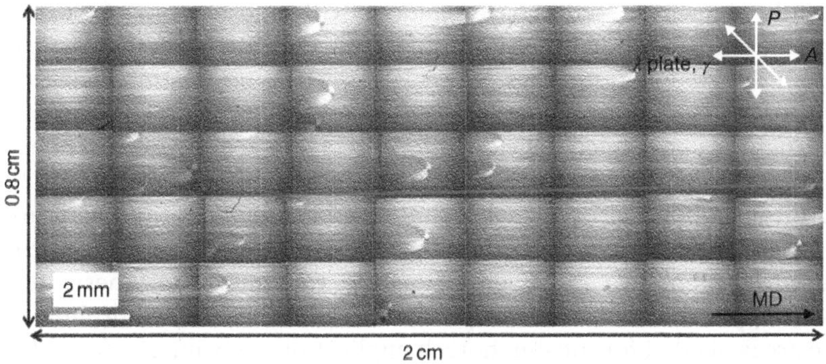


**Figure 12.22** Mosaic image of uniaxial crystallization of PEO crystals through directional solidification observed under cross-polarizers with lambda plate (scale bar: 2 mm) [21]. © John Wiley & Sons.

The film was then set in motion at a speed of 25 μm/s with a calculated moving gradient of 60°C/cm [21].

After directional crystallization, the sample was characterized using polarized optical microscope with lambda plate; a composite of multiple images is shown in Figure 12.22. Elongated crystals were observed with extended sectors of spherulites that start from the nucleation points. The elongated crystals were visible by naked eye, and some of the elongated crystals ran through the whole length of the sample.

### 12.5.2 Block Copolymer Oriented with Thermal Gradient

BCPs undergo spontaneous phase separation at the nanoscale, forming a diversity of self-assembled structures with functional architectures, such as lamellae, cylinders, and gyroids. With BCPs it is possible to attain nanostructures that have periodicity in the range of 5–100 nm, making them highly attractive for alternative route to nano-patterning where the present limit using photolithography is 22 nm half pitch *via* 193 nm light [47]. Several researchers have demonstrated that it is indeed possible to obtain nano-patterns based on highly ordered BCPs [48–50], but translating these to manufacturable large area patterns with minimal defects is a major challenge requiring facile-directed assembly methods that can be adopted by industry. Such methods will open avenues for exciting areas of application of BCPs in organic photovoltaic devices (OPVs) where it is expected that the very high interfacial area of the blocks with approximately 10–20 nm domain spacing would be useful for exciton separation or serve as templates for directed assembly of nanoparticles as has been demonstrated by several researchers; that may even have interesting implications for plasmonics and optoelectronics.

More recently, pioneering studies have explored BCP nanostructures for nanoporous membrane formation for ultrafiltration concepts, yet again these would benefit from directed alignment of channels in vertical orientations favoring fluid transport.

Various methods have been developed for directed assembly of BCP phase orientation including shear alignment [51], electric field alignment [52], solvent annealing [53], and thermal techniques [45] including temperature gradients [54], zone casting, topographic and chemical patterning, gradient substrate surface energy, and even nanoscale roughness. Of these, few methods are applicable to continuous production of desired BCP morphology over large areas needed for most nanotech applications, a primary reason for the disconnect between lab-scale experiments and scale-up to intermediate and large-scale device manufacturing processes. Thus the practical reality of achieving these ordered BCP structures with scalable continuous processing methods on flexible substrates as needed for nanomanufacturing still remains a formidable challenge.

Here, the viability of applying a dynamic gradient thermal annealing process to vertically ordering BCPs, with natural extensionability to established R2R technology was demonstrated, providing a facile route toward continuous large area fabrication of highly ordered nanomaterials. Moreover, the use of flexible substrates makes this processing method immediately applicable to target applications in flexible organic electronics and photovoltaics.

Zone annealing is a well-established method predominantly utilized for metallurgical and semiconductor purification process, where recrystallization and oriented grain growth occur on the planar front formed by the cooling edge of the zone. In contrast to "hot" zone annealing of BCPs where maximum temperature "$T_{max}$" of the thermal zone is above the order–disorder transition temperature "$T_{ODT}$," causing the two blocks to intermix completely; the zone annealing in the "cold" regime was performed where $T_{max} < T_{ODT}$, but above the glass transition temperature "$T_g$" of either blocks, and hence the term cold zone annealing (CZA). This process was previously applied to create highly ordered BCP cylinders that are parallel to the substrate with desirable features of orientational control, long-range order, and faster ordering kinetics than oven annealing. Although directed thermal zones have been utilized for ordering organic systems, including BCPs and organic electronics, their potential to vertically order BCPs has not been investigated to date despite its simplicity and viability as an industrially adaptable method [46].

CZA-sharp(S) results could be accomplished under realistic manufacturing conditions, and the results are presented in Figure 12.23a–c for the R2R conditions, utilizing only a single CZA unit. The resulting temperature gradient profile is shown in Figure 12.23g. The $T_{max}$ and $\nabla T_{max}$ are 210 and 27°C/mm, respectively. The CZA-S annealing velocity is maintained at 25 µm/s.

When the thickness of the PS-b-PMMA film was above 60 nm, approximately 80% of perpendicular cylinders were observed, whereas when the thickness of

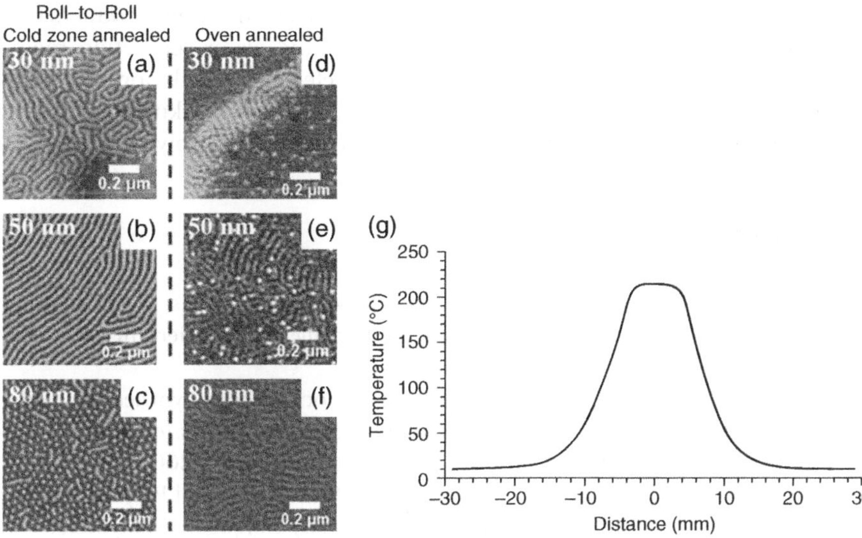


**Figure 12.23** Large-scale R2R fabrication of vertically oriented BCP thin films. (a–c) AFM morphologies of the R2R CZA-S annealed PS-PMMA thin films at the velocity of 25 μm/s and (d–f) AFM images of oven annealed samples, (g) Temperature gradient curve is shown, $T_{max}$ and $\nabla T_{max}$ are 210 and 27°C/mm, respectively. Size of scale bar is 200 nm [46]. © American Chemical Society.

the film was below 50 nm or around $L_0$, mostly parallel cylinders were observed. A morphology transition from parallel to perpendicular cylinders within a thickness variation of 10 nm from 50 to 60 nm was indicative of BCP film thickness commensurability near substrate playing an important role of ability to orient these structures. Identical samples were annealed in a vacuum oven at 210°C for 24 h as shown in Figure 12.23d–f. Oven-annealed samples show a mixed morphology containing both parallel and perpendicular cylinders at all three thicknesses. In principle, morphology obtained via previously described prototype R2R zone annealed films can used to fabricate ultrahigh dense silicon oxide dots via a procedure identical to the one described before. Thus continuous production of ultradense silicon oxide dots can be realized easily in the near future, which can then be used for further applications [46].

## 12.6 Conclusions

A novel R2R process was designed to produce continuous high value added nanostructured products with the aid of several external fields. These include use of electric field, magnetic field, and thermal gradient to promote defect-free "Z-direction" oriented flexible films that can be achieved by using one or combination of different capabilities built into this R2R processing line.

Hence using the R2R capability, we can make flexible, transparent products that can lead to supercapacitors, heat spreaders, gas and water separation membranes, fuel cell membranes, thermal insulators, electrical insulation, fuel cells, and so on, as these applications require enhanced through-thickness properties that can be achieved by alignment of particles or phases in a polymer matrix.

## References


1 Varadan, V.; Pillai, A.; Mukherji, D.; Dwivedi, M.; Chen, L., Eds. Back Matter. In *Nanoscience and Nanotechnology in Engineering*; World Scientific: Singapore, **2010**; pp. 275–301.

2 Nicolais, L.; Carotenuto, G. Metal-Polymer Nanocomposite Synthesis: Novel ex situ and in situ Approaches. In *Metal-Polymer Nanocomposites*; Nicolais, L.; Carotenuto, G., Eds.; John Wiley & Sons, Inc.: Hoboken, NJ, **2004**.

3 Xu, P.; Han, X.; Zhang, B.; Du, Y.; Wang, H.-L. Multifunctional Polymer-Metal Nanocomposites via Direct Chemical Reduction by Conjugated Polymers. *Chem. Soc. Rev.* **2014**, 43, 1349–1360.

4 Sharma, B.; Mahajan, S.; Chhibber, R.; Mehta, R. Glass Fiber Reinforced Polymer-Clay Nanocomposites: Processing, Structure and Hygrothermal Effects on Mechanical Properties. *Procedia Chem.* **2012**, 4, 39–46.

5 Chirdon, W. M.; O'Brien, W. J.; Robertson, R. E. Diffuse Reflectance of Short-Fiber-Reinforced Composites Aligned by an Electric Field. *Dent. Mater.* **2006**, 22, 57–62.

6 Patsidis, A. C.; Psarras, G. C. Structural Transition, Dielectric Properties and Functionality in Epoxy Resin–Barium Titanate Nanocomposites. *Smart Mater. Struct.* **2013**, 22, 115006.

7 Liu, C.-Y.; Bard, A. J. In-Situ Regrowth and Purification by Zone Melting of Organic Single-Crystal Thin Films Yielding Significantly Enhanced Optoelectronic Properties. *Chem. Mater.* **2000**, 12, 2353–2362.

8 Gao, F. *Advances in Polymer Nanocomposites*; Woodhead Publishing Limited, Cambridge, **2012**.

9 Newnham, R. E.; Skinner, D. P.; Cross, L. E. Connectivity and Piezoelectric-Pyroelectric Composites. *Mater. Res. Bull.* **1978**, 13, 525–536.

10 Ionescu, E.; Kleebe, H.-J.; Riedel, R. Silicon-Containing Polymer-Derived Ceramic Nanocomposites (PDC-NCs): Preparative Approaches and Properties. *Chem. Soc. Rev.* **2012**, 41, 5032–5052.

11 Kunz, D. A.; Schmid, J.; Feicht, P.; Erath, J.; Fery, A.; Breu, J. Clay-Based Nanocomposite Coating for Flexible Optoelectronics Applying Commercial Polymers. *ACS Nano* **2013**, 7, 4275–4280.

12 Marconnet, A. M.; Yamamoto, N.; Panzer, M. A.; Wardle, B. L.; Goodson, K. E. Thermal Conduction in Aligned Carbon Nanotube-Polymer Nanocomposites with High Packing Density. *ACS Nano* **2011**, 5, 4818–4825.

**13** Vaughan, B. R.; Marand, E. Transport Properties of Polymer–aluminophosphate Nano-Composites Prepared by Simple Mixing. *J. Membr. Sci.* **2008**, 310, 197–207.

**14** Lin, H.; Li, L.; Ren, J.; Cai, Z.; Qiu, L.; Yang, Z.; Peng, H. Conducting Polymer Composite Film Incorporated with Aligned Carbon Nanotubes for Transparent, Flexible and Efficient Supercapacitor. *Sci. Rep.* **2013**, 3, 1353.

**15** Zhang, Y.; Wang, Y.; Deng, Y.; Guo, Y.; Bi, W.; Li, M.; Luo, Y.; Bai, J. Excellent Dielectric Properties of Anisotropic Polymer Composites Filled with Parallel Aligned Zinc Flakes. *Appl. Phys. Lett.* **2012**, 101, 192904.

**16** Yuan, D.; Liu, Z.; Tay, S. W.; Fan, X.; Zhang, X.; He, C. An Amphiphilic-like Fluoroalkyl Modified SiO2 nanoparticle@Nafion Proton Exchange Membrane with Excellent Fuel Cell Performance. *Chem. Commun. (Camb.)* **2013**, 49, 9639–9641.

**17** Huynh, W.; Dittmer, J.; Teclemariam, N.; Milliron, D.; Alivisatos, A.; Barnham, K. Charge Transport in Hybrid Nanorod-Polymer Composite Photovoltaic Cells. *Phys. Rev. B* **2003**, 67, 115326.

**18** Lin, T.-J.; Chen, C.-C.; Lee, W.; Cheng, S.; Chen, Y.-F. Electrical Manipulation of Magnetic Anisotropy in the Composite of Liquid Crystals and Ferromagnetic Nanorods. *Appl. Phys. Lett.* **2008**, 93, 13108.

**19** Darling, S. B. Directing the Self-Assembly of Block Copolymers. *Prog. Polym. Sci.* **2007**, 32, 1152–1204.

**20** Mita, K.; Tanaka, H.; Saijo, K.; Takenaka, M.; Hashimoto, T. Ordering of Cylindrical Domains of Block Copolymers under Moving Temperature Gradient: Separation of $\nabla$T-Induced Ordering from Surface-Induced Ordering. *Macromolecules* **2008**, 41, 6787–6792.

**21** Cakmak, M.; Batra, S.; Yalcin, B. Field Assisted Self-Assembly for Preferential Through Thickness ("z-Direction") Alignment of Particles and Phases by Electric, Magnetic, and Thermal Fields Using a Novel Roll-to-Roll Processing Line. *Polym. Eng. Sci.* **2015**, 55, 34–46.

**22** Besra, L.; Liu, M. A Review on Fundamentals and Applications of Electrophoretic Deposition (EPD). *Prog. Mater. Sci.* **2007**, 52, 1–61.

**23** Ghosal, S. Fluid Mechanics of Electroosmotic Flow and Its Effect on Band Broadening in Capillary Electrophoresis. *Electrophoresis* **2004**, 25, 214–228.

**24** Pethig, R. Dielectrophoresis: Status of the Theory, Technology, and Applications. *Biomicrofluidics* **2010**, 4, 22811.

**25** Gutierrez, C. P.; Mosley, J. R.; Wallace, T. C. Electrophoretic Deposition: A Versatile Coating Method. *J. Electrochem. Soc.* **1962**, 109, 923.

**26** Velev, O. D.; Gangwal, S.; Petsev, D. N. Particle-Localized AC and DC Manipulation and Electrokinetics. *Annu. Rep. Sect. C Phys. Chem.* **2009**, 105, 213.

**27** Lyklema, J. Water at Interfaces: A Colloid-Chemical Approach. *J. Colloid Interface Sci.* **1977**, 58, 242–250.

**28** Velev, O. D.; Bhatt, K. H. On-Chip Micromanipulation and Assembly of Colloidal Particles by Electric Fields. *Soft Matter* **2006**, 2, 738.

29 Pohl, H. *Dielectrophoresis: The Behavior of Neutral Matter in Nonuniform Electric Fields*; Cambridge university Press: Cambridge, **1978**.
30 Amarandei, G.; Beltrame, P.; Clancy, I.; O'Dwyer, C.; Arshak, A.; Steiner, U.; Corcoran, D.; Thiele, U. Pattern Formation Induced by an Electric Field in a Polymer–air–polymer Thin Film System. *Soft Matter* **2012**, 8, 6333.
31 Verma, R.; Sharma, A.; Kargupta, K.; Bhaumik, J. Electric Field Induced Instability and Pattern Formation in Thin Liquid Films. *Langmuir* **2005**, 21, 3710–3721.
32 Batra, S.; Unsal, E.; Cakmak, M. Directed Electric Field Z-Alignment Kinetics of Anisotropic Nanoparticles for Enhanced Ionic Conductivity. *Adv. Funct. Mater.* **2014**, 24, 7698–7708.
33 David, E.; Fréchette, M.; Zazoum, B.; Daran-Daneau, C.; Ngô, A. D.; Couderc, H. Dielectric Properties of PE/Clay Nanocomposites. *J. Nanomater.* **2013**, 2013, 1–11.
34 Aranda, P.; Ruiz-Hitzky, E. Poly(ethylene Oxide)-Silicate Intercalation Materials. *Chem. Mater.* **1992**, 4, 1395–1403.
35 Hutchison, J. C.; Bissessur, R.; Shriver, D. F. Conductivity Anisotropy of Polyphosphazene–Montmorillonite Composite Electrolytes. *Chem. Mater.* **1996**, 8, 1597–1599.
36 Kitajima, S.; Matsuda, M.; Yamato, M.; Tominaga, Y. Anisotropic Ionic Conduction in Composite Polymer Electrolytes Filled with Clays Oriented by a Strong Magnetic Field. *Polym. J.* **2012**, 45, 738–743.
37 Batra, S.; Cakmak, M. Ultra-Capacitor Flexible Films with Tailored Dielectric Constant Using Electric Field Assisted Assembly of Nanoparticles. *Nanoscale* **2015**, 7, 20571–20583.
38 Louh, R.-F.; Hsu, Y.-H. Fabrication of Barium Titanate Ferroelectric Layers by Electrophoretic Deposition Technique. *Mater. Chem. Phys.* **2003**, 79, 226–229.
39 Wang, J. Q.; Kuwabara, M. Electrophoretic Deposition of BaTiO3 Films on a Si Substrate Coated with Conducting Polyaniline Layers. *J. Eur. Ceram. Soc.* **2008**, 28, 101–108.
40 Hermanson, K. D.; Lumsdon, S. O.; Williams, J. P.; Kaler, E. W.; Velev, O. D. Dielectrophoretic Assembly of Electrically Functional Microwires from Nanoparticle Suspensions. *Science* **2001**, 294, 1082–1086.
41 Chen, Y.; Guo, Y.; Batra, S.; Wang, E.; Wang, Y.; Liu, X.; Wang, Y.; Cakmak, M. Transparent and Through Thickness Conductive Polystyrene Films Using External Magnetic Fields for "Z" Alignment of Nickel Nanoparticles. *Nanoscale* **2015**, 7, 14636–14642.
42 Lovinger, A. J.; Wang, T. Investigation of the Properties of Directionally Solidified Poly(vinylidene Fluoride). *Polymer (Guildf)* **1979**, 20, 725–732.
43 Lovinger, A. J.; Gryte, C. C. The Morphology of Directionally Solidified Poly(ethylene Oxide) Spherulites. *Macromolecules* **1976**, 9, 247–253.

**44** Hashimoto, T.; Bodycomb, J.; Funaki, Y.; Kimishima, K. The Effect of Temperature Gradient on the Microdomain Orientation of Diblock Copolymers Undergoing an Order-Disorder Transition. *Macromolecules* **1999**, 32, 952–954.

**45** Mita, K.; Tanaka, H.; Saijo, K.; Takenaka, M.; Hashimoto, T. Cylindrical Domains of Block Copolymers Developed via Ordering under Moving Temperature Gradient. *Macromolecules* **2007**, 40, 5923–5933.

**46** Singh, G.; Batra, S.; Zhang, R.; Yuan, H.; Yager, K. G.; Cakmak, M.; Berry, B.; Karim, A. Large-Scale Roll-to-Roll Fabrication of Vertically Oriented Block Copolymer Thin Films. *ACS Nano* **2013**, 7, 5291–5299.

**47** Bosse, A.; Douglas, J.; Berry, B.; Jones, R.; Karim, A. Block-Copolymer Ordering with a Spatiotemporally Heterogeneous Mobility. *Phys. Rev. Lett.* **2007**, 99, 23–26.

**48** Sakurai, S. Progress in Control of Microdomain Orientation in Block Copolymers: Efficiencies of Various External Fields. *Polymer (Guildf)* **2008**, 49, 2781–2796.

**49** Grigorova, T.; Pispas, S.; Hadjichristidis, N.; Thurn-Albrecht, T. Magnetic Field Induced Orientation in Diblock Copolymers with One Crystallizable Block. *Macromolecules* **2005**, 38, 7430–7433.

**50** Albert, J. N. L.; Epps, T. H., III Self-Assembly of Block Copolymer Thin Films. *Mater. Today* **2010**, 13, 24–33.

**51** Morrison, F.; Mays, J.; Muthukumar, M. Shear-Induced Morphological Structures in Triblock Copolymers. *Macromolecules* **1993**, 26, 5271–5273.

**52** Xu, T.; Zvelindovsky, A.; Sevink, G.; Lyakhova, K.; Jinnai, H.; Russell, T. P. Electric Field Alignment of Asymmetric Diblock Copolymer Thin Films. *Macromolecules* **2005**, 38, 10788–10798.

**53** Kim, T. H.; Hwang, J.; Hwang, W. S.; Huh, J.; Kim, H.-C.; Kim, S. H.; Hong, J. M.; Thomas, E. L.; Park, C. Hierarchical Ordering of Block Copolymer Nanostructures by Solvent Annealing Combined with Controlled Dewetting. *Adv. Mater.* **2008**, 20, 522–527.

**54** Berry, B. C.; Bosse, A. W.; Douglas, J. F.; Jones, R. L.; Karim, A. Orientational Order in Block Copolymer Films Zone Annealed below the Order: Disorder Transition Temperature. *Nano Lett.* **2007**, 7, 2789–2794.

# Index



*Roll-to-Roll Manufacturing: Process Elements and Recent Advances*, First Edition.
Edited by Jehuda Greener, Glen Pearson, and Miko Cakmak.

### C

## d



## e

## t

Printed and bound by CPI Group (UK) Ltd, Croydon, CR0 4YY
17/09/2026
14964288-0001